2002

Chemical Ecology of Plants: Allelopathy in Aquatic and Terrestrial Ecosystems

Edited by Inderjit and Azim U. Mallik

Birkhäuser Verlag
Basel · Boston · Berlin

Editors:

Prof. Azim U. Mallik
Biology Department
Lakehead University
Thunder Bay
Ontario P7B 5E1
Canada

Dr. Inderjit
Department of Botany
University of Delhi
Delhi 110007
India

Library of Congress Cataloging-in-Publication Data

Chemical ecology of plants : allelopathy in aquatic and terrestrial ecosystems / edited by Inderjit and
Azim U. Mallik
     p. cm
   Includes bibliographical references and index.
   ISBN 3764365358 (alk. paper)
   1. Allelochemicals. 2. Allelopathic agents. 3. Allelopathy. 4. Plant chemical ecology.
   I. Inderjit, 1963- II. Mallik, A. U. (Azim U.)

QK898.A43 C425 2002
571.9'2--dc21   2002074636

Deutsche Bibliothek Cataloging-in-Publication Data

Chemical ecology of plants : allelopathy in aquatic and terrestrial ecosystems / ed. by Inderjit and
     Azim U. Mallik. - Basel ; Boston ; Berlin : Birkhäuser, 2002
   ISBN 3-7643-6535-8

ISBN 3-7643-6535-8 Birkhäuser Verlag, Basel - Boston - Berlin

© 2002 Birkhäuser Verlag, P.O. Box 133, CH-4010 Basel, Switzerland
Member of the BertelsmannSpringer Publishing Group
Printed on acid-free paper produced from chlorine-free pulp. TCF ∞
Cover design: General view of a corn field infested with common lamb's-quarters (see page 188).
Small pictures: Right: Assay for discovery and evaluation of active antifungal agents (see page 24).
Above: Inhibition of morning glory seedling biomass (see page 103).

Printed in Germany
ISBN 3-7643-6535-8

9 8 7 6 5 4 3 2 1   http://www.birkhauser.ch

Contents

List of contributors VII

Preface IX

Azim U. Mallik and Inderjit
Problems and prospects in the study of plant allelochemicals:
a brief introduction 1

Dale G. Nagle and David E. Wedge
Antifungal properties of cyanobacteria and algae: ecological
and agricultural implications 7

Dale G. Nagle and Inderjit
The chemistry and chemical ecology of biologically active
cyanobacterial metabolites 33

Maria T. Gallardo-Williams and Dean F. Martin
Phytotoxic compounds from *Typha domingensis* P. 57

Francisco A. Macías, Rosa M. Varela, Ascensión Torres,
José L.G. Galindo, and José M.G. Molinillo
Allelochemicals from sunflowers: chemistry, bioactivity
and applications 73

X. Carlos Souto and François Pellissier
Feedback mechanism in the chemical ecology of plants: role of
soil microorganisms 89

Harleen Kaur, Inderjit, and K.I. Keating
Do allelochemicals operate independent of substratum factors? 99

Erik Tallak Nilsen
Ecological relevance of allelopathy: some considerations related to
Mediterranean, subtropical, temperate, and boreal forest shrubs 109

Azim U. Mallik
Linking ecosystem disturbance with changes in keystone species,
humus properties and soil chemical ecology: implications for
conifer regeneration with ericaceous understory 131

Shibu Jose
Black walnut allelopathy: current state of the science 149

Stephen R. Gliessman
Allelopathy and agroecology 173

Inderjit and Prasanta C. Bhowmik
Allelochemicals phytotoxicity in explaining weed invasiveness and their
function as herbicide analogues 187

Inderjit and Harsh Nayyar
Shift in allelochemical functioning with selected abiotic stress factors . 199

Julia Koricheva and Anna Shevtsova
Pitfalls in interpretation of allelochemical data in ecological studies:
implications for plant-herbivore and allelopathic research 219

Stephen D. Murphy
Biochemical and physiological aspects of pollen allelopathy 245

Index 261

List of contributors

Prasanta C. Bhowmik, Department of Plant and Soil Sciences, Stockbridge Hall, University of Massachusetts, Amherst, MS 01003-7245, USA; e-mail: pbhowmik@pssci.umass.edu

José L.G. Galindo, Departamento de Química Orgánica, Facultad de Ciencias, Universidad de Cádiz, Apdo. 40, 11510 Puerto Real, Cádiz, Spain; e-mail: joseluis.garciagalindo@alum.uca.es

Maria T. Gallardo-Williams, Laboratory of Molecular Toxicology, National Institute of Environmental Health Sciences, P.O. Box 1233, Mail drop E1-06, Research Triangle Park, NC 27709, USA; e-mail: Gallardo@niehs.nih.gov

Institute for Environmental Studies, Department of Chemistry–SCA 400, University of South Florida, 4202 East Fowler Avenue, Tampa, FL 33620-5202, Florida, USA; e-mail: @

Stephen R. Gliessman, Center for Agroecology, Department of Environmental Studies, University of California Santa Cruz, Santa Cruz, California 95064, USA; e-mail: gliess@zzyx.ucsc.edu

Inderjit, Department of Botany, University of Delhi, Delhi 110007, India; e-mail: allelopathy@satyam.net.in

Shibu Jose, School of Forest Resources and Conservation, University of Florida, 5988 Hwy 90, Building 4900, Milton, FL 32583, USA; e-mail: sjose@ufl.edu

Harleen Kaur, Department of Botany, Panjab University, Chandigarh 160 014, Sector 14, India; e-mail: harleen786@rediffmail.com

K.I. Keating, Department of Environmental Sciences, Rutgers University of New Jersey, Cook Campus, New Brunswick, New Jersey 08903, USA; e-mail: KKeating@rci.rutgers.edu

Julia Koricheva, Section of Ecology, Department of Biology, University of Turku, FIN-20014 Turku, Finland; e-mail: julkoric@utu.fi

Francisco A. Macías, Departamento de Química Orgánica, Facultad de Ciencias, Universidad de Cádiz, Apdo. 40, 11510 Puerto Real, Cádiz, Spain; e-mail: famacias@uca.es

Azim U. Mallik, Department of Biology, Lakehead University, Thunder Bay, Ontario P7B 5E1, Canada; e-mail: azim.mallik@lakeheadu.ca

Dean F. Martin, Institute for Environmental Studies, Department of Chemistry –SCA 400, University of South Florida, 4202 East Fowler Avenue, Tampa, FL 33620-5202, Florida, USA; e-mail: dmartin@chuma1.cas.usf.edu

José M.G. Molinillo, Departamento de Química Orgánica, Facultad de Ciencias, Universidad de Cádiz, Apdo. 40, 11510 Puerto Real, Cádiz, Spain; e-mail: chema.gonzalez@uca.es

Stephen D. Murphy, Department of Environment and Resource Studies, University of Waterloo, Waterloo, Ontario, N2L 3G1, Canada; e-mail: sd2murph@fes.uwaterloo.ca

Dale G. Nagle, Department of Pharmacognosy and National Center for Natural Products Research, Research Institute of Pharmaceutical Sciences, School of Pharmacy, University of Mississippi, University, MS 38677-1848, USA; e-mail: dnagle@olemiss.edu

Harsh Nayyar, Department of Botany, Panjab University, Chandigarh 160 014, India; e-mail: harshnayyar@hotmail.com

Erik Tallak Nilsen, Biology Department, Virginia Polytechnic Institute and State University, 2119 Derring Hall, Blacksburg, VA 24061, USA; e-mail: enilsen@vt.edu

François Pellissier, Dynamics of Altitude Ecosystems Laboratory, University of Savoie, 73376 Le Bourget-du-Lac, France; e-mail: pellissier@univ-savoie.fr

Anna Shevtsova, Department of Forest Vegetation Ecology, Swedish University of Agricultural Sciences, S-90183 Umeå, Sweden; e-mail: Anna.Shevtsova@svek.slu.se

X. Carlos Souto, Departamento de Ingeniería de los Recursos Naturales y Medio Ambiente, E.U.E.T. Forestal, Universidade de Vigo, 36005 Pontevedra, Spain; e-mail: csouto@uvigo.es

Ascensión Torres, Departamento de Química Orgánica, Facultad de Ciencias, Universidad de Cádiz, Apdo. 40, 11510 Puerto Real, Cádiz, Spain; e-mail: chon.torres@uca.es

Rosa M. Varela, Departamento de Química Orgánica, Facultad de Ciencias, Universidad de Cádiz, Apdo. 40, 11510 Puerto Real, Cádiz, Spain; e-mail: rosa.varela@uca.es

David E. Wedge, United States Department of Agriculture, Agricultural Research Service, Natural Products Utilization Research Service, Thad Cochran National Center for Natural Products Research, University, MS 38677-8048, USA; e-mail: dwedge@olemiss.edu

Preface

There has been significant progress in plant allelochemical research with the recent advancement of conceptual understanding of ecological processes and breakthroughs in biotechnological and molecular techniques. Allelochemicals play a major role in influencing community structure, nutrient dynamics, substratum and mycorrhizal ecology, and resource competition. There is a wide gap between ecologists and natural product chemists in understanding the significance of an allelochemical in the natural environment. While natural product chemistry is an integral part of allelochemical research, ecological understanding of allelochemicals is required to generate ecologically relevant data. We feel that merely isolating allelochemicals from a plant or its habitat is not enough to argue the occurrence of allelochemical interactions in field conditions. We must unfold the chain of events initiated with the release of allelochemicals in the environment to the manifestation of their impact in community organization.

This book is devoted to highlighting the recent findings in allelochemical research from an ecological standpoint. The aim of the book is to provide an up-to-date insight into examples of allelochemical research from aquatic, forest, and agricultural ecosystems. Chapter 1 summarizes the problems and prospects in the study of allelopathy and serves as brief introduction for the book. Chapters 2 and 3 specifically discuss algal allelochemicals having ecological roles in plant defense and community structure. Chapter 4 discusses the role of allelochemicals in the aquatic weed *Typha domingesis*. Chapter 5 discusses in-depth chemical and applied aspects of sunflower allelochemicals. The significance of abiotic and biotic soil factor in determining allelochemical response is discussed in Chapters 6 and 7. The role of allelochemicals in forest and agroecosystems is reviewed in Chapters 8 through 12. Plants generally have to cope with abiotic stresses. Chapter 13 raises several questions on shifts in allelochemical functioning that are due to selected abiotic stresses. Chapter 14 argues that if the focus of a study is on the mechanisms rather than the consequences of variation in plant allelochemical composition, other methods of chemical data analysis and interpretation should be used. Chapter 15 highlights the update information of pollen allelochemicals. All in all, the book illustrates the different aspects of allelochemical research in aquatic and terrestrial ecosystems.

We are grateful to the authors for submitting their valuable work for this book within the given time lines. We are indebted to the following referees for their critical and timely reviews: Hans Lambers, Jeffrey Weidenhamer, John

Romeo, Stella Elakovich, Erik Nilsen, Steve Duke, Chester Foy, Dave Orcutt, Rod Heisey, Francois Pellissier, Steve Murphy, Dean Martin, Maria Olofsdotter, Robert Thacker, William Gerwick, Shmuel Carmeli, Gabriele König, Kathleen I. Keating, Arthur Zangerl, Jim Pratley, Geoffrey Smith, Udo Blum, David Nagle, Steve Gliessman, and Tony Hooper. Last but not least, Inderjit sincerely thanks Professor C.R. Babu, the Pro Vice Chancellor, University of Delhi, who helped him in various ways. We appreciate the help and cooperation of Dr. Hans Detlef Klüber, Acquisitions Editor, Birkhäuser Verlag. Editorial help of Miss Harleen Kaur is gratefully acknowledged. It is our hope that the information presented in this book will serve the scientific community well.

Azim U. Mallik and Inderjit
January 2002

Problems and prospects in the study of plant allelochemicals: a brief introduction

Azim U. Mallik[1] and Inderjit[2]

[1] *Department of Biology, Lakehead University, Thunder Bay, Ontario P7B 5E1, Canada*
[2] *Department of Botany, University of Delhi, Delhi 110007, India*

A series of simplistic experiments followed by a broad and bold generalization by Hans Molisch in the late 1930s regarding the influence of one plant on another through release of chemicals into the environment initiated the foundation of allelopathy [1]. However, the phenomenon of a plant's growth suppression by chemicals released from plants in its vicinity has been known since ancient times [2]. Molisch articulated the phenomenon by defining and coining the term allelopathy. As a plant physiologist, Molisch drew his conclusions from controlled experiments, mostly by exposing his test plant(s) or plant parts to ethylene emitting from apples. Around the same time, F. Boas conducted controlled experiments with aquatic extracts of leaves of buttercup (*Ranunculus* spp.), camomile, and several conifers such as fir, pine, and juniper and found inhibitory as well as stimulatory effects on test plant(s) depending on the type of extract. From these limited tests, he developed far-reaching conclusions of plant community structuring [3, 4]. Our intention is not to dismiss or undermine the contributions of these pioneers of allelopathy research but to emphasize the fact that a complex phenomenon like allelopathy cannot be demonstrated by such a simplified experimental approach [5]. Many researchers in the past, and to some extent present, have taken this bad lesson of using inappropriate and simplified approaches draw unjustifiable conclusions and extrapolation in the study of allelopathy [6–9]. The discipline has suffered a great deal because of it [10].

There have been legitimate concerns regarding the definition of allelopathy and allelochemicals. The all-encompassing definition of allelopathy by E.L. Rice [11] "*as any direct or indirect harmful or beneficial effect by one plant (including microorganisms) on another through production of chemical compounds that escape into the environment*" has been criticized for being "boundless and therefore meaningless" as a definition [12]. The working definition of the International Allelopathy Society [13], "*any process involving secondary metabolites produced by plants, algae, bacteria and viruses that influence the growth and development of agricultural and biological systems; a study of the functions of secondary metabolites, their significance in biological organiza-*

tion, their evolutionary origin and elucidation of the mechanisms involving plant–plant, plant–microorganisms, plant–virus, plant–insect, plant–soil–plant interactions," also suffers from the similar problem of all-inclusiveness and boundlessness. However, the fact remains that the phenomenon of plant–plant interaction through the release of secondary compounds in nature is undeniable. Even one of the most influential ecologists and a vocal critics of allelopathy of our time, J.L. Harper, said, *"Demonstrating allelopathy has proved extraordinarily difficult—it is logically impossible that it does not happen and perhaps nearly impossible to prove that it does"* [10]. Therefore, the concern is the burden of proving the existence of allelopathy in natural ecosystem amongst all other interacting phenomena rather than the existence of allelopathy. Terminology and trends in allelopathy research were convincingly reviewed by Willis [14]. He suggested that allelopathy should be considered under plant ecology and should be kept as sub-discipline of allelochemicals [14].

The other challenge is in the definition of allelochemicals. There are several questions which need to be addressed. For example, do allelochemicals include only secondary metabolites or also primary metabolites? Are allelochemicals necessarily a specialized group of compounds that are exudates or volatilized or degraded products of non-allelopathic compounds at certain concentration and/or in certain mixtures influencing the growth of the neighboring plants? We all know that many usual secondary metabolites of plants such as phenolics are considered allelochemicals under certain concentrations or in certain mixtures. Are there non-allelopathic compounds that become allelopathic under certain circumstances? When does a compound become allelopathic? These are the sorts of questions we have yet to resolve in order to develop a solid conceptual basis of allelochemical research.

Having highlighted some of the academic questions that we have yet to answer in order to better articulate the premise of allelopathy, there is no denying that we have experienced significant progress in allelopathy research in the last four decades. Development of new and sophisticated methods of isolation and identification of allelochemicals and development of innovative experimental designs for the study of allelopathy have contributed significantly to this success. Better methods are able to detect the effects of dominant chemical and environmental factors that play important roles in the manifestation of allelopathy. Although some researchers were able to demonstrate the separation of allelopathic effects from competition under controlled conditions [15–17], others [18] advocated against the separation of allelopathy from resource competition. Inderjit and Del Moral [18] opined that nature is too complex to be explained by one factor. Several factors—allelopathy, resource competition, nutrient immobilization, mycorrhizae, substratum—operate sequentially and/or simultaneously and influence community structure. We can separate allelopathy from other interference factors under controlled condition, but the question is: Does this actually happen in nature? By using sophisticated, modern equipment and techniques, the mode of action and fate of allelochemicals in soil and their effects on microorganisms and nutrient availability have been investigated [19, 20]. A methodological impasse in allelopathy research has been further elevated by the recent advances in molecular biology and digital technology. As a result, the discipline has been diversified from molecular genetics, to pollen allelopathy, to weed management in agriculture and forestry. Scientists with many different specializations are working independently and in collaboration to gain a better understanding of the workings of allelochemicals and allelopathic phenomenon. Breakthroughs in allelopathic research in molecular biology will play a leading role in developing practical methods for sustainable agriculture and forestry. Digital imaging techniques and information technology will play an increasingly important role in in-depth study of allelochemical origin and function and will elucidate the phenomenon of allelopathy [21].

Recently, ecosystem-level understanding of allelopathy has been explored in agriculture and forestry by several researchers [16, 22]. The concept of keystone species and their role as ecosystem engineers has been linked to substrate ecology and allelopathy, which can preempt competition in structuring plant communities (Mallik, this volume).

Research areas need in-depth investigations

While many unanswered questions remain in the study of allelopathy, new discoveries on the behavior of secondary compounds as allelochemicals, new tools for sophisticated chemical analyses, and renewed understanding of ecosystem function together provide an exciting opportunity to resolve the challenging fundamental questions [23]. On the application side, there is a bright prospect of applying allelopathic principles in sustainable agriculture and forestry by refining agronomic and silvicultural techniques, allelopathic gene transfer, and biological weed control. The future lies in understanding the ecosystem perspective of allelopathic phenomena and allelochemical effects on soil chemistry. Demonstration of the subtlety of the phenomenon and its interconnectedness with other ecological processes that influence ecosystem functioning will make a significant contribution to science and natural resources management. We feel that the following areas can be used as examples for understanding the diversity and complexity of allelochemical functioning in ecosystems.

1. Significance of algal metabolites and their importance in pharmaceutical and community structure (Nagle and Wedge, this volume; Nagle and Inderjit, this volume).
2. Compared to terrestrial ecosystems, little attention has been paid to allelochemical interactions in aquatic ecosystems. Gallardo-Williams and Martin (this volume) discuss allelochemicals from *Typha domningesis* and their ecological significance.

3. Sunflower has long been a source of interest to chemist and agroecologists. Macías et al. (this volume) provide an update on the chemistry, bioactivity, and application of sunflower allelochemicals.

4. An in-depth insight on ecological relevance of allelochemicals in forest and agroecosystems can contribute significantly to developing sustainable land management, and these are discussed by several authors in this book (e.g., Nilsen, Mallik, Jose, Inderjit and Bhowmik, and Gliessman).

5. The functioning of allelochemicals is greatly influenced by abiotic stress factors in field situations. Inderjit and Nayyar (this volume) raise several questions about allelochemical functioning and abiotic stress.

6. Koricheva and Shevtsova (this volume) argue that if the focus of a study is on the mechanisms rather than on the consequences of variation in plant allelochemical composition, presentation of data in terms of concentrations, contents, and plant biomass should be encouraged.

7. Pollen allelopathy is of great interest to eco-physiologists and weed ecologists. Recent information on the biochemical and physiological aspects of pollen allelopathy is reviewed by Murphy (this volume).

Topics covered in this volume provide information not only on the recent developments in allelochemical research in relation to sustainable agriculture and forestry but also on our latest understanding about the complexity of interaction between allelochemicals and various biotic and abiotic factors.

References

1 Molisch H (1937) *Der Einfluß einer Pflanze auf die andere, Allelopathie.* Verlag von Gustav Fischer, Jena, Germany. English translation (2001) LJ LaFleur, MAB Mallik (translators): *Influence of one plant on another*, Scientific Publisher, Jodhpur

2 Theophrastus (ca. 300 B.C.) *Enquiry into plants and minor works on odorous and weither signs.* 2 Vols. Transl. to English by A Hort, Loeb Classical Libr., W. Heinemann, London (1916)

3 Boas F (1934) Beiträge zur Wirkungsphysiologie einheimischer Pflanzen, I [Contributions to the metabolic physiology of indigenous plants, I]. *Berichte Deutsch Bot Gesell* 52: 126–131

4 Boas F (1935) Vergleichende Untersuchungen über Wachstumsanreger in einheimischen Pflanzen, Beiträge zur Wirkungsphysiologie einheimischer Pflanzen, II [Comparative investigations of growth stimulants in indigenous plants: contributions to the metabolic physiology of indigenous plants, II]. *Berichte Deutsch Bot Gesell* 53: 495–511

5 Inderjit, Weiner J (2001) Plant allelochemical interference or soil chemical ecology? *Perspect Plant Ecol Evol Syst* 4: 3–12

6 Willis RJ (1985) The historical basis of the concept of allelopathy. *J Hist Biol* 18: 71–102

7 Inderjit, Dakshini KMM (1995) On laboratory bioassays in allelopathy. *Bot Rev* 61: 28–44

8 Inderjit, Weston LA (2000) Are laboratory bioassays for allelopathy suitable for prediction of field responses? *J Chem Ecol* 26: 2111–2118

9 Romeo J (2000) Raising the beam: moving beyond phytotoxicity. *J Chem Ecol* 26: 2011–2014

10 Harper JL (1975) Allelopathy. *Q Rev Biol* 50: 493–495

11 Rice EL (1974) *Allelopathy.* Academic Press, New York

12 Watkinson AR (1998) Reply from A.R. Watkinson. *Trends Ecol Evol* 13: 407

13 International Allelopathy Society (1998) Allelopathy: is this the definition we want? *In:* R Willis (ed): *IAS Newsletter 2: 5*, University of Melbourne, Australia

14 Willis RJ (1994) Terminology and trends in allelopathy. *Allelo J* 1: 6–28

15 Weidenhamer JD, Hartnett DC, Romeo JT (1989) Density-dependent phytotoxicity: distinguish-

ing resource competition and allelopathic interference in plants. *J Appl Ecol* 26: 613–624

16 Zackrisson O, Nilsson M-C (1992) Allelopathic effects by *Empetrum hermaphroditum* on seed germination of two boreal tree species. *Can J For Res* 22: 1310–1319

17 Nilsson MC (1994) Separation of allelopathy and resource competition by the boreal dwarf shrub *Empetrum hermaphroditum* Hagerup. *Oecologia* 98: 1–7

18 Inderjit, Del Moral R (1997) Is separating resource competition from allelopathy realistic? *Bot Rev* 63: 221–230

19 Einhellig FA (2001) An integrated view of phytochemicals amid multiple stresses. *In*: Inderjit, KMM Dakshini, CL Foy (eds): *Principles and practices in plant ecology: allelochemical interactions*. CRC Press, Boca Raton, 479–494

20 Blum U, Shafer SR, Lehman ME (1999) Evidence of inhibitory allelopathic interactions involving phenolic acids in field soils: concepts *vs.* an experimental model. *Crit Rev Plant Sci* 18: 673–693

21 Weston LA, Czarnota MA (2001) Activity and persistence of sorgoleone, a long-chain hydroquinone produced by *Sorghum bicolor*. *J Crop Prod* 4: 363–377

22 Wardle DA, Nilsson MC, Gallet C (1998) An ecosystem-level perspective of allelopathy. *Biol Rev* 73: 305–319

23 Mallik AU (2000) Challenges and opportunities in allelopathy research: a brief overview. *J Chem Ecol* 26: 2007–2009

Chemical Ecology of Plants: Allelopathy in Aquatic and Terrestrial Ecosystems
ed. by Inderjit and Azim U. Mallik

Antifungal properties of cyanobacteria and algae: ecological and agricultural implications

Dale G. Nagle[1] and David E. Wedge[2]

[1] Department of Pharmacognosy and National Center for Natural Products Research, Research Institute of Pharmaceutical Sciences, School of Pharmacy, University of Mississippi, University, MS 38677-1848, USA
[2] United States Department of Agriculture, Agricultural Research Service, Natural Products Utilization Research Service, Thad Cochran National Center for Natural Products Research, University, MS 38677-8048, USA

Introduction

Scientific reports of fungi associated with marine algae date back to the late nineteenth century [1, 2]. Since then, many species of fungi have been identified from marine macroalgae [3]. As many as one-third of the described species of marine higher fungi and many species of lower fungi are associated with marine algae [4, 5]. Similarly, aquatic fungi are often found in association with freshwater algae and microscopic cyanobacteria [6]. While most reports are primarily taxonomic, the ecological nature of the interactions between these associations is diverse. Saprophytic fungi are common, and parasitic fungal species can seriously affect vast populations of macroalgae raised in commercial sea-farming enterprises [7].

It is often difficult to establish the nature of the algal-fungal relationship without careful observation of the ecological process of fungal attack on the alga. The overall health of the alga before fungal infestation must be considered. The intertidal brown marine algae *Pelvetia canaliculata* (L.) Dcne et Thur. and *Ascophyllum nodosum* (L.) Le Jol. are reported to be permanently and systemically infected by the ascomycete fungus *Mycosphaerella ascophylli* Cotton [8, 9]. While much of the work has focused on *A. nodosum*, a mutualistic interrelationship has been suggested to exist between these organisms [10, 11]. More recent studies have failed to find any significant benefit that the algae derive from the fungus and have questioned the implied mutualistic nature of this relationship [12].

In terrestrial systems, algae and fungi appear to exist in pathogenic, competitive, and symbiotic relationships, as observed in lichens. In much the same way, evidence suggests that freshwater and marine algae are intimately associated with fungi. It is our intention neither to exhaustively examine the interactions between algae and fungi nor to describe all aspects of algal defense strategies. However, this review does briefly examine the pathogenic nature of the relationship between these two distinctly different groups of organisms that somehow seem inexorably bound. Significant research efforts have probed the chemical arsenal produced by cyanobacteria and algae as new sources of biologically active natural products for use in both agriculture and medicine. Numerous reports exist concerning the production of antifungal compounds by cyanobacteria and algae [13–17]. However, most reports of antifungal activity by compounds from algae and cyanobacteria concern their evaluation for pharmaceutical activity. There are substantially fewer reports on the evaluation of these organisms for antifungal activity against economically important plant pathogenic fungi. Very little is known with regard to the chemical ecology of algae and fungi associations in marine and aquatic systems. In order to spur new research efforts and suggest possible avenues for future chemical ecology research, we have elected to specifically discuss the substances produced by cyanobacteria and algae with clearly documented antimicrobial activities against only those species of fungi that are recognized as pathogenic to either plants or algae. In addition, a discussion of antifungal algae natural products in agricultural pest control and the approaches that we are currently using to discover antifungal compounds from algae and cyanobacteria is presented.

Fungal growth and parasitism on algae

Ecological interactions between parasitic fungi and freshwater algae are well documented [6]. Marine algae are raised commercially throughout the world as both food and commercial sources of polysaccharides such as carrageenin, alginic acid, and agar. Several Oomycete species of fungi have been directly implicated in pathological infestations of maricultured red algae. *Petersenia pollagaster* (Petersen) Sparrow has been found to infect a cultured strain of *Chondrus crispus* Stackh. [18]. This polysaccharide-rich strain of *C. crispus*, termed "T4", is cultured in Nova Scotia as a major commercial source of κ-carrageenin.

Nori (*Porphyra yezoensis* Ueda, *Porphyra tenera* Kjellum) mariculture is a multi-billion-dollar food industry in Japan and throughout coastal Asia [19]. The Oomycete *Pythium porphyrae* Takahashi et Sasaki commonly infects *Porphyra* sp. growing in cultivation farms. This process is known as red rot disease or "aka-gusare" [20]. Red rot disease is reported to have nearly destroyed entire crops of *Porphyra* sp. in Japan and Korea within only a few weeks [7, 21, 22]. *Pythium porphyrae* has since been found to infect other species including *Porphyra miniata* (Lyngbye) C. Agardh [23]. Oomycete species have been observed to infect other red algae, brown algae, and green algae [24]. The Oomycete *Ectrogella perforans* Petersen acts as an endobiotic parasite of marine diatoms and is reported to cause epidemic infections of *Licmorphora* and other diatom species [25].

The large marine kelp *Laminaria saccharina* (L.) Lam. is subject to parasitic infection by the ascomycete *Phycomelaina laminariae* (Rostrup) Kohlm [26]. *Phycomelaina laminariae* infects the meristematic tissue of the algal stipe, forming black patches, and appears to increase algal susceptibility to invasion by otherwise non-pathogenic species of fungi. Deuteromycete fungi are known to infect several species of large marine algae. *Sargassum* sp. and *Cystoseira* sp. are parasitized by *Spaceloma cecidii* Kohlm. [27]. In addition, numerous species of fungi have recently been isolated and cultured from the surfaces of marine algae as part of modern microbial drug-discovery programs [28–33].

Disease cycle

Although considerable research has focused on the process of fungal invasion of terrestrial plant tissues [34], far less is known about the process of fungal infection in algae. An overview of the fungus disease cycle in plants is provided. However, considerable research is needed before the true nature of the algae infection process and the role(s) of antifungal chemical defenses in the control of fungal invasion of algae are clear. A plant or alga becomes diseased in most cases when it is attacked by a pathogen. In order for successful infection to occur, each of the three following components must be present: (1) the pathogen must be virulent, (2) environmental conditions must be present that favor infection, and (3) the host must be susceptible. In every infectious disease, whether in plants or animals, a series of distinct events occurs in succession that leads to disease development: (1) attachment of the pathogen to the host; (2) recognition between the host and pathogen; (3) penetration/invasion by the pathogen; (4) growth and reproduction of the pathogen within the host; and (5) subsequent survival of the pathogen in the absence of the host [34].

All plant pathogens, in their vegetative state, can initiate infection. Eight thousand fungal species are known to cause plant diseases, compared with only 100 fungal species that cause human mycosis. Fungal spores are most often the initial infectious propagule. Spore germination is often stimulated by nutrient diffusion from the plant surface, and the subsequent development of germ tube, appressorium, and penetration peg is influenced by the presence or absence of numerous plant chemicals.

Almost all plant pathogens are mobile and are transported to an immobile host. This situation is in contrast to human diseases, in which the host is mobile and most pathogens are immobile. Before a pathogen can colonize its host, it must first become attached to the host surface. Pathogen propagules have evolved sophisticated processes to locate, attach to, and recognize suitable hosts. Most pathogens have on their cellular surface mucilaginous substances, glycoproteins, and/or fibrillar materials that aid in adherence to the host surface [35]. Successful invasion, infection, and disease development occur when the pathogen successfully overcomes the physical and biochemical defenses of the host. In the process of invasion, numerous enzyme types (e.g., cutinases, pectinases, cellulases, hemicellulases, amylases, and lipases) can degrade the host cell into chemical fragments, which in turn may elicit the biochemical defenses of the host organism [36, 37]. Research suggests that a disease cycle, similar to that observed in fungal infection of plants, is responsible for the fungal infection process observed in marine algae [38]. The pathogenic nature of the marine fungus *Pythium porphyrae* that infects red algae (*Porphyra* spp.) has been the subject of considerable research [38]. *Pythium porphyrae* infects only a relatively small number of algae species. Studies have demonstrated that *Pythium porphyrae* will attach to and encyst on many species of red algae [38]. However, *P. porphyrae* was shown to infect *Porphyra* spp. and *Bangia atropurpurea* (Roth) only in laboratory culture. Research has implicated porphyran, a sulfated galactan polysaccharide, as a surface determinant on the thallus surface of the marine red alga *Porphyra* spp. that is recognized by zoospores of the pathogenic marine fungus *Pythium porphyrae* [38]. Under laboratory conditions, *P. porphyrae* zoospores attach, encyst, germinate, and form appressoria on thin films of porphyran in suspension. Films composed of agar or agarose media also induced *P. porphyrae* zoospore attachment and cyst formation. However, no appressoria were formed. Films composed of agarose media and either water-soluble extracts of *Porphyra yezoensis* or purified porphyran polysaccharide induced *Pythium porphyrae* appressoria formation [38].

Plants and algae defend themselves against pathogens by a combination of structural characteristics that act as physical barriers to inhibit the invasion and colonization of a host by the pathogen. Biochemical reactions take place in the host tissues to produce substances that either are directly toxic to the pathogen or indirectly create conditions that inhibit infection and further growth and development of the pathogen. It is becoming increasingly clear that plant resistance to pathogen attack depends more on the production of biochemical metabolites in cells before or shortly after infection than on preexisting physical or structural barriers [35].

Antibiotics, antineoplastics, herbicides, and insecticides often originate from plant and microbial chemical defense mechanisms [39]. Secondary metabolites, once considered unimportant products, are now thought to mediate chemical defense mechanisms by providing chemical barriers against animal and microbial predators [35, 39]. This chemical warfare between algae and their pathogens shows promise to provide new natural product leads for biomedical research and agricultural pest control.

Direct acting defense chemicals

Since the discovery of the vinca alkaloids in 1963, many of the major known antitubulin agents used in today's cancer chemotherapy arsenal are products of secondary metabolism. These "natural products" are probably defense chemicals that target and inhibit cell division in invading pathogens [39]. Other phytochemicals such as resveratrol [40], ellagic acid, beta-carotene, and vitamin E may possess anti-mutagenic and cancer-preventive activity [41, 42]. Therefore, it is reasonable to hypothesize that plants and algae produce chemicals that act in defense directly, by inhibiting pathogen proliferation, or indirectly, by disrupting chemical signal processes related to growth and development of pathogens or herbivores.

Indirect acting defense chemicals

Plant resistance to pathogens is considered to be systemically induced by some endogenous signal molecule produced at the infection site that is then translocated to other parts of the plant [43]. Search and identification of the putative signal is of great interest to many plant scientists because such molecules have possible uses as "natural product" disease-control agents. However, research indicates that there is not a single compound but rather a complex signal-transduction pathway in plants that can be mediated by a number of compounds and that appears to influence octadecanoid metabolism. In response to wounding or pathogen attack, fatty acids of the jasmonate cascade are formed from membrane-bound α-linolenic acid by lipoxygenase-mediated peroxidation [44]. Analogous to the prostaglandin cascade in mammals, α-linolenic acid is thought to participate in a lipid-based signaling system where jasmonates induce the synthesis of a family of wound-inducible defensive proteinase inhibitors [45] and low- and high-molecular-weight phytoalexins such as flavonoids, alkaloids, and terpenoids [46, 47].

Fatty acids are known to play an important role in signal-transduction pathways *via* the inositol phosphate mechanism in both plants and animals. In animals, several polyunsaturated fatty acids, such as linolenic acid, are precursors for hormones. Interruption of fatty acid metabolism produces a complex cascade of effects that are difficult to separate independently. In response to hormones, stress, infection, inflammation, and other stimuli, a specific phospholipase present in most mammalian cells acts upon membrane phospholipids, releasing arachidonate. Arachidonic acid is parent to a family of very potent biological signaling molecules that act as short-range messengers, affecting tissues near the cells that produce them. The role of various phytochemicals and their ability to disrupt arachidonic acid metabolism in mammalian systems by inhibiting cyclooxygenase-mediated pathways are of major pharmacological importance.

Eicosanoids, which include prostaglandins, prostacyclin, thromboxane A2, and leukotrienes, are a family of very potent autocoid signaling molecules that act as chemical messengers with a wide variety of biological activities in various tissues of vertebrate animals. It was not until the general structure of prostaglandins was determined, a 20-carbon unsaturated carboxylic acid with a cyclopentane ring, that the role of fatty acids was realized. Eicosanoids are formed *via* a cascade pathway in which the 20-carbon polyunsaturated fatty acid arachidonic acid is rapidly metabolized to oxygenated products by several enzyme systems including cyclooxygenases [48], lipoxygenases [49, 50], or cytochrome P450s [51] (Fig. 1). The eicosanoids maintain this 20-carbon scaffold, often with cyclopentane-ring, (prostaglandins), double cyclopentane-ring (prostacyclin), or oxirane-ring (thromboxanes) modifications. The first enzyme in the prostaglandin synthetic pathway is prostaglandin endoperoxide synthase, or fatty acid cyclooxygenase. This enzyme converts arachidonic acid to unstable prostaglandin endoperoxide intermediates. Aspirin, derived from salicylic acid in plants, irreversibly inactivates prostaglandin endoperoxide synthase by acetylating an essential serine residue on the enzyme, thus producing anti-inflammatory and anti-clotting actions [52].

Jasmonic acid (Fig. 2), a 12-carbon pentacyclic polyunsaturated fatty acid derived from α-linolenic acid, plays a role in plants similar to that of arachidonic acid metabolites [53] and has a structure similar to the prostaglandins. It is synthesized in plants from α-linolenic acid by an oxidative pathway analo-

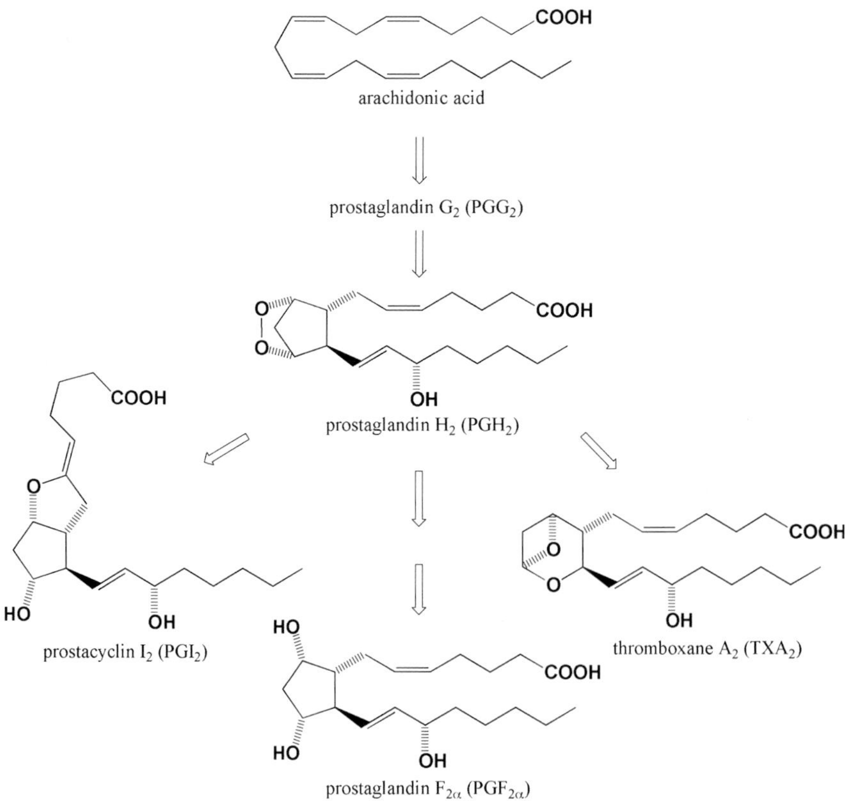

Figure 1. Cyclooxygenase-catalyzed metabolism of arachidonic acid to eicosanoids.

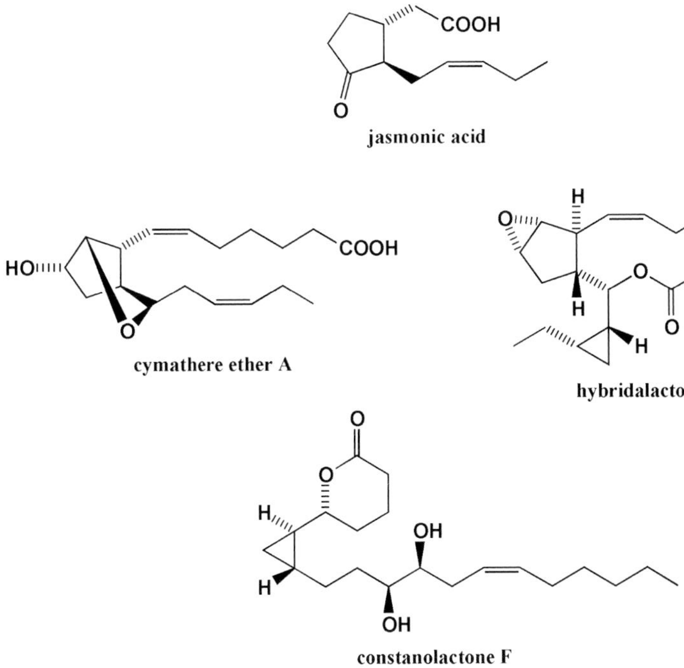

Figure 2. Jasmonic acid and several algal oxylipins.

gous to the hydroxyeicosanoids in animals. In animals, eicosanoid synthesis is triggered by release of arachidonic acid from membrane lipids into the cytoplasm, where it is converted into secondary messenger molecules. Conversion of α-linolenic acid to jasmonic acid through several steps is perhaps a mechanism analogous to that of arachidonate, which allows the plant to respond to wounding or pathogen attack [54]. Linolenic acid is released from precursor lipids by action of lipase and subsequently undergoes oxidation to jasmonic acid. Apparently, jasmonic acid and its octadecanoid precursors in the jasmonate cascade are an integral part of a general signal-transduction system that must be present between the elicitor-receptor complex and the gene-activation process responsible for induction of enzyme synthesis [45, 47, 54]. Closely related fatty acids that are not jasmonate precursors are ineffective at activating the signal transduction of wound-induced proteinase inhibitor genes [45]. Arachidonic acid, eicosapentaenoic acid, and other unsaturated fatty acids (e.g., linoleic acid, linolenic acid, and oleic acid) are also known elicitors for sesquiterpenoid phytoalexins and induce systemic resistance against *Phytophthora infestans* in the potato [55].

Increasing evidence suggests that plant cellular defenses may be analogous to the "natural" immune response of vertebrates and insects. The biosynthesis of phytoalexins after microbial attack appears to be similar to the phenomenon of immune reaction in vertebrates [56]. In addition to cell structural similarities, plant and mammalian defense responses share functional similarities. In mammals, natural immunity is characterized by the rapid induction of gene expression after microbial invasion. A characteristic feature of plant disease resistance is the rapid induction of a hypersensitive response in which a small area of cells containing the pathogen is killed. Other aspects of plant defense include an oxidative burst leading to the production of reactive oxygen intermediates (ROIs), expression of defense-related genes, alteration of membrane potentials, increase in lipoxygenase activity, modification of the cell wall, and production of antimicrobial compounds such as phytoalexins [57].

In the mammalian immune response, ROIs induce acute-phase response genes by activating the transcription factors NFκB and AP-1 [58, 59]. In plants, ROIs and salicylic acid regulate pathogen resistance through transcription of resistance gene-mediated defenses. Functional and structural similarities among evolutionarily divergent organisms suggest that the mammalian immune response and the plant pathogen defense pathways may be built from a common template [60]. Wedge and Camper [39] believe that similar biosynthetic processes involved in signaling pathogen invasion and stress in plants and animals may account for the physiological cross-activity of various fatty acid intermediates and other pharmacologically active phytochemicals.

Algal oxylipins

Algae, especially temperate marine macroalgae, produce a wide variety of eicosanoids and eicosanoid-like metabolites [61]. These chemically complex fatty acid metabolites are often highly oxidized, and many contain unusual carbocyclic ring systems. As in jasmonic acid, many of these metabolites are derived from fatty acid precursors other than the 20-carbon compound arachidonic acid [61, 62] and are, therefore, technically not eicosanoids. These structurally unique natural products are now commonly referred to as "oxylipin". Algae contain oxylipins that are structurally identical to the 20-carbon eicosanoids produced by animals (Fig. 2) [63]. Like plants, algae also produce low-molecular-weight oxylipin metabolites, such as jasmonic acid [64]. In addition, algae produce a wide variety of structurally novel oxylipins found nowhere else in nature [61–63, 65]. Algal oxylipins are thought to be formed through biosynthetic pathways that involve an initial lipoxygenase-mediated fatty acid peroxidation step, often followed by a series of relatively unprecedented epoxidation, carbocyclization, and lactonization reactions [66, 67]. Gerwick [68] has suggested that many marine oxylipin metabolites may be formed through the central intermediacy of a proposed epoxy allylic carbocation-type precursor. While the biogenetic origins of algal oxylipins are still the subject of ongoing research, several have been shown to be biosynthetic precursors to the reproductive pheromones used by many species of algae [69]. However, the actual biochemical functions of most algal oxylipins remain unknown [63, 65]. It is conceivable that these oxylipins serve a biochemical function similar to that of oxylipins produced by plants. Algal oxylipins may act as second messengers that regulate chemical communication pathways in response to environmental stress, microbial infections, or in defense against herbivory.

Antifungal compounds

Algae appear to be able to produce a wide array of antifungal natural products. Studies suggest that histidine treatment of *Porphyra* sp. infected with *Pythium porphyrae* induces an antifungal chemical defense by stimulating the algae to produce fatty acids and phenylacetic acid [70, 71].

Marine cyanobacteria produce a wide diversity of secondary metabolites. The chemistry of macrophytic algae is predominately of terpene or fatty acid origin, whereas most secondary metabolites isolated from marine cyanobacteria contain nitrogen and appear to derive from amino acid or mixed biosynthetic pathways [72, 73].

Since the early 1970s, agriculture has struggled with the evolution of pathogen resistance to antimicrobial disease-control agents. Increased necessity for repeated chemical applications, development of pesticide cross-resistance, and disease-resistance management strategies have overshadowed the use of agricultural chemicals. Scientists are currently attempting to control agricultural pests with fewer effective chemical controls. In addition, the desire for safer pesticides with less environmental toxicity is a major public concern. Particularly desirable is the discovery of totally novel, prototype pesticidal agents representing new chemical classes with different toxicities that operate by different modes of action and, consequently, lack cross-resistance with currently used chemicals. In this respect, evaluating natural products and extracts to identify potential new pesticides offers an approach to discover new chemical entities that have never been synthesized by chemists.

Although plants offer great potential for the discovery of new pesticides, other virtually untapped sources are eukaryotic algae and cyanobacteria (blue-green algae) that together comprise most of the earth's biomass [74]. Algae are prolific producers of secondary metabolites used for defense (e.g., toxins), competition (e.g., allelochemicals), and for other yet unknown purposes. Compounds produced by phytoplankton that permit them to compete for light and nutrients are of particular interest as potential algal- and weed-control agents because these compounds may inhibit photosynthesis as a mechanism of toxicity.

Several compounds isolated from freshwater species of cyanobacteria have been identified as having potent algicidal properties. The secondary metabolites fischerellin A [75] and fischerellin B (Fig. 3) [76] are produced by the cyanobacterium *Fischerella musciola* Thuret. Fischerellin A has been demonstrated to inhibit photosynthesis with the site of action at photosystem II (PSII) [77]. In addition, fischerellin A was found to be more toxic to cyanobacteria

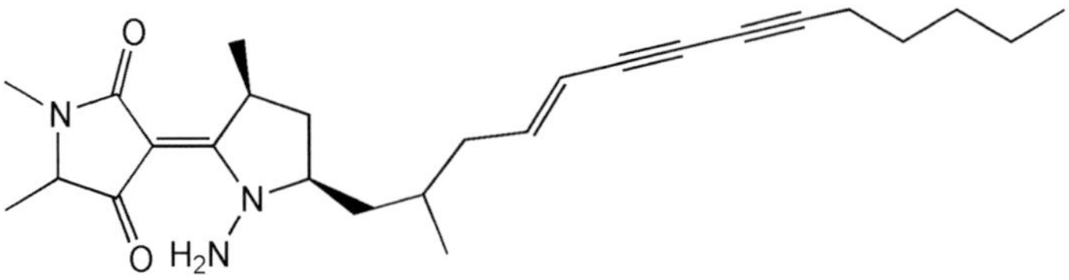

fischerellin A

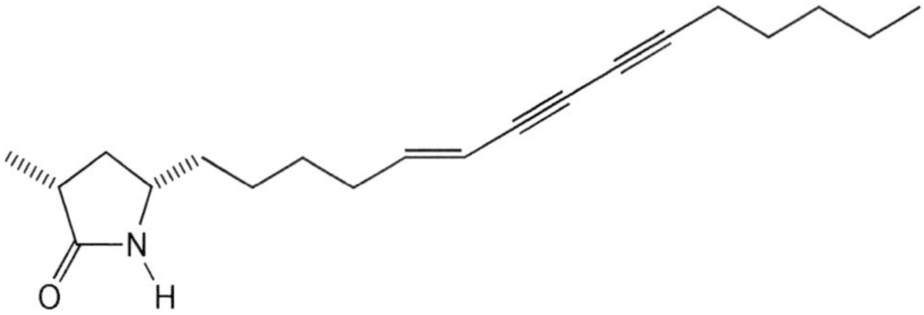

fischerellin B

Figure 3. Fischerellins from the cyanobacterium *Fischerella musciola*.

than to the species of green algae screened [75, 78] and, therefore, may be of potential use as a selective algicide in catfish aquaculture.

Fischerellin A has also been reported to have potential for use as a fungicide. Hagmann and Jüttner [75] found that fischerellin A had differential antifungal activity against several plant pathogenic fungi. Fischerellin A showed 100% inhibition of *Uromyces appendiculatus* (brown rust) at 250 $\mu g \cdot mL^{-1}$. At 1000 $\mu g \cdot mL^{-1}$, fischerellin A caused 100% inhibition of *Erysiphe graminis* (powdery mildew), 80% inhibition of *Phytophthora infestans* and *Pyricularia oryzae* (rice blast), and was much less active against *Monilinia fructigena* (brown rot) and *Pseudocercosporella herpotrichoides* (stem break).

Extracts from the cyanobacterium *Nostoc muscorum* Agardh have been found to inhibit the *in vitro* growth of the fungal plant pathogens *Sclerotinia sclerotiorum* (Cottony rot of vegetables and flowers) and *Rhizoctonia solani* (root and stem rots [15]). In addition, marine macroalgae extracts have been reported to reduce the incidence of gray mold incited by *Botrytis cinerea* on strawberry plants [79]. The active compounds in these extracts were not identified. So far, most of the cyanobacterial extracts reported to exhibit antifungal activity have not had significant antibacterial activity [16].

Other studies have actually identified the active compounds in antifungal cyanobacterial extracts and determined the structures of active compounds. For example, several cyclic peptides found in antifungal extracts from various species of cyanobacteria have been identified as the active components. Crude extracts from *Anabaena laxa* FK-1-2 contain laxaphycins (Fig. 4), a large group of cyclic undecapeptides and dodecapeptides, which act synergistically

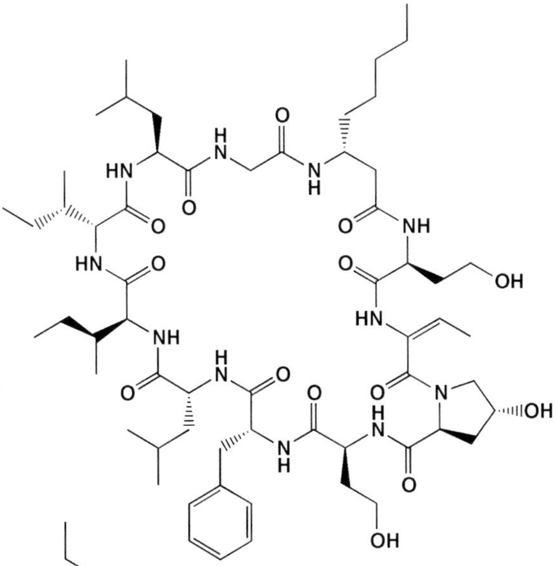

laxaphycin A

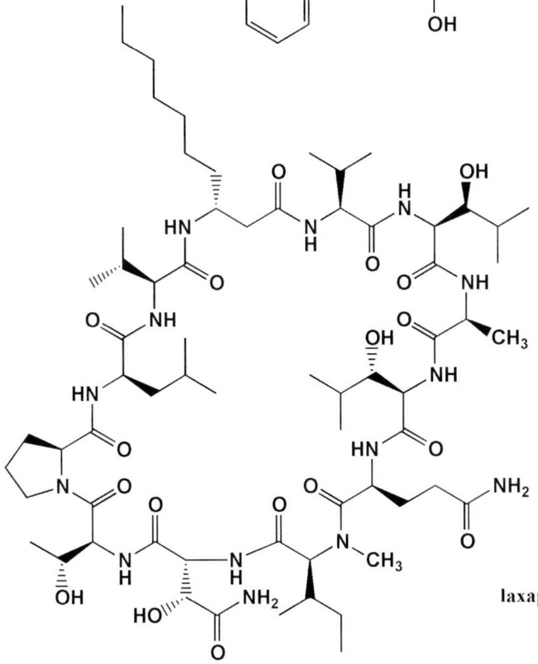

laxaphycin B

Figure 4. Laxaphycins from the cyanobacterium *Anabaena laxa*.

to inhibit fungal growth [16]. Calophycin (Fig. 5), a cyclic decapeptide, is the antifungal component of crude extracts from *Calothrix fusca* EU-10-1 [80]. The cyclic undecapeptide scytonemin A (Fig. 5), produced by a *Scytonema* sp. [81], and the cyclic undecapeptide schizotrin A (Fig. 6), produced by a

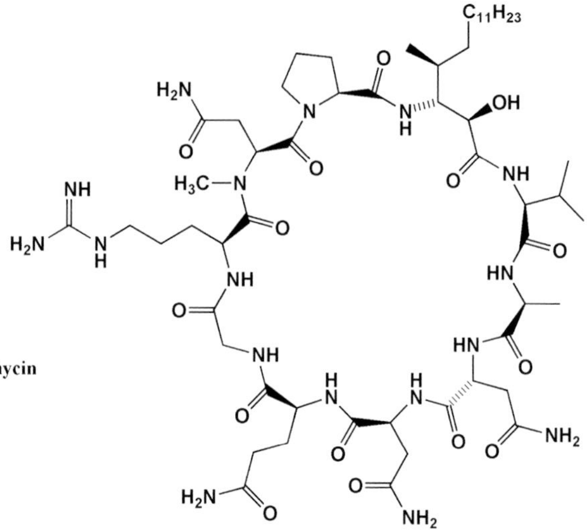

calophycin

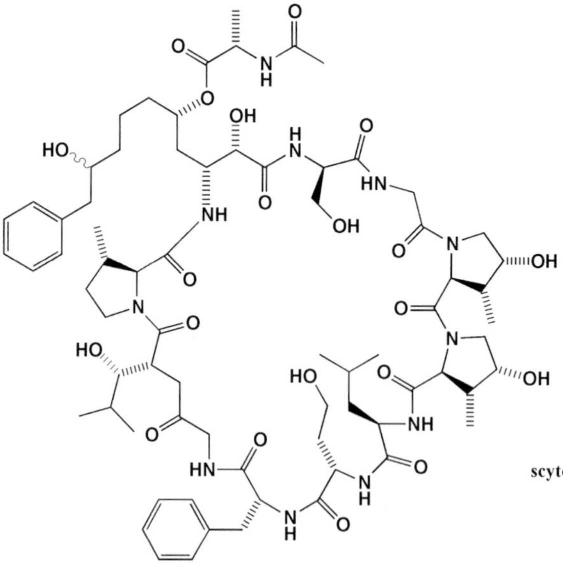

scytonemin A

Figure 5. Antifungal cyclic peptides from aquatic cyanobacteria.

Schizothrix sp., have moderate antifungal activity against *Sclerotium rolfsii* (Southern blight), *R. solani*, *Fusarium oxysporum* (vascular wilts, root and stem rots), and *Colletotrichum gloeosporioides* (anthracnose) [82]. Carter et al. [83] isolated and identified majusculamide C (Fig. 6), a cyclic depsipeptide from *Lyngbya majuscula*. Majusculamide C was shown to have activity against

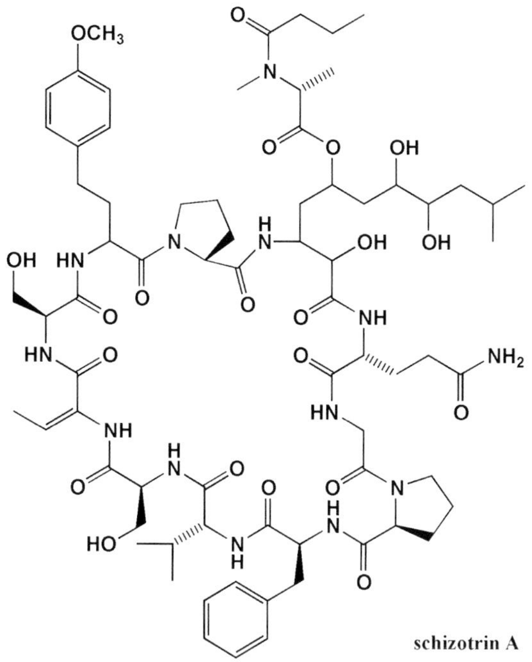

schizotrin A

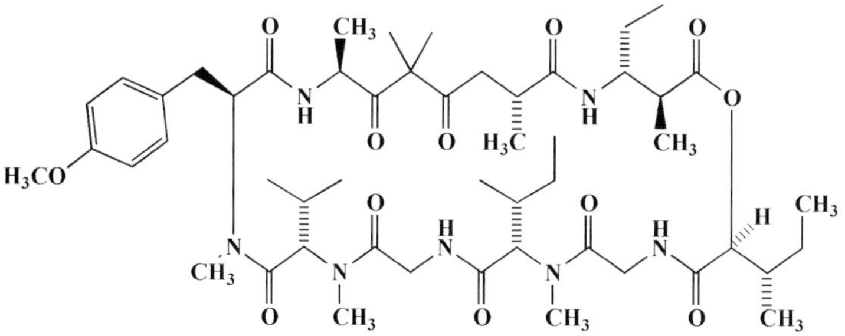

majusculamide C

Figure 6. Antifungal cyclic peptides from marine cyanobacteria.

two important Oomycetes, *Phytophthora infestans* (potato late blight) and *Plasmopara viticola* (grape downy mildew). The potent antifungal activity of majusculamide C appears to be due to its ability to depolymerize microfilaments, and it may have agricultural applications in the disease management of fungicide-resistant plant pathogens [16]. Tjipanazoles (Fig. 7), N-glycosides of indol-[2, 3-a]-carbozoles isolated from *Tolypothrix tjipanasensis*, exhibited antifungal activity in tests against unidentified phytopathogenic fungi [84].

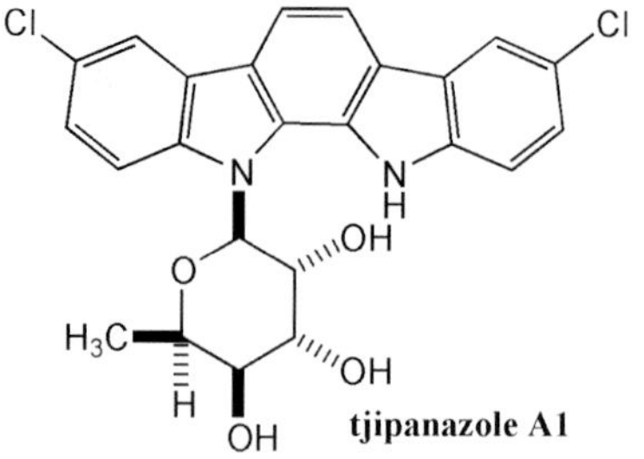

tjipanazole A1

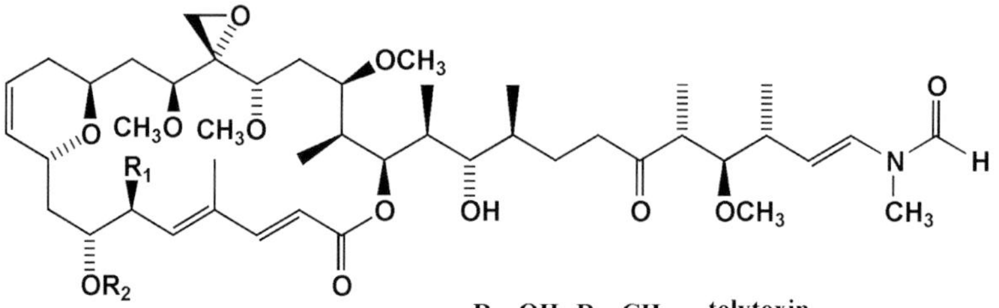

R_1=OH; R_2=CH_3	**tolytoxin**
R_1=H; R_2=H	**scytophycin B**

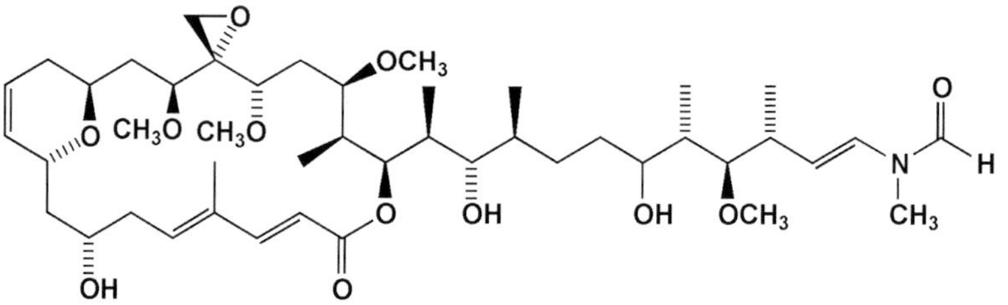

scytophycin A

Figure 7. Antifungal substances produced by *Tolypothrix* spp. and *Scytonema* spp.

Tolytoxin and scytophycins (Fig. 7) are a class of antifungal and highly cytotoxic compounds, most of which have been isolated from terrestrial cyanobacteria in the genera *Scytonema* and *Tolypothrix*. Scytophycins A and B were obtained and identified from *Scytonema pseudohofmanni* collected from heavily forested areas of Oahu, Hawaii [85]. Scytophycins A and B have chemical and pharmacological properties similar to those of tolytoxin, originally isolated from *Tolypothrix conglutinata* var. *colorata* [86]. Carmeli et al. [85] obtained tolytoxin and three scytophycins (6-hydroxyscytophycin B, 19-*O*-demethylscytophycin C, and 6-hydroxy-7-*O*-methylscytophycin E) from cultured strains of *S. burmanicum, S. mirabile*, and *S. ocellatum.* Jung et al. [87] isolated several scytophycins from *Cylindrospermum muscicola*, an unrelated genus that belongs to the family Nostocales.

The antifungal activity of tolytoxin and scytophycins toward specific phytopathogens is reported in several patent documents [88, 89]. Tolytoxin showed minimum inhibitory concentrations (MIC) in the nanomolar range against *Alternaria alternata* (leaf spots and blights), *Aspergillus oryzae* (seed decay), *Bipolaris incurvata* (leaf spot of grasses), *Colletotrichum coccodes* (anthracnose), *Phyllosticta capitalensis* (anamorph of *Guignardia bidwelli* causing black rot of grape), *S. rolfsii, Phytophthora nicotianae* (root rots), and *Thielaviopsis paradoxa* (vegetable root and lower stem rots). Less antifungal activity was shown by scytophycins B, C, and E to *S. rolfsii, P. nicotianae,* and *T. paradoxa* [89]. While tolytoxin and scytophycins have shown growth inhibition of fungal phytopathogens, their cytotoxicity towards nontarget organisms and their unknown mode of action may limit their use in agriculture. Information on the phytotoxicity of this group of antifungal agents would help clarify their potential for use as plant protectants.

Brown algae are known for their production of a large number of prenylated quinones, hydroquinones [13], and polycyclic terpenoids with ichthyotoxic and cytotoxic activity, but only a few species have been reported to produce compounds with antifungal activity towards phytopathogens. *Stypopodium zonale* produces at least three pentacyclic sesquiterpenoids with biological activity, but only taondiol (Fig. 8) had antifungal activity against *Eurotium repens* (telemorph of *Aspergillus* causing seed decays and producer of aflatoxins) in an agar diffusion assay [90]. Two fungitoxic hydroquinones, zonarol and isozonarol (Fig. 8), that were isolated from *Dictopteris undulata* had moderate activity against *Phytophthora cinnamomi* (root rots of numerous crops), *R. solani, S. sclerotiorum,* and *S. rolfsii* [91–93]. Fucosterol (Fig. 8) obtained from *Jolyna laminarioides* showed activity against *Curvularia lunata* (leaf spots of grasses) at 250 $\mu L \cdot mL^{-1}$ [94]. Methoxybifurcarenone (Fig. 8), a meroditerpenoid from *Cystoseira tamariscifolia*, possesses antifungal activity toward *Botrytis cinerea, F. oxysporum* sp. *lycopersici* (tomato wilt), and *Verticillium alboatrum* (verticilium wilt) [95].

A green algal terpenoid isolated from several members of the Udoteaceae showed mycelial growth inhibition of *Leptosphaeria* sp. (anamorph of *Phoma* sp. causing cabbage rot) and *Alternaria* sp. (many leaf spots and blights) [96].

Marine red algae of the genus *Laurencia* are rich in sesquiterpenes and C_{15}-acetogenins and produce secondary metabolites characterized by a high degree of halogenation [13]. Allolaurinterol (Fig. 9) obtained from *Laurenica obtusa* showed inhibitory activity against *Ustilago violacea* (Ustomycetes), *Mycotypha microspora* (Zygomycetes), *Eurotium repens* (Ascomycetes), and *Fusarium oxysporum* (Deuteromycetes) at 50 μg, but no significant activity at 5 μg [97]. Also, a terpenoid from *L. obtusa* was shown to have inhibitory activity toward *P. infestans* [98]. From the temperate marine red alga *L. rigidia*, König and Wright [99] isolated two sesquiterpenes, elatol and deschloroelatol (Fig. 9), that showed antifungal activity. Growth inhibition of *M. microspora* and *E. repens* occurred at 25 μg of deschloroelatol and 135 μg of elatol. Elatol (135 μg) was also inhibitory to *U. violacea* (smuts) and *F. oxysporum*; however, activity at

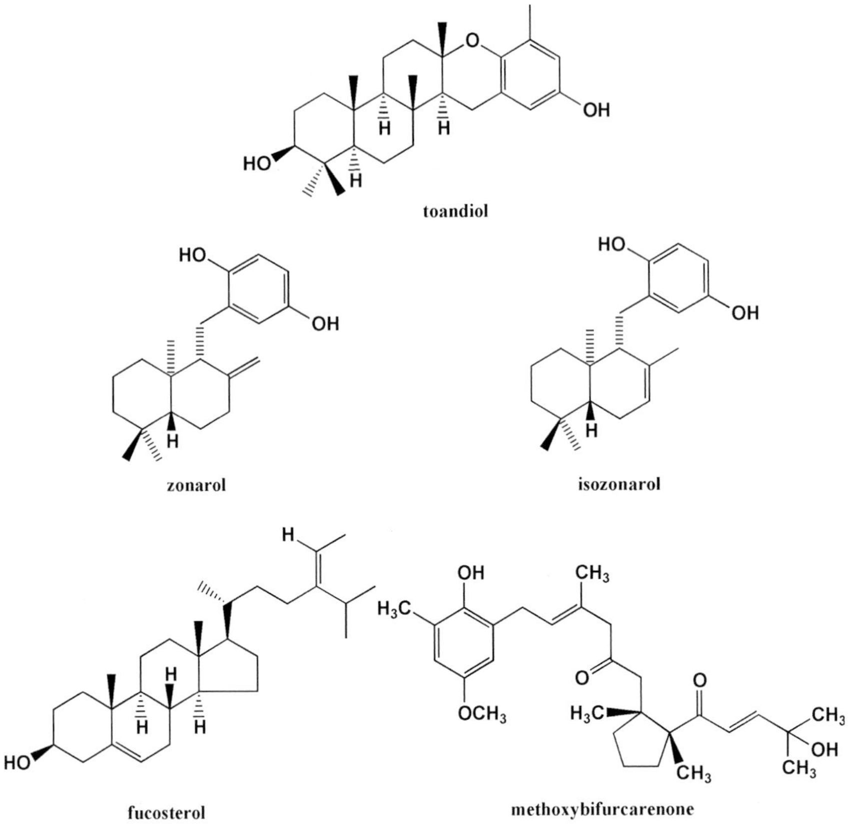

Figure 8. Antifungal substances produced by brown algae.

this level is not sufficient for agricultural applications. A polyhalogenated monoterpene artifact isolated from a tropical marine species, *Plocamium hamatum,* showed weak antifungal activity against *U. violacea, F. oxysporum,* and *E. repens* [100]. Extracts from several marine algae (*Centroceras clavulatum, Gastroclonium clavatum, L. obtusa,* and *Lophocladia lallemandii*) from eastern Sicily have shown 70% to 80% inhibition of *Phoma tracheiphila* pycnidiospores (cause of citrus malsecco disease) [101].

Biological evaluation of antifungal activity

Many experimental approaches have been used to screen compounds and extracts from plants and algae in order to discover new antifungal compounds. However, our focus in this chapter is the bioassay methods that we are currently using in our laboratories to examine the antifungal potential of com-

allolaurinterol

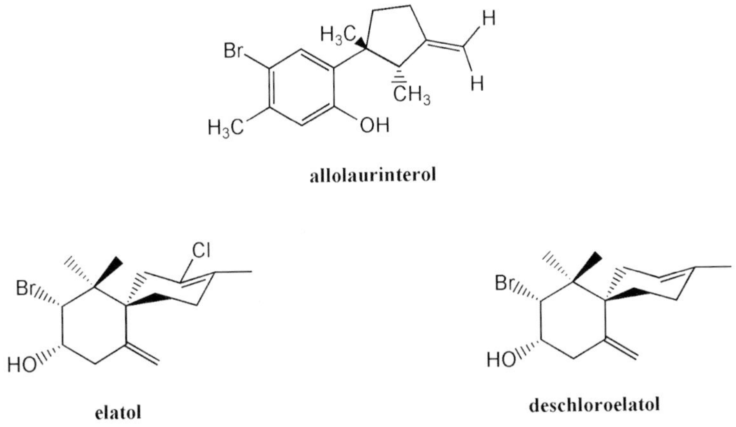

elatol

deschloroelatol

Figure 9. Antifungal substances produced by red algae.

pounds produced by marine and freshwater cyanobacteria and algae. A unique-ly synergistic collaboration involving the discovery of new fungicides from marine natural products is in place between the United States Department of Agriculture, Agricultural Research Service, Natural Products Utilization Research Unit and scientists from the University of Mississippi Department of Pharmacognosy and the Thad Cochran National Center for Natural Products Research. The collaborative team has evaluated numerous extracts. Evaluation of the "hit-rates" for antifungal activity between terrestrial plant extracts and extracts from aquatic organisms reveals that aquatic organisms appear to have evolved more antifungal chemical defenses, most likely to survive in complex microbe-rich ecosystems.

As part of a program to discover natural product-based fungicides with low environmental and mammalian toxicity, two sensitive detection systems have been developed for the evaluation of antifungal agents (Fig. 10). A two-dimen-sional direct bioautography assay is used to detect antifungals effective against *Colletotrichum* spp. [102]. A standardized high-throughput 96-well microbias-say is used to identify antifungal agents effective against *Colletotrichum* spp., *Botrytis* sp., and *Fusarium* sp. [103, 104].

Failure to control *Colletotrichum*, *Botrytis*, and *Fusarium* species can result in serious economic losses to agriculture worldwide. Anthracnose diseases, especially those caused by *Colletotrichum* sp. (*Gloeosporium* sp.), are com-mon and destructive to numerous crops worldwide [34], especially in subtrop-ical and tropical regions [105]. *Colletotrichum* spp. cause significant diseases of cereals, grasses, legumes, vegetables, perennial crops, and tree fruits.

Gray molds or gray mold rots (*Botrytis cinerea*) can infest almost all types of fresh fruit, vegetables, bulbs in the field and in storage, and greenhouse-grown ornamentals [37]. Flower and fruit blights caused by *Botrytis cinerea* are considered to be the most serious diseases of fruit and plant production in

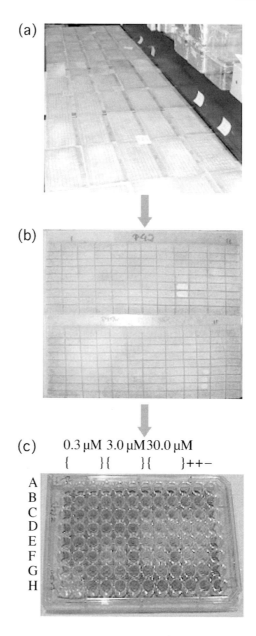

(a)

(b)

(c)   0.3 µM 3.0 µM 30.0 µM
        {   }{   }{   }++−
    A
    B
    C
    D
    E
    F
    G
    H

Figure 10. Combination of bioautography and microtiter assays provides an efficient protocol for discovery and evaluation of natural products. (a) and (b) Bioautography allows the investigator to visually follow active antifungal agents through the extraction process, can be used to identify the number of active compounds in an extract, and has been adapted to a high-throughput screening format. (c) The 96-well microtiter assay provides the investigator with high-throughput capability, the ability to perform 3- or 6-point dose-response studies, compare internal fungicide standards and control strains of fungi, and perform complex mode-of-action studies in a chemically defined and buffered medium.

many areas of the world, especially in strawberries and greenhouse-grown tomatoes [106, 107]. Few fungicides are now available for effective control of *Botrytis* disease, and new approaches for the control of *B. cinerea* are necessary as the effectiveness and availability of commercial fungicides decrease.

Several *Fusarium* spp., especially *F. oxysporum* and *F. solani*, are becoming serious fungal pathogens, often uncontrolled by commercial fungicides. *Fusarium* spp. cause vascular wilts of vegetables, flowers, herbaceous perennial ornamentals, field crops (e.g., cotton, tobacco), plantation crops (e.g., banana, plantain, coffee, and sugarcane) and a few shade trees [34]. Therefore, the discovery of new natural product-based disease-control agents for *Colletotrichum, Botrytis,* and *Fusarium* species are major research objectives of our collaboration team.

These studies have already revealed that a relatively high proportion of aquatic algae and marine cyanobacteria produce natural products with activity against several *Colletotrichum* spp. Significant "hit-rates" of antifungal activity from aquatic algae and cyanobacteria against *Colletotrichum* spp. have been higher than those observed for a diverse sampling of extracts from tropical rainforest plant species (12% to 30% compared to only 1.5% for terrestrial plants).

A number of bioautography techniques are used to detect antifungal agents in crude extracts, including a two-dimensional thin-layer chromatography (2D-TLC) direct bioautography method, where the TLC plates are developed once with a polar solvent, turned 90°, and then developed a second time with a nonpolar solvent system [102]. This method takes advantage of the resolving power of 2D-TLC to separate chemically diverse mixtures found in extracts of algae and cyanobacteria. Two-dimensional TLC bioautography is well suited to resolving extracts containing lipophilic natural products that are difficult to separate by single-elution TLC. Antifungal metabolites can be readily located on the plates by visually observing clear zones where the active compounds inhibit fungal growth [108, 109]. This method eliminates the need for the development of large numbers of plates in multiple solvent systems, reduces the amount of waste solvents for disposal, and substantially reduces the time required to identifying active compounds.

Results from these studies suggest that chemically different populations of cyanobacteria can easily be distinguished by their characteristic "2D-TLC fingerprint" and antifungal zone patterns. Thus, this system represents a very useful technique for the identification and selection of chemically unique chemotypes from microbial isolates [110]. Chromatographic properties (relative polarity, UV absorbance, chemical reactivity, etc.) associated with each active metabolite provide valuable information that allows for rapid dereplication of known or nuisance compounds. When strains of different phytopathogenic fungi with dissimilar fungicide resistance profiles are inoculated onto replicate bioautography plates prepared from any given extract containing active metabolites, it is possible to visually observe distinct differences in sensitivity of each fungal pathogen to single metabolites. These differences in pathogen sensitivity (fungicide resistance) can be observed by direct comparison of inhibition zone dimensions produced by active metabolites and control standards against each pathogenic strain tested. Chemical profiles can provide valuable information for rapidly selecting specific antifungal metabolites with unique activity against fungicide-resistant pathogens and identifying new compounds with novel mechanisms of action.

Discovery, evaluation, and development of natural product fungicides are totally dependent upon the availability of miniaturized antifungal bioassays. A reference method (M27-A from the National Committee for Clinical Laboratory Standards) for broth-dilution antifungal susceptibility testing of yeast was adapted for evaluation of antifungal compounds against sporulating filamentous fungi [103]. This 96-well microtiter assay is used to determine and compare the sensitivity of fungal plant pathogens with natural and synthetic compounds with known fungicidal standards [104]. Each fungal species is challenged in a dose-response format using test compounds with final treatment concentrations of 0.3, 3.0, and 30.0 µM. Fungal growth is evaluated by measuring absorbance of each well at 620 nm at 0, 24, 48, and 72 hours. Mean absorbance values and standard errors are used to evaluate fungal growth. Differences in growth or sensitivity to particular concentrations of a compound indicate fungistatic or fungicidal effects. This microbioassay allows several compounds to be tested in a dose-response format. Standardization of inoculum allows for meaningful comparison of growth inhibition between different fungal pathogens and test compounds. The 96-well microbioassay is accurate and sensitive; as little as 0.1 µM of test compound permits discrimination between germination and mycelial growth inhibitors. The assay utilizes a chemically defined liquid medium with a zwitterion buffer that limits chemical interaction with test compounds. This new standardized method provides high-throughput capability with the capacity to study a chemical compound in detail, to perform mode-of-action studies, and to determine fungicide-resistance profiles for specific fungal pathogens.

Conclusions

Bioassay-directed screening of compounds and extracts is the initial step in the discovery process for new agrochemicals and pesticides. Efficacy testing ultimately helps to determine the potential usefulness of compounds as pest-control agents. Marine and freshwater algae and cyanobacteria provide a relatively unexplored group of organisms in nature that produce toxic metabolites with the potential for use as new pest-control agents. There is great potential for finding new bioactive agents from algae and cyanobacteria for use as pesticides, as indicated by the discoveries made so far. To maximize the detection of natural products, high-throughput bioassay techniques must target significant agricultural pests, include relevant standards (commercial pesticides), and adhere to sound statistical principles. Ultimately, the discovery of natural products with low mammalian and environmental toxicity for the control of pests and diseases in agriculture will provide for a safe and dependable food supply.

The use of natural product-based agrochemicals provides an opportunity for better natural resource management by reducing dependence on synthetic chemicals. The results of this type of research will help to ensure food safety and enhance the international sustainability of agriculture.

Most of the research discussed places considerable emphasis on the discovery of new antifungal substances produced by cyanobacteria and algae as potential means of agricultural pest control (summarized in Table 1). Fungi

Table 1. Summary of *in vitro* antifungal activity for selected antimicrobial natural products towards agricultural pathogens

	Chemical class	Susceptible pathogen(s)
Cyanobacteria		
Fischerellin A	Enediyne-pyrrolidinedione	*Uromyces appendiculatus, Erysiphe graminis, Phytophthora infestans, Pyricularia oryzae*
Scytonemin A	Cyclic undecapeptide	*Sclerotium rolfsii, Rhizoctonia solani, Fusarium oxysporum, Colletotrichum gloeosporioides*
Majusculamide C	Cyclic depsi-octapeptide	*Phytophthora infestans, Plasmopara viticola*
Tolytoxin	Polyketide macrolide	*Alternaria alternata, Aspergillus oryzae, Bipolaris incurvata, Colletotrichum coccodes, Phyllosticta capitalensis, Guignardia bidwelli, Sclerotium rolfsii, Phytophthora nicotianae, Thielaviopsis paradoxa*
Brown Algae		
Taondiol	Aromatic diterpene	*Eurotium repens*
Zonarol/isozonarol	Sesquiterpene Hydroquinone	*Phytophthora cinnamomi, Sclerotium rolfsii, Sclerotium sclerotiorum, Rhizoctonia solani*
Fucosterol	Sterol	*Curvularia lunata*
Methoxybifurcarenone	Meroditerpenoid	*Botrytis cinerea, Fusarium oxysporum* sp. *mycopersici, Verticilium alboatrum*
Red Algae		
Allolaurinterol	Halogenated sesquiterpene	*Ustilago violacea, Mycotypha microspora, Eurotium repens, Fusarium oxysporum*
Elatol/deschloroelatol	Halogenated sesquiterpene	*Ustilago violacea, Mycotypha microspora, Eurotium repens, Fusarium oxysporum*

and algae are closely associated in aquatic and marine systems, and cyanobacteria and algae produce an extraordinarily high number of antifungal chemical defenses. It seems reasonable to suggest that these compounds play a role in the regulation of the chemical-ecological balance of competition and parasitism between these organisms. However, future research efforts must directly examine the effects of algal defenses on the particular species of fungi that are routinely found in competition with these specific cyanobacteria and algae. The ecological relevance of field-based competition experiments and pathogenic infection studies should be emphasized as a necessary extension of laboratory-based model systems. New research into the chemical ecology of algae and fungi will likely be spurred as a direct consequence of an increased appreciation for the enormous economic impact of fungal disease on major algal food crops and polysaccharide production in Asia and elsewhere.

Acknowledgements
This work was supported in part by the USDA, Agricultural Research Service Specific Cooperative Agreement # 58-6408-2-0009.

References

1 Church AH (1893) A marine fungus. *Ann Bot* 7: 399–400
2 Jones HL (1898) A new species of Pyrenomycetes parasitic on an alga. *Bull Oberlin Coll Lab* 9: 3–4
3 Jones EBG (1976) Lignicolous and algicolous fungi. *In*: EBG Jones (ed): *Recent advances in aquatic mycology*. Elek Science, London, 1–49
4 Johnson TW, Sparrow FK Jr, (1961) *Fungi in oceans and estuaries*. J. Cramer, Weinheim
5 Kohlmeyer J (1974) Higher fungi as parasites and symbionts of algae. *Veröff Inst Meeresf Bremerhaven* Supp 5: 339–356
6 Masters MJ (1976) Freshwater phycomycetes on algae. *In*: EBG Jones (ed): *Recent advances in aquatic mycology*. Elek Science, London, 489–512
7 Fujita Y (1990) Diseases of cultivated *Porphyra* in Japan. *In*: I Akatsuka (ed): *Introduction to applied phycology*. SPB Academic Publishing, The Hague, 177–190
8 Drew EA (1969) Uptake and metabolism of exogenously supplied sugars by brown algae. *New Phytol* 68: 35–43
9 Kremer BP (1973) Untersuchungen zur Physiologie von Volemitol in der Marinen Braunalge *Pelvetia canaliculata*. *Mar Biol* 22: 31–35
10 Webber FC (1967) Observations on the structure, life history and biology of *Mycosphaerella ascophylli*. *Trans Br Mycol Soc* 50: 583–601
11 Kohlmeyer J, Kohlmeyer E (1972) Is *Mycosphaerella ascophylli* lichenized? *Bot Mar* 15: 109–112
12 Kingham DL, Evans LV (1986) The *Pelvitia–Mycosphaerella* interrelationship. *In*: ST Moss (ed): *The biology of marine fungi*. Cambridge University Press, Cambridge, 177–187
13 Bhakuni DS (1998) Some aspects of bioactive marine natural products. *J Indian Chem Soc* 75: 191–205
14 Dixon GK (1996) Biologically active compounds from algae. *Crit Rep Appl Chem* 35: 114–216
15 Kulik MM (1995) The potential for using cyanobacteria (blue-green algae) and algae in the biological control of plant pathogenic bacteria and fungi. *Eur J Plant Path* 101: 585–599
16 Moore RE (1996) Cyclic peptides and depsipeptides from cyanobacteria: a review. *J Ind Microbiol* 16: 134–143
17 Siddhanta AK, Shanmugam M (1999) Metabolites of tropical marine algae of the family Codiaceae (Chlorophyta): chemistry and bioactivity. *J Indian Chem Soc* 76: 323–334

18 Molina FI (1986) *Petersenia pollagaster* (oomycetes): An invasive fungal pathogen of *Chondrus crispus* (Rhodophyceae). *In*: ST Moss (ed): *The biology of marine fungi*. Cambridge University Press, Cambridge, 165–175

19 Ohno M, Largo DB (1998) The seaweed resources of Japan. *In*: AT Critchey, M Ohno (eds): *Seaweed resources of the world*. International Cooperation Agency, Japan, 1–15

20 Kazama FY, Fuller MS (1979) Colonization of *Porphyra perforata* thallus discs by *Pythium marinum*, a marine facultative parasite. *Mycologia* 69: 246–254

21 Fujita Y, Zenitani B (1970) Studies on the pathogenic *Pythium* of laver red rot disease in Ariake sea farm. I. general mycological characteristic. *Nippon Sisan Gakkaishi* 42: 1183–1188

22 Park CS, Sakaguchi K, Kakinuma M, Amano H (2000) Comparison of the morphological and physiological features of the red rot disease fungus *Pythium* sp. isolated from *Porphyra yezoensis* from Korea and Japan. *Fish Sci* 66: 261–269

23 Fuller MS, Lewis B, Cook P (1966) Occurrence of *Pythium* sp. on the marine alga *Porphyra*. *Mycologia* 58: 313–318

24 Porter D (1986) Mycosis of marine organisms: an overview of pathogenic fungi. *In*: ST Moss (ed): *The biology of marine fungi*. Cambridge University Press, Cambridge, 141–153

25 Sparrow FK (1960) *Aquatic phycomycetes*, 2nd ed. University of Michigan Press, Ann Arbor

26 Schatz S (1983) The developmental morphology and life history of *Phycomelaina laminariae*. *Mycologia* 75: 762–772

27 Kohlmeyer J, Kohlmeyer E (1979) *Marine mycology: the higher fungi*. Academic Press, New York

28 Takahashi C, Numata A, Yamada T, Minoura K, Enomoto S, Konishi K, Nakai M, Matuda C, Nomoto K (1996) Penostatins, novel cytotoxic metabolites from a *Penicillium* species separated from a green alga. *Tetrahedron Lett* 37: 655–658

29 Chen C, Imamura N, Nishijima M, Adachi K, Sakai M, Sano H (1996) Halymecins, new antimicroalgal substances produced by fungi isolated from marine algae. *J Antibiot* 49: 998–1005

30 Iamoto C, Minoura K, Hagashi S, Nomoto K, Numata A (1998) Penostatins F-I, novel cytotoxic metabolites from a *Penicillium* species separated from an *Enteromorpha* marine alga. *J Chem Soc Perkin Trans* 13: 449–456

31 Amagata T, Minoura K, Numata A (1998) Cytotoxic metabolites produced by a fungal strain from a *Sargassum* alga. *J Antibiot* 51: 432–434

32 Jenkins KM, Toske SG, Jensen PR, Fenical W (1998) Solanapyrones E-G, antialgal metabolites produced by a marine fungus. *Phytochemistry* 49: 2299–2304

33 Iwamoto C, Minoura K, Oka T, Ohta T, Hagishita S, Numata A (1999) Absolute stereostructures of novel cytotoxic metabolites, penostatins A-E, from a *Penicillium* species separated from an *Enteromorpha* alga. *Tetrahedon* 55: 14353–14368

34 Agrios GN (1997) Parasitism and disease development. *In*: GN Agrios: *Plant pathology*. Academic Press, San Diego, 43–62

35 Agrios GN (1997) How pathogens attack plants. *In*: GN Agrios: *Plant pathology*. Academic Press, San Diego, 63–82

36 Manners JG (1993) Entry into the Host Plant. *In*: JG Manners: *Principles of plant pathology*, 2nd ed. Cambridge University Press, Cambridge, 66–78

37 Walton JD (1997) Biochemical plant pathology. *In*: PM Dey, JB Harborne (eds): *Plant biochemistry*. Academic Press, San Diego, 487–502

38 Uppalapati SR, Fujita Y (2000) Carbohydrate regulation of attachment, encystment, and appressorium formation by *Pythium porphyrae* (Oomycota) zoospores on *Porphyra yezoensis* (Rhodophyta). *J Phycol* 36: 359–366

39 Wedge DE, Camper ND (2000) Connections between agrochemicals and pharmaceuticals. *In*: HG Cutler, SJ Cutler (eds): *Biologically active natural products: agrochemicals and pharmaceuticals*. ACS Symposium Series, American Chemical Society, Washington, 1–15

40 Jang M, Cai L, Udeani GO, Slowing KV, Thomas CF, Beecher CWW, Fong HHS, Farnsworth NR, Kinghorn AD, Mehta RG et al (1997) Cancer chemopreventative activity of resveratrol, a natural product of grape. *Science* 275: 218–220

41 Balandrin MF, Kinghorn AD, Farnsworth NR (1993) Plant-dervied natural products in drug discovery and development: an overview. *In*: AD Kinghorn, MF Balandrin (eds): *Human medicinal agents from plants*. ACS Symposium Series No. 534, American Chemical Society, Washington, 2–12

42 Pezzuto JM (1993) Cancer chemopreventative agents: from plant materials to clinical intervention trials. *In*: AD Kinghorn, MF Balandrin (eds): *Human medicinal agents from plants*. ACS

Symposium Series No. 534, American Chemical Society, Washington, 205–215

43 Oku H (1994) *Plant pathogensis and disease control.* CRC Press, Cleveland

44 Vick BA, Zimmerman DC (1984) Biosynthesis of jasmonic acid by several plant species. *Plant Physiol* 75: 458–461

45 Farmer EE, Ryan CA (1992) Octadecanoid precursors of jasmonic acid activate the synthesis of wound-inducible proteinase inhibitors. *Plant Cell* 4: 129–134

46 Gundlach H, Muller MJ, Kutchan TM, Zenk MH (1992) Jasmonic acid is a signal transducer in elicitor-induced plant cell cultures. *Proc Natl Acad Sci USA* 89: 2389–2393

47 Mueller MJ, Brodschelm W, Spannagl E, Zenk MH (1993) Signaling in the elicitation process is mediated through the octadecanoid pathway leading to jasmonic acid. *Proc Natl Acad Sci USA* 90: 7490–7494

48 Smith WL (1992) Prostanoid biosynthesis and mechanisms of action. *Am J Physiol* 268: F181–F191

49 Samuelsson B (1983) Leukotrienes: mediators of immediate hypersensitivity reactions and inflammation. *Science* 20: 568–575

50 Needleman P, Turk J, Jakschik BA, Morrison AR, Lefkowith JB (1986) Arachidonic acid metabolism. *Annu Rev Biochem* 55: 69–102

51 Fitzpatrick FA, Murphy RC (1989) Cytochrome P-450 metabolism of arachidonic acid: formation and biological actions of epoxygenase-derived eicosanoids. *Pharmacol Rev* 40: 229–241

52 Insel PA (1996) Analgesic-antipyretic and antiinflammatory agents and drugs employed in the treatment of gout. *In*: JG Hardmand, LE Limbird, PB Molinoff, RW Ruddon, AG Gilman (eds): *The pharmacological basis of therapeutics.* McGraw-Hill, New York, 617–657

53 Davies PJ (1995) *Plant hormones: physiology, biochemistry and molecular biology.* Kluwer Academic Publishers, Boston

54 Farmer E, Ryan C (1990) Interplant communication: airborne methyl jasmonate induces synthesis of proteinase inhibitors in plant leaves. *Proc Natl Acad Sci USA* 87: 7713–7716

55 Cohen Y, Gisi U, Mosinger E (1991) Systemic resistance of potato plants against *Phytophthora infestans* induced by unsaturated fatty acids. *Physiol Mol Plant Path* 38: 255–263

56 Oku H (1997) Resistance of plants against pathogens. *In*: *Plant Pathogenesis and Disease Control.* Lewis Publishers, Boca Raton, 64

57 Dixon RA, Harrison MJ, Lamb CJ (1994) Early events in the activation of plant defense responses. *Annu Rev Phytopathol* 32: 479–501

58 Schreck R, Baeuerle PA (1991) A role for oxygen radicals as second messengers. *Trends Cell Biol* 1: 39–42

59 Kopp E, Ghosh S (1994) Inhibition of NF-kappa B by sodium salicylate and aspirin. *Science* 265: 956–959

60 Baker B, Zambryski P, Staskawicz B, Dinesh-Kumar SP (1997) Signaling in plant-microbe interactions. *Science* 276: 726–733

61 Gerwick WH (1999) Eicosanoids in nonmammals. *In*: U Sankawa, DHR Barton, K Nakanishi, O Meth-Cohn (eds): *Comprehensive natural products chemistry, Volume 1.* Elsevier, New York, 207–254

62 Gerwick WH, Bernart MW, Jiang ZD, Moghaddam MF, Nagle DG, Proteau PJ, Wise ML, Hamberg M (1993) Patterns of oxylipin metabolism in marine organisms. *Dev Indust Microbiol* 1: 369–378

63 Gerwick WH, Proteau PJ, Nagle DG, Wise ML, Jiang ZD, Bernart MW, Hamberg M (1993) Biologically active oxylipins from seaweeds. *Hydrobiologia* 260/261: 653–665

64 Fujii S, Yamamoto R, Miyamoto K, Ueda J (1997) Occurrence of jasmonic acid in *Dunaliella* (Dunaliellales, Chlorophyta). *Phycol Res* 45: 223–226

65 Gerwick WH (1993) Carbocyclic oxylipins of marine origin. *Chem Rev* 93: 1807–1823

66 Nagle DG, Gerwick WH (1994) Structure and stereochemistry of constanolactones A-G, lactonized cyclopropyl oxylipins from the marine red alga *Constantinea simplex. J Org Chem* 59: 7227–7237

67 Jiang Z-D, Gerwick WH (1997) Novel oxylipins from the temperate red alga *Polyneura latissima*: evidence for an arachidonate 9(*S*)-lipoxygenase. *Lipids* 32: 231–235

68 Gerwick WH (1993) Structure and biosynthesis of marine algal oxylipins. *Biochim Biophys Acta* 1211: 243–255

69 Boland W (1995) The chemistry of gamate attraction: Chemical structures, biosynthesis, and (a)biotic degradation of algal pheromones. *Proc Natl Acad Sci USA* 92: 37–43

70 Noda H, Amano H, Kano S, Ohta F (1983) Unsaturated fatty acids as antifungal substances effective against "akagusare" disease of the laver *Porphyra* sp. *Bull Jpn Soc Sci Fish* 49: 1583–1586

71 Noda H, Amano H, Kano S, Ohta F (1983) Isolation and identification of antifungal substances from "akagusare"-infected laver *Porphyra* sp. *Bull Jpn Soc Sci Fish* 49: 1587–1590

72 Ireland CM, Roll DM, Molinski TF, Mckee TC, Zabriskie TM, Swersey JC (1998) Uniqueness of the marine chemical environment: categories of marine natural products from invertebrates. *In:* DG Fautin (ed): *Biomedical importance of marine organisms.* California Academy of Sciences, San Francisco, 41–57

73 Gerwick WH, Roberts MA, Proteau PJ, Chen J-L (1994) Screening cultured marine microalgae for anticancer-type activity. *J Appl Phycol* 6: 143–149

74 Cannell RJP (1993) Algae as a source of biologically active products. *Pestic Sci* 39: 147–153

75 Hagmann L, Jüttner F (1996) Fischerellin A, a novel photosystem-II-inhibiting allelochemical of the cyanobacterium *Fischerella muscicola* with antifungal and herbicidal activity. *Tetrahedron Lett* 37: 6539–6542

76 Papke U, Gross EM, Francke W (1997) Isolation, identification, and determination of the absolute configuration of fischerellin B: a new algicide from the freshwater cyanobacterium *Fischerella muscicola* (Thuret). *Tetrahedron Lett* 38: 379–382

77 Srivastava A, Jüttner F, Strasser RJ (1998) Action of the allelochemical, fischerellin A, on photosystem II. *Biochim Biophys Acta* 1364: 326–336

78 Gross EM, Wolk CP, Jüttner F (1991) Fischerellin, a new allelochemical from the freshwater cyanobacterium *Fischerella muscicola*. *J Phycol* 27: 686–692

79 Stephenson WM (1966) The effect of hydrolysed seaweed on certain plant pests and diseases. *In:* EC Young, JL McLachlan (eds): *Proceedings 5th international seaweed symposium,* Pergamon Press, Oxford, 405–415

80 Moon SJ, Chen L, Moore RE, Patterson GML (1992) Calophycin, a fungicidal cyclic decapeptide from the terrestrial blue-green alga *Calothrix fusca. J Org Chem* 57: 1097–1103

81 Helms GL, Moore RE, Niemczura WP, Patterson GML (1988) Scytonemin A, a novel calcium antagonist from a blue-green alga. *J Org Chem* 53: 1298–1307

82 Pergament I, Carmeli S (1994) Schizotrin A; a novel antimicrobial cyclic peptide from a cyanobacterium. *Tetrahedron Lett* 35: 8473–8476

83 Carter DC, Moore RE, Mynderse JS, Niemczura WP, Todd JS (1984) Structure of majusculamide C, a cyclic depsipeptide from *Lyngbya majuscula. J Org Chem* 49: 236–241

84 Patterson GML, Larsen LK, Moore RE (1994) Bioactive natural products from blue-green algae. *J Appl Phycol* 6: 151–157

85 Carmeli S, Moore RE, Patterson GML (1990) Tolytoxin and new scytophycins from three species of *Scytonema. J Nat Prod* 53: 1533–1542

86 Moore RE, Patterson GML, Mynderse JS, Barchi J (1986) Toxins from cyanophytes belonging to the Scytonemataceae. *Pure Appl Chem* 58: 263–271

87 Jung JH, Moore RE, Patterson GML (1991) Scytophycins from a blue-green alga belonging to the Nostocaceae. *Phytochemistry* 30: 3615–3616

88 Moore RE, Furusawa E, Norton TR, Patterson GML, Mynderse JS (1989) Scytophycins. U.S. Patent No. 4,863,955

89 Patterson GML, Moore RE, Carmeli S, Smith CD, Kimura LH (1995) Scytophycin compounds, compositions and methods for their production and use. U.S. Patent No. 5,493,933

90 Wessels M, König GM, Wright AD (1999) A new tyrosine kinase inhibitor from the marine brown alga *Stypopodium zonale. J Nat Prod* 62: 927–930

91 Fenical W, Sims JJ (1973) Zonarol and isozonarol, fungitoxic hydroquinones from the brown seaweed *Dictyopteris zonarioides. J Org Chem* 38: 2383–2386

92 Welch SC, Rao ASCP (1977) Stereoselective total synthesis of the fungitoxic hydroquinones (±) zonarol and (±) isozonarol. *J Org Chem* 43: 1957–1961

93 Welch SC, Rao ASCP (1977) Stereoselective total synthesis of (±) zonarol and (±) isozonarol. *Tetrahedron Lett* 6: 505–508

94 Atta-ur-Rahman Choudhary MI, Shabbir MM, Ghani U, Shameel M (1997) A succinylanthranilic acid ester and other bioactive constituents of *Jolyna laminarioides. Phytochemistry* 46: 1215–1218

95 Bennamara A, Abourriche A, Berrada M, Chaarrouf M, Chaib N, Boudouma M, Garneau FX (1999) Methoxybifurcarenone: an antifungal and antibacterial meroditerpenoid from the brown alga *Cystoseira tamariscifolia. Phytochemistry* 52: 37–40

96 Fenical W, Paul VJ (1984) Antimicrobial and cytotoxic terpenoids from tropical green algae of the family Udoteaceae. *Hydrobiologia* 116/117: 135–140

97 König GM, Wright AD (1997) Sesquiterpene content of the antibacterial dichloromethane extract of the marine red alga *Laurencia obtusa. Planta Med* 63: 186–187

98 Fenical W (1983) Investigation of benthic marine algae as a resource for new pharmaceuticals and agricultural chemicals. *In*: CK Tseng (ed): *Proceedings of the joint China-US Phycology Symposium,* Qindao, Science Press, Beijing, 497–521

99 König GM, Wright AD (1997) *Laurencia rigida*: chemical investigations of its antifouling dichloromethane extract. *J Nat Prod* 60: 967–970

100 König GM, Wright AD, Linden A (1999) *Plocamium hamatum* and its monoterpenes: chemical and biological investigations of the tropical marine red alga. *Phytochemistry* 52: 1047–1053

101 Caccamese S, Azzolina R, Furnari G, Cormaci M, Grasso S (1981) Antimicrobial and antiviral activities of some marine algae from Eastern Sicily. *Bot Mar* 24: 365–367

102 Wedge DE, Nagle DG (2000) A new 2D-TLC bioautography method for the discovery of novel antifungal agents to control plant pathogens. *J Nat Prod* 63: 1050–1054

103 Wedge DE, Kuhajek JM (1998) A microbioassay for fungicide discovery. *SAAS Bull Biochem Biotech* 11: 1–7

104 Wedge DE, Galindo JCG, Marcias FA (2000) Fungicidal activity of natural and synthetic sesquiterpene lactone analogs. *Phytochemisty* 53: 747–757

105 Bailey JA, Jeger MJ (1992) *Colletotrichum: biology, pathology and control.* CAB International, Wallingford

106 Maas JL (1998) *Compendium of strawberry diseases.* APS Press, St. Paul

107 Maas JL, Palm ME (1997) Occurrence of anthracnose irregular leafspot, caused by *Colletotrichum acutatum*, on strawberry in Maryland. *Adv Strawberry Res* 16: 68–70

108 Tellez MR, Dayan FE, Schrader KK, Wedge DE, Duke SO (2000) Composition and some biological activities of the essential oil of *Callicarpa americana* (L.). *J Agric Food Chem* 48: 3008–3012

109 Vincent A, Dayan FE, Maas JL, Wedge DE (1999) Detection and isolation of antifungal compounds in strawberry inhibitory to *Colletotrichum fragariae. Adv Strawberry Res* 18: 28–36

110 Nagle DG, Paul VJ (1999) Production of secondary metabolites by filamentous tropical marine cyanobacteria: ecological functions of the compounds. *J Phycol* 35 (Suppl. 607): 1412–1421

Chemical Ecology of Plants: Allelopathy in Aquatic and Terrestrial Ecosystems
ed. by Inderjit and Azim U. Mallik
© 2002 Birkhäuser Verlag/Switzerland

The chemistry and chemical ecology of biologically active cyanobacterial metabolites

Dale G. Nagle[1] and Inderjit[2]

[1] Department of Pharmacognosy, School of Pharmacy, University of Mississippi, University, MS 38677-1848, USA
[2] Department of Botany, University of Delhi, Delhi 110007, India

Introduction

In recent years, two major developments have spurred a rapid increase in cyanobacteria chemical and ecological research. Traditionally, the interest in isolating compounds from cyanobacteria has focused on identifying their antifungal, antiviral, antibacterial, antimitotic, antihelminthic, anticoagulating, hemagglutinating, and toxic metabolites [1–13]. Over the past decade, cyanobacteria have become a major source of important new classes of pharmacologically active natural products with potential therapeutic applications in the treatment of cancer [14–16], and blooms of marine cyanobacteria are becoming an increasingly familiar occurrence around tropical reef systems [17, 18]. These growing areas of intense research are discussed in light of bioactive cyanobacterial natural products chemistry with respect to recent evidence of chemical defenses in bloom phenomena.

Cytotoxins and pharmacologically active compounds

Bioactive organic compounds isolated from members of the phylum Cyanobacteria have been grouped and discussed under the general term "cyanotoxins" [3, 5]. Cyanotoxins include cytotoxins and biotoxins, depending upon the assay system used to test for them. Carmichael [5], based on the bioassay to screen activity, grouped secondary algal metabolites into (1) cytotoxins and (2) biotoxins.

Cytotoxins

Cytotoxins exhibit activity against algae, fungi, bacteria, and mammalian cells, but have generally not been shown to be extremely lethal to animals.

Cytotoxins are generally assayed with cultured mammalian cells, mainly tumor cells, while biotoxins are assayed with small animals such as mice and aquatic invertebrates. Two phenolic compounds, 2,3-dibromo-4,5-dihydroxy-benzyl alcohol (lanosol) and 3,5-dibromo-*p*-hydroxybenzyl alcohol, have been reported from axenic cultures of *Calothrix brevissima* [19]. Lunularic acid, a natural dihydrostilbene, was isolated from *Anabaena* spp. [20]. However, identification of these compounds was only presumptive, as they did not involve unequivocal identification, e.g., through gas chromatography/mass spectrometry (GC/MS). Hapalindole A, an indole alkaloid, and its oxidation product, anhydrohaploxindole, were isolated from *Hapalosiphon fontinalis* [21, 22]. Later, two minor alkaloids, dichlorofontonamide, anhydrohapaloxindoles B and M, and hapalonamide G, H, and V were isolated from this alga [23]. Cytotoxins include compounds such as acutiphycin, indolcarbazoles, mirabilene isonitriles, paracyclophanes, scytophycins, tantazoles, tolytoxin, toyocamycin, and tubercidin [24]. Cytotoxins were found to be produced by members of the Nostocales and Stigonematales families.

Two macrolides, acutiphycin and 20,21-didehydroacutiphycin, were isolated from the freshwater cyanobacterium *Oscillatoria acutissima* [25]. Cyanobacterin, a chlorinated secondary metabolite possessing γ-ylidene and γ-butyrolactone moieties was reported from the freshwater cyanobacterium *Scytonema hofmanni* [26–28]. Schwartz et al. [29] isolated two unusual cyclopropane-containing hapalindolinones from cultured, cyanobacteria *Fischerella* (ATCC 53558). From the marine alga *Rivularia firma*, rivularin D_3 was reported [30].

Many metabolites produced by marine cyanobacteria are toxic or pharmacologically active [31–33]. Members of the Oscillatoriaceae family, especially the filamentous cyanobacterium *Lyngbya majuscula* Gomont, have proven to be rich sources of biologically active natural products. The *L. majuscula* lipid metabolite malyngolide has been shown to have antimicrobial activity [34]. Tanikolide, a lipid structurally related to malyngolide, has recently been isolated from a Madagascan collection of *L. majuscula* and has been shown to be ichthyotoxic to goldfish (*Carassius auratus*), molluscicidal to *Biomphalaria glambrata*, and antifungal to *Candida albicans* [35].

Lyngbyatoxin A, a cyclic depsipeptide produced by *L. majuscula*, is a potent phorbol-ester-type tumor promoter and has been linked to at least one severe human poisoning [36–38]. Debromoaplysiatoxin is cytotoxic to P-388 lymphocytic leukemic cells [39] and is a potent chemical irritant [40, 41]. Consumption of the red alga *Gracilaria coronopifolia* J. Agardh (Gracilariaceae) has sporadically resulted in human poisonings in Hawaii [42, 43]. It is now believed that the source of toxins responsible for these toxic episodes was likely due to filaments of *L. majuscula* simply becoming entangled in the branches of *G. coronopifolia* [43, 44]. Debromoaplysiatoxin, aplysiatoxin, and other metabolites, produced by *L. majuscula,* have been found in toxic collections of *G. coronopifolia*. The lipopeptides malyngamide H [45] and I [46], kalkitoxin [47], and the depsipeptides antillatoxin [48] and antillatoxin B [49] are ichthyotoxic to goldfish (*Carassius auratus*). Kalkitoxin and the antillatoxins appear to act as potent neurotoxins that function through N-methyl-D-aspartate (NMDA) receptor-mediated mechanisms. Malyngamides J, K, and L [50], hermitamides A and B [51], and the γ-pyrone-containing compound kalkipyrone [52] are toxic to both brine shrimp (*Artemia salina*) and goldfish. Brine shrimp assays are used to find natural products that are toxic to fast growing cells and thereby serve as a biological model to detect cytostatic and cytotoxic agents with potential in the treatment of cancer [53, 54]. Recently, the isomalyngamides A and B (structural isomers of malyngamides A and B) have been isolated from Hawaiian collections of *L. majuscula* and shown to be toxic to crayfish (*Procambarus clarkii*) [55].

The cytotoxic depsipeptides lyngbyabellin A and B have been isolated from collections of *L. majuscula* obtained from both Guam and Florida [56–58]. Lyngbyabellin B has also been shown to be toxic to brine shrimp and antifungal to *C. albicans*. Curacin A is a new antimitotic agent with potent brine shrimp toxicity [59]. Curacin A binds to the colchicine binding site on tubulin and potently inhibits the growth of tumor cells ($IC_{50} = 38$ nM in the MCF-7 breast tumor cell line) [60, 61]. Two cyclopropyl-containing fatty acid metabolites, grenadadiene and grenadamide, have recently been isolated from Grenada collections of *L. majuscula* [62]. Grenadadiene is cytotoxic to tumor cells, and grenadamide is toxic to brine shrimp and has modest cannabinoid receptor-binding activity. The cytotoxic cyclic depsipeptides apratoxin A [63], pitipeptolide A, and pitipeptolide B [64] have recently been isolated from collections of *L. majuscula* obtained in Guam. Apratoxin A showed remarkably potent *in vitro* cytotoxicity against the human cell lines at nM levels. However, apratoxin proved highly toxic *in vivo*, and its antitumor activity was not significant at sublethal doses [63]. Jimémez and Scheuer isolated a structurally novel acyclic lipopeptide called dragonomide and a series of small polychlorinated peptides (pseudodysidenin, dysidenamide, and nordysidenin) from a red-colored sample of *L. majuscula* collected from Panama [65]. Pseudodysidenin, dysidenamide, and nordysidenin are closely related to the sponge metabolites dysidenin and isodysidenin, previously isolated from South Pacific collections of the tropical marine sponge *Dysidea herbacea*. Pseudodysidenin and dragonamide were cytotoxic but showed no specific tumor selectivity [65]. Barbamide, an unusual polychlorinated metabolite isolated from a Curaçao collection of *L. majuscula,* is potently molluscicidal to the snail *Biomphalaria glambrata* [66]. The unusual lipopeptides carmabin A and B were also isolated from the same collection of *L. majuscula* [67]. Microcolins A and B are cytotoxic and immunosuppressive at nanomolar concentrations [68]. A subsequent study identified apramides A–G from a sample of *L. majuscula* obtained from Guam [69]. The apramides are lipopeptides that are structurally related to the microcolins. The lipopeptides tumonic acids A–C and methyl tumonate A and B were isolated from the same mixed assemblage of *L. majuscula* and *Schizothrix calcicola* collected from Guam that yielded the potent antitumor agent lyngbyastatin 1 [70].

Recent studies have found that a small indanone isolated from a Micronesian collection of *L. majuscula* inhibits hypoxia-induced transcriptional activation of the vascular endothelial growth factor (VEGF) gene promoter in Hep3B human liver tumor cells, *in vitro* [71]. VEGF is a critical angiogenic factor produced by tumor cells to promote new blood vessel formation and facilitate tumor growth.

Majusculamide C, a depsipeptide isolated from *L. majuscula,* is both cytotoxic and has a broad spectrum of antifungal activity against pathogenic plant fungi [72, 73]. The isolation of majusculamide C from marine cyanobacteria provided the first indication that the structurally related dolastatins, first isolated from sea hares, may have been of dietary origin. Dolastatin 10 and dolastatin analogs (symplostatin 1) have recently been isolated from the marine cyanobacteria *Symploca* sp. [16, 74]. The cyclic peptides dolastatin 3, homodolastatin 3, kororamide, lyngbyastatin 2, and norlyngbyastatin 2 have been isolated from collections of *Lyngbya majuscula* obtained in Micronesia [75, 76]. Lyngbyastatin 2 and norlyngbyastatin 2 were shown to be cytotoxic analogs of dolastatin G in biological assays used to discover new anticancer agents [75]. Lyngbyastatin 2 exhibited potent cytotoxicity against human nasopharyngeal carcinoma cells (IC_{50} = 20 ng/mL in KB cells) and human colon adenocarcinoma cells (IC_{50} = 14 ng/mL in LoVo cells). In addition, dolastatin 3 has shown activity against HIV-integrase [76]. Dolastatin 12 and lyngbyastatin 1 have been isolated from mixed cyanobacterial assemblages obtained in Micronesia [77], and the somamides A and B, cyclic depsipeptides structurally related to dolastatin 13, have been identified from mixed assemblages of *L. majuscula* and *Schizothrix* sp. collected in the Fijian Islands [78]. However, the dolastatin 13 analog symplostatin 2 was isolated from *Symploca hydnoides* collections from Guam [79]. Dolastatins and dolastatin analogs are potent cytotoxins, three of which are currently in clinical evaluation as potential new anticancer drugs [80]. Dolastatins were originally isolated as extremely minor constituents from large collections of the marine opisthobranch mollusk *Dollabela auricularia* Lightfoot [81]. The discovery of dolastatins in cyanobacteria suggests that their occurrence in sea hares may be the result of dietary consumption. Therefore, the production of these anticancer agents by cyanobacteria represents a potentially valuable new marine source of these antitumor agents.

Additional species of marine cyanobacteria also produce unusual secondary metabolites. Hormothamnin A, a cyclic peptide isolated from *Hormothamnion enteromorphoides* Grunow (Nostocaceae), is both antibiotic and cytotoxic [82]. The nakienones A–C and nakitriol are cytotoxic cyclic C_{11} metabolites that have been isolated from a *Synechocystis* sp. found overgrowing coral (*Porites* sp.) in Okinawa [83].

Cryptophycins, cyclic depsipeptides first isolated from the freshwater cyanobacterium *Nostoc* sp., are potent antimicrotubule agents that have been shown to inhibit the growth of tumors in animal models [15, 84]. Cryptophycins competitively inhibit the binding of dolastatins to tubulin and are up to a thousand times more potent than other antitumor agents such as paclitaxel and vinblastine [85]. Cryptophycin 1 has even been shown to be active against Taxol-resistant breast tumors in mice [86].

Biotoxins

Biotoxins, on the other hand, are found in surface supplies of freshwater and have been shown to be produced by species of *Anabaena, Microcystis, Nostoc, Nodularia, Plektonema, Oscillatoria*, and *Aphanizomenon* [5, 87]. Two groups of biotoxins, neurotoxins (alkaloid) and hepatotoxins (cyclic peptides), have frequently been detected in cyanobacteria blooms. Neurotoxins such as ana-toxin-A (secondary amine alkaloid) were isolated from *Anabaena flos-aquae* strain NRC-44-1; anatoxin-A(s) (N-hydroxyguanidine methyl-phosphate ester), from *A. flos-aquae* strain NRC-527-17; and aphanatoxin (saxitoxin and neosaxitoxin), from *Aphanizomenon flos-aquae* strain NH-5 [2, 87]. Similarly, there are hundreds of reports of cyclic peptide hepatotoxins called micro-cystins from *Anabaena, Microcystis, Nostoc,* and *Oscillatoria* or nodularins from *Nodularia spumigena* [2, 87, 88]. Watanabe et al. detected microcystins in *Microcystis aeruginosa* and found that microcystin concentrations varied with season and site [89]. The amount of microcystin was 230 to 950 $\mu g \cdot g$ dry wt^{-1} at one sampling station and 160 to 746 $\mu g \cdot g$ dry wt^{-1} at the other sampling station. Further, the highest amount of microcystins (819 to 952 $\mu g \cdot g$ dry wt^{-1}) at sampling station 1 was observed from September through early October, while at sampling station 2 the amount was highest from July to early September [89]. Azevedo et al. isolated microcystins from a Brazilian cyanobacterium, *Microcystis aeruginosa* [90]. Biotoxins also have been found in marine cyanobacteria, e.g., lyngbyatoxin A (indole alkaloid), debro-moaplysiatoxin and an aplysiatoxin (phenolic bislactone) from *Lyngbya majuscula*, debromoaplysiatoxin (phenolic bislactone) from *Schizothrix calci-cola*, and oscillatoxin A (phenolic bislactone) from *Oscillatoria nigroviridis* [2, 91]. Effects of antibiotics from *Oscillatoria late-virens* were studied on the growth and physiological parameters of *Microcystis aeruginosa* and higher plants [92, 93].

Cyanobacterin, a secondary metabolite from the freshwater cyanobacterium *Scytonema hofmanni*, interfered with electron transport associated with photo-system II [26]. Similarly, fischerellin from *Fischerella muscicola* inhibited photosynthesis in Chlorophyceae and Cyanophyceae, and its mode of action is on photosystem II [94].

Kohlhase and Pohl reported 24-ethylcholest-5,22-dien-3β-ol, 24-ethyl-cholest-5-en-3β-ol, and 24-ethylcholestan-3β-ol in the sterol of *Anabaena, Nodularia*, and *Nostoc* [95]. Ikawa et al. reported that lipids of cyanobacteri-um *Aphanizomenon flos-aquae* inhibited the growth of green alga *Chlorella pyrenodosa* [96]. They identified the sterol as poriferasterol.

Chemical defense

Dense cyanobacterial blooms often wash ashore and accumulate in mass. Yet little is known of the interactions between algal grazers and the occurrence of marine cyanobacterial blooms. Benthic bloom-forming marine cyanobacteria are rich producers of secondary metabolites [14, 31, 32]. Most secondary metabolites isolated from marine cyanobacteria contain nitrogen and appear to derive from amino acid or mixed biosynthetic pathways, in contrast to the large numbers of terpenes and fatty acid-derived compounds found in macrophytic algae [59, 97]. However, mat-forming cyanobacteria typically have been grouped with turf algae in ecological studies [98].

Until only recently, little experimental evidence has existed to establish the ecological significance of most cyanobacterial metabolites [99]. Compounds produced by marine cyanobacteria have been shown to deter feeding by several species of herbivorous fishes. The *Lyngbya majuscula* lipid malyngolide [36] has been shown to act as a feeding deterrent to juvenile rabbitfishes (*Siganus argenteus* Quoy and Gaimard and *Siganus spinus* Linnaeus) and the parrotfish (*Scarus schlegeli* Bleeker) [100]. The lipopeptides malyngamide A and malyngamide B and majusculamides deter feeding by herbivorous fishes [100–102]. Moreover, a mixed cyanobacteria bloom composed of *Schizothrix calcicola* (Ag.) Gomont with sparsely distributed strands of *Lyngbya majuscula* occurred simultaneously with a massive die-off of juvenile rabbitfishes in Guam [103]. These events appear to be linked to a broadly acting, feeding-deterrent secondary metabolite produced by the cyanobacteria bloom. This cyanobacteria assemblage produces ypaoamide, a previously unknown lipopeptide, which deters feeding by herbivorous reef, fishes, and sea urchins [104]. Another mixed assemblage of *L. majuscula* and *Schizothrix* sp. from Yanuca Island in Fiji was found to produce the lipophilic depsipeptides yanucamides A and B, which are toxic to brine shrimp [105].

Initial studies examined how cyanobacterial secondary metabolites interact with macrograzers (fishes, sea urchins) and mesograzers (sea hares, amphipods, crabs) in tropical reef habitats [100, 104, 106–108]. More recently, Nagle and Paul combined fish feeding-deterrence assays with chemical separation methods such as two-dimensional thin layer chromatography (2D-TLC) to evaluate the ability of chemically distinct cyanobacterial "chemotypes" to deter grazing by generalist species of reef fishes [109]. Some cyanobacterial "chemotypes" produce a single major secondary metabolite [100], while other populations can produce an assortment of secondary metabolites in different concentrations and in a number of possible combinations [45, 48, 62, 66, 100, 109]. Purified secondary metabolites isolated from *Lyngbya majuscula* [100], *Hormothamnion enteromorphoides* [107], and mixed cyanobacteria assemblages [104] have been examined in similar experiments. Malyngolide, malyngamides, ypaoamide, laxaphycins (hormothamnins), and other purified cyanobacteria metabolites strongly deter grazing by *S. schlegeli*.

Cyanobacteria assemblages, even those composed of identical species combinations, produce remarkably different secondary metabolites [81, 103, 110]. Ypaoamide, the broadly acting feeding-deterrent compound isolated from a bloom of *L. majuscula* mixed with *S. calcicola* [108], has not been found in any other bloom (2D-TLC analysis of extracts). Nagle and Paul evaluated chemically distinct crude extracts of a variety of different bloom-forming cyanobacteria from Micronesia for their ability to deter grazing of herbivorous fish [109]. Cyanobacteria extracts were incorporated into artificial diets and tested in laboratory feeding-preference assays to generalist grazing reef fish. Remarkably, the extracts of every species of cyanobacteria examined effectively deterred herbivore grazing at ecologically relevant concentrations. Therefore, it appears likely that the production of deterrent metabolites is vital to cyanobacterial survival in herbivore-rich coral reef environments.

Herbivore specialization

The opisthobranch sea hare *Stylocheilus longicauda* is a reef grazer that specializes on *L. majuscula* and sequesters cyanobacteria metabolites, which may function in defense against predators [101, 102, 111]. It has been suggested that *Stylocheilus* dietary selection might be regulated by the concentration of specific chemical cues produced by *L. majuscula*. Preliminary studies were unable to show a connection between cyanobacteria chemistry and feeding specialization of *S. longicauda* on *L. majuscula* [106]. These experiments tested *L. majuscula* compounds at relatively high concentrations (2% of dry mass). In more recent studies, the cyanobacteria compounds microcolin B, ypaoamide, and malyngolide were shown to deter feeding by *S. longicauda* at ecologically relevant concentrations [108]. Nagle and Paul also suggested that a finely tuned chemical association may exist because *S. longicauda* feeding-preference experiments, using artificial diets treated in narrow increments with purified cyanobacteria metabolites, produced a range of sea hare feeding responses [108, 109]. In contrast to generalist herbivores, *S. longicauda* was shown to respond to artificial food treated with pure *L. majuscula* compounds with more than one type of behavior. *Lyngbya* metabolites incorporated into sea hare food strips acted as feeding deterrents or feeding stimulants to *S. longicauda*. Remarkably, the molluscicide barbamide stimulated sea hare feeding. Malyngamides A and B and majusculamides A and B, the most common natural products found in Guam samples of *L. majuscula*, increase sea hare feeding at low concentrations and inhibit feeding at the slightly higher concentrations that occur in some populations of *L. majuscula*.

Ecological significance of chemical diversity

Chemotypes of *L. majuscula* and other filamentous marine cyanobacteria produce a wide range of secondary metabolites at various concentrations. Little is known of the genetic differences between these chemically distinct populations of cyanobacteria. However, several adaptive advantages may be envisioned by the production of a suite of different defensive natural products rather than a single antifeedent compound. Benthic filamentous cyanobacteria are subject to grazing by an assortment of different generalist and specialist feeders [107, 112, 113]. Production of deterrent metabolites may be vital to large-scale mat and bloom formation, especially in the presence of grazers. Different species of reef fishes respond differently to natural products produced by marine cyanobacteria [100, 107]. Compounds that deter grazing by one species of reef fish may have little or no effect on another species. In addition, metabolites that deter herbivory by reef fishes can act to stimulate feeding by sea hares [104, 108]. A diverse combination of natural products produced by benthic filamentous cyanobacteria is essential for survival in a complex grazer community structure. Chemical defenses produced by marine cyanobacteria probably allow these organisms to reach and maintain bloom levels, even in the presence of generalist herbivores

Biomedical importance of toxic cyanobacterial metabolites

It has been demonstrated that severe erythematous and a papulovesicular dermatitis in humans, known as "swimmer itch", are caused by the marine cyanobacterium *Lyngbya majuscula* [114]. Lyngbyatoxin, aplysiatoxin, and oscillatoxins isolated from species of *Lyngbya* have been found to be inflammatory and vesiculatory skin irritants as well as tumor promotors. Biotoxins produced by certain algae are responsible for sickness and death of small mammals such as mice and aquatic invertebrates [2, 3, 86]. Macrolides from *Oscillatoria acutissima* possess potent cytotoxic and antineoplastic activity [25].

Allelopathic importance of cyanobacterial secondary metabolites

Harder [115] was the first to observe algal allelopathy. According to Rice [116], the phenomenon was first recognized as an influence in natural systems by Akehurst in 1931 in his explaining of algal succession [117]. Inderjit and Dakshini [118] reviewed the existing literature on algal allelopathy with special reference to allelopathic interactions among and between algae, autotoxicity, toxins, bioassays, and implications of algal allelopathy. In an extensive study carried out by Lefevre et al. [119], filtrates from a canal infested with *Aphanizomenon gracile* inhibited growth of several algal forms such as *Pediastrum boryanum, P. clathratum, Cosmarium lundellii*, and *Phormidium uncinatum.* A bloom of *Oscillatoria planktonica* appeared later in the same canal and its filtrate inhibited *Chlorella pyrenoidosa, Cosmarium lundellii, C. obtusatum, Pediastrum boryanum, Phormidium uncinatum,* and *Scenedesmus quadricauda.* These *Oscillatoria* blooms disappeared after two months and previously present alga appeared again. In a study on Linsley Lakes for five years in the 1970s, Keating [120, 121] demonstrated the role of allelopathy in determining algal bloom sequence, tying laboratory and *in situ* events. Five years of bloom-dominant organisms were isolated in culture. Both cell-free filtrates of natural waters collected before, during, and after blooms and cell-free filtrates of each cultured dominant had only positive or neutral effects on succeeding and only negative or neutral effects on preceding, dominant species. No exceptions were observed. This strongly suggests that both positive and negative allelochemical incidents are involved in the control of freshwater bloom sequence [120]. Additionally [121], in all cases cyanobacterial filtrates inhibited diatom growth, suggesting that replacement of diatom blooms by cyanobacterial blooms during cultural eutrophication is also allelochemically mediated. In 1999 Schlegel et al. [122] isolated cyanobacterial strains from diverse habitats. Selected cyanobacterial strains showed antibiotic activity against green algae such as *Coelastrum microsporum, Monoraphidium convolutum,* and *Scenedesmus acutus.* They identified three strains of *Fischerella,* seven strains of *Nostoc* species, and three strains of *Calothrix* species that showed antibiotic activities.

In late 1990s Gross [123] reported that two species of *Fischerella (F. muscicola* and *F. ambigua*) produce allelochemicals having strong phytotoxic effects against other cyanobacteria and lesser effects against green algae and diatoms. Fischerellin A, the major allelochemical, inhibits cyanobacteria and chlorophytes in the agar-diffusion medium assay. Its mode of action is PSII [94]. Fischerellin B, the minor metabolite, also exhibits inhibition in the agar-diffusion assay [124]. Srivastava et al., also working with fischerellin A [125], reported that the IC_{50} for 10-minute incubation with *Anabaena* P9 thylakoids is 200 nM. A broad array of additional cyanobacterial allelochemical effects have been reported: a minimum lethal dose of ether extract of *Oscillatoria latevirens* to *Microcystis* Pasteur culture collection (PCC) strain was found to be 16 to 20 $\mu g \cdot mL^{-1}$ [92]; *Nodularia harveyana,* a Mediterranean cyanobacterium, was reported to possess lipophilic bioactive substances [126]; and an acetone extract of *N. harveyana* has high activity against cyanobacteria (*Anabaena variabilis, A. azollae, Spirulina platensis*), eubacteria (*Enterococcus faecalis, Streptococcus pneumonia*), rotifer (*Brachionus calyciflorus*), and crustacea (*Thamnocephalus platyurus*). Additionally, Flores and Wolk [127] found that *Nostoc* species is reported to produce the anticyanobacterial compound nostocyclamide [128, 129]. The minimum inhibition concentration (MIC) of nostocyclamide is 0.1 μM against several cyanobacteria and 1 μM against chlorophycean species such as *Nannochloris, Ankistrodesmus, Scenedesmus obliquus,* and *S. subspicatus.* Significantly, it appears that allelochemicals from cyanobacteria such as fischerellin A and B (*Fischerella muscicola*), hapalindole A (*Hapalosiphon fontinalis*), cyanobacterin (*Scytonema hofmanni*), cyanobacterin LU-12 (*Nostoc* sp), nostocyclamide (*Nostoc* sp 31), and an unidentified compound ($C_{29}H_{48}O_3$ from *Oscillatoria late-virens*) all influence photosynthetic electron transport [130]. From species of *Fischerella* and *Calothrix*, Doan et al. [131] reported 12-*epi*-hapalindole E isonitrile and the indolophenanthridine calothrixin A, respectively. These alkaloids adversely affect gram-negative bacteria (*Escherichia coli, Proteus mirabilis, Klebsiella pneumoniae*), gram-positive bacteria (*Bacillus subtilis 168, Bacillus cereus, Staphylococcus cerevisiae*), cyanobacteria (*Anabaena doliolum*), and green algae (*Monoraphidium convolutum, Coelastrum Microsporum, Scenedesmus acutus*).

Toxic effects of cyanobacteria may be affected by different abiotic factors. Utkilen and Gjølme [132] found that light intensity had significant effects on both the toxicity and the peptide toxin production of *Microcystis aeruginosa*; however, light quality had little effect. The often toxic species *Microcystis* accompanies increasing eutrophication and responds to light intensity, nutrients, and temperature as well as nutrient substrates [133, 134]. The content of cyanobacterial toxin is generally lower under conditions of N and P deficiency. For example, N deficiencies in the medium significantly decrease fischerellin content. Gross [123, 135] suggested that such a decrease in the fischerellin content resulting from N depletion does not adversely affect the competitive ability of *Fischerella* in planktonic environment because chlorophytes, diatoms, and other algae cannot fix N and thus would still be less able to compete against cyanobacteria under N-deficient conditions. In benthic environments, where the main competitors of *Fischerella* would be N-fixing cyanobacteria, light is the main limiting factor. While *Fischerella* is favored in terms of production of fischerellin, that production severely decreases under acute low-light conditions. Low amounts of fischerellin under acute N and light conditions may cause a disadvantage for the producer *Fischerella* strains. However, the mode of transfer of fischerellin is *via* cell-cell contact [135]. Because of such an effective mode of allelochemical transfer, *Fischerella* may still be able to suppress the growth of other benthic cyanobacteria.

Cyanotoxins have been employed widely for their biotechnological utilization, e.g., in the pharmaceutical industry [136]. However, their role in natural and agroecosystems is not fully explored. Aside from playing an important role in influencing algal growth and dominance, allelochemicals from cyanobacteria have the potential to influence higher plants. Gleason and Case [26] reported that the secondary metabolite cyanobacterin, produced by *Scytonema hofmanni*, inhibited the growth of aquatic and terrestrial angiosperms, e.g., *Setaria viridis, Avena fatua, Rumax crispus, Polygonium convolvulus, Zea mays, Pisum sativum*, and *Lemna gibba*. Allelochemicals from *Oscillatoria* species also have been shown to suppress the growth of aquatic angiosperms such as *Spirodela polyrhiza* and terrestrial angiosperms such as wheat (*Triticum aestivum*) [93]. These examples of inhibition must be of concern because cyanobacterial inocula are used to enhance the N supply in paddy fields [137]. Such inocula include different strains of cyanobacteria such as *Aulosira fertilissima, Tolypothrix tenuis, Nostoc muscorum*, and *Anabaena variabilis*. When Inderjit and Dakshini [137] studied the effect of cyanobacterial inoculum on soil chemistry and growth of wheat (*Triticum aestivum*) and rice (*Oryza sativa*), they found that soil amended with high amounts of cyanobacterial inoculum had significantly different values for pH, electrical conductivity, total phenolics, N, and other inorganic ions. In spite of higher N, seedling growth of cereals was suppressed because of altered soil characteristics and high values of phenolics. Based on these findings, and the known allelochemical production of many Cyanophyta, these authors suggest that prior to preparing inocula, the allelopathic potential of cyanobacterial strains involved should be assessed.

Compared to the possible interactions of higher plants, less research has been done to understand cyanobacterial allelopathic interactions. Research is needed to understand the role of cyanobacterial allelochemicals in (1) algal succession and bloom formation, (2) their physical and chemical effects on soils, and (3) their role as bioherbicides sources. Cyanobacteria can be excellent experimental organisms to demonstrate allelopathy. Gross et al. [135], for example, found that *Fischerella* does not release allelochemicals in medium but directly transfers the bioactive substance to the target species by cell-cell contact. This is a good model for demonstrating allelopathy because the transfer of chemicals from donor to target species is through cell-to-cell contact. If a species releases chemicals in the medium, the chemicals get diluted and chances of degradation (physical, chemical, or biological) are greater. Demonstrating allelopathy in nature has proven difficult, particularly in terrestrial systems [138]. In aquatic systems, the phenomenon can be proven convincingly, as is evident from the work of Keating [120, 121, 139] and Gross [123]. Using labeled ^{14}C compound, Pflugmacher et al. [140] demonstrated the uptake of the cyanobacterial toxin microcystin-LR in the emergent reed plant, *Phragmites australis*. The compound was found to be stable for three days, and no degradation was detected. Uptake of the same compound was also demonstrated in other aquatic macrophytes, e.g., *Ceratophyllum demorsum, Elodea canadensis*, and *Vesicularia dubyana* [141]. Such examples are helpful in demonstrating allelopathy in field situations.

Acknowledgements
We sincerely thank Professors Dean Martin, Shmuel Carmeli, Geoffry Smith, Kathleen Keating, and Elisabeth Gross for their constructive comments. This work was supported in part by the USDA, Agricultural Research Service Specific Cooperative Agreement #58-6408-2-0009.

Appendix

Structures of biologically active cyanobacterial metabolites mentioned in the text. (In alphabetical order.)

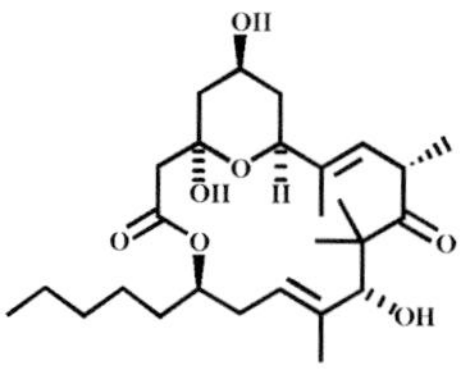

Acutiphycin

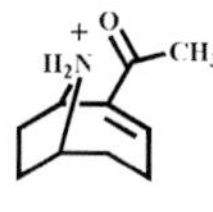

Anatoxin-a

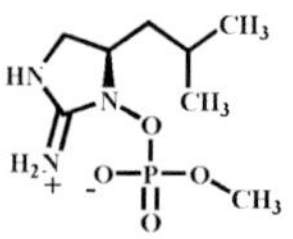

Anatoxin-a(s)

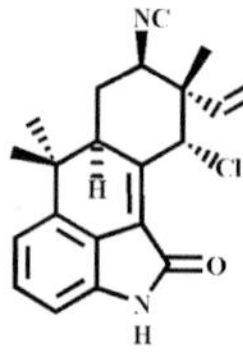

Anhydrohapalindole A

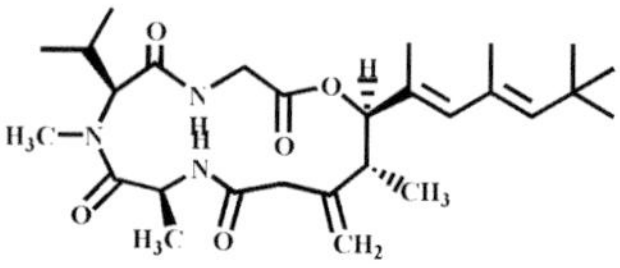

Antillatoxin

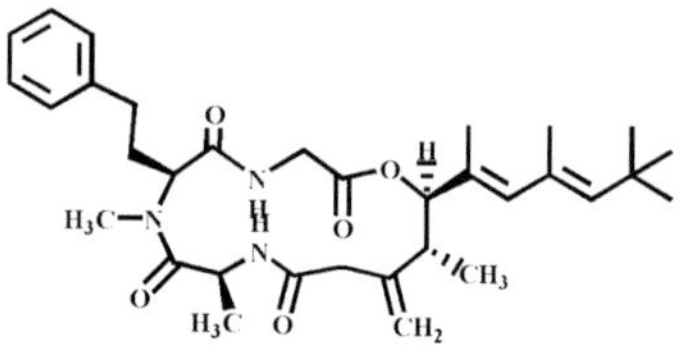

Antillatoxin B

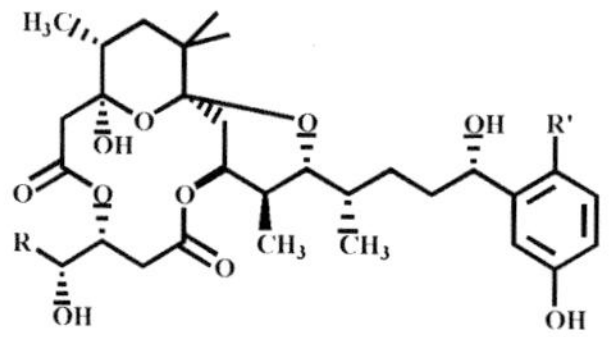

Aplysiatoxin : R=CH₃ R'=Br
Debromoaplysiatoxin : R=CH₃ R'=H
Oscillatoxin A : R=H   R'=Br

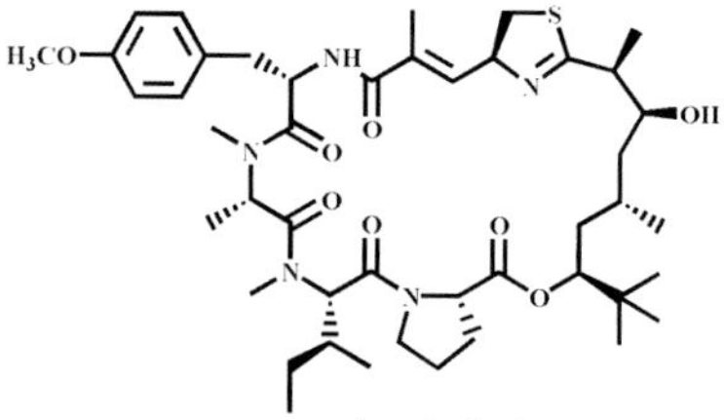

Apratoxin A

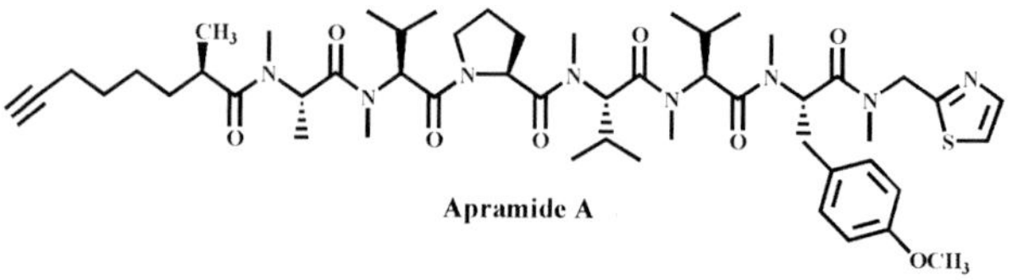

Apramide A

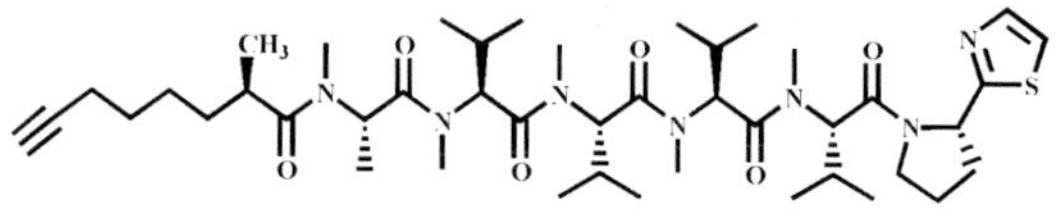

Apramide G

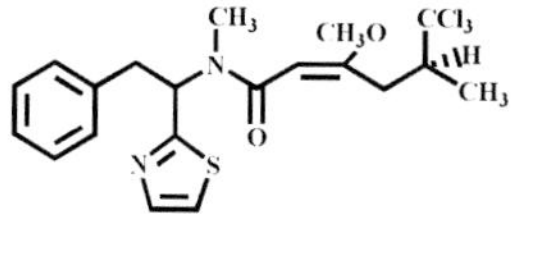

Barbamide

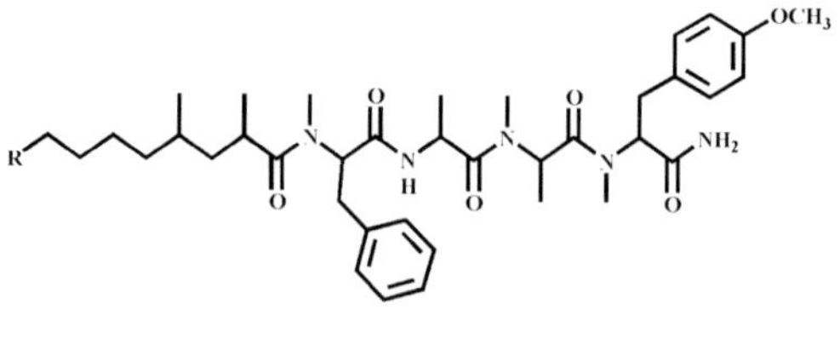

Carmabin A: R= C≡CH
Carmabin B: R= COCH₃

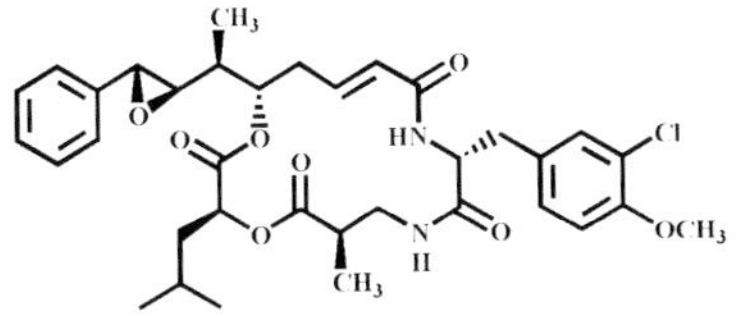

Cryptophycin 1

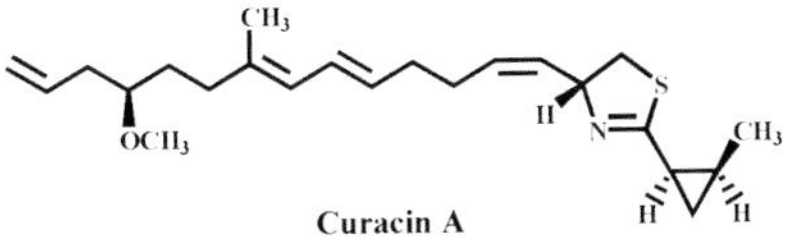

Curacin A

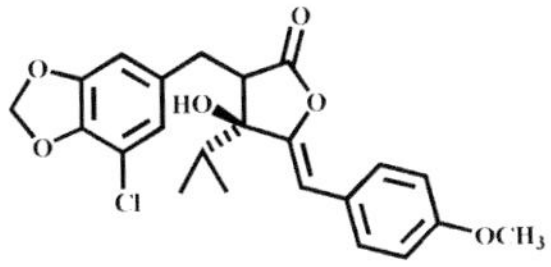

Cyanobacterin

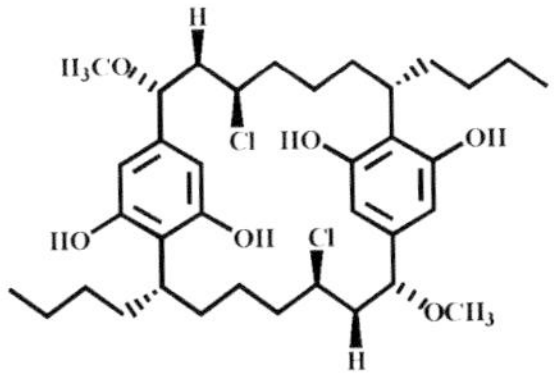

Cylindrocyclophane

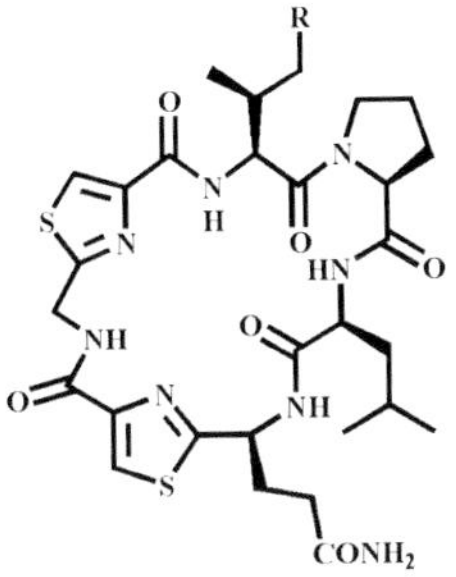

Dolastatin 3: R=H
Homdolastatin 3: R=CH₃

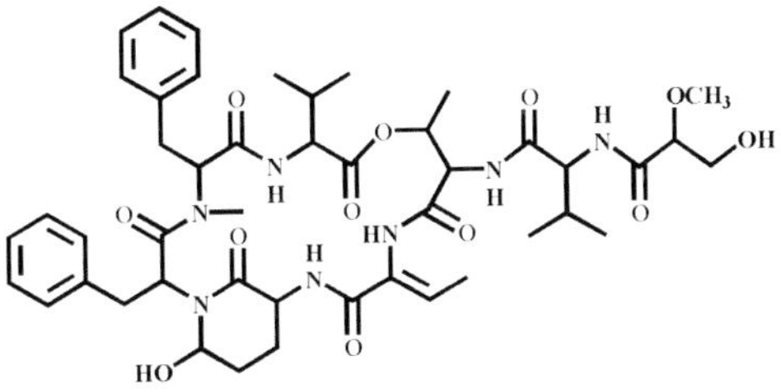

Dolastatin 13

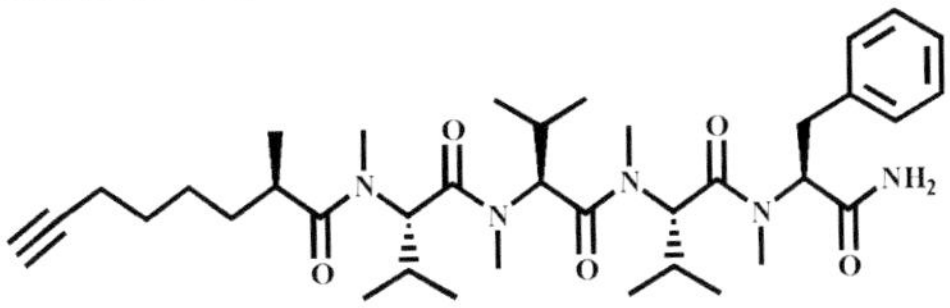

Dragonamide

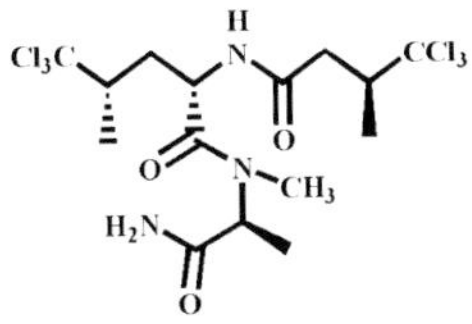

Dysidenamide

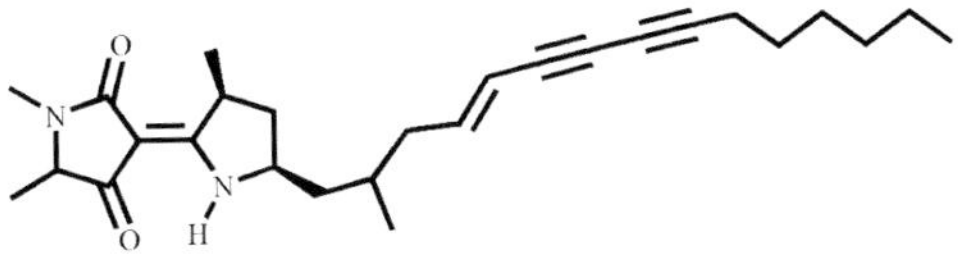

Fischerellin A

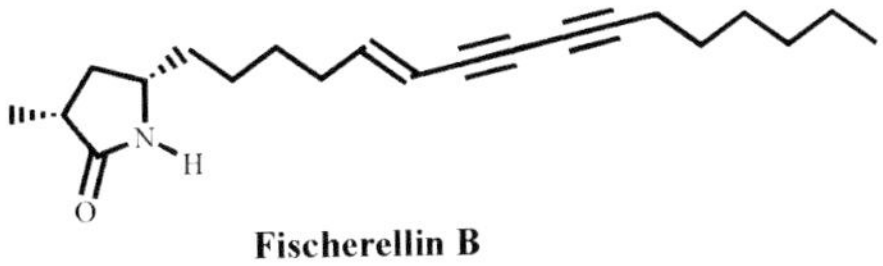

Fischerellin B

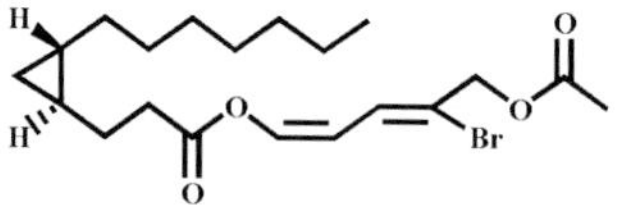

Granadadiene

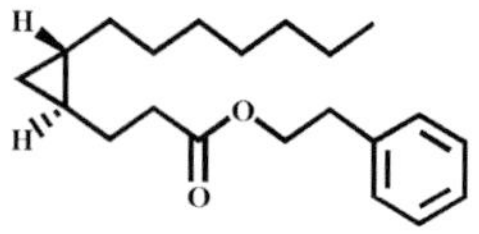

Granadamide

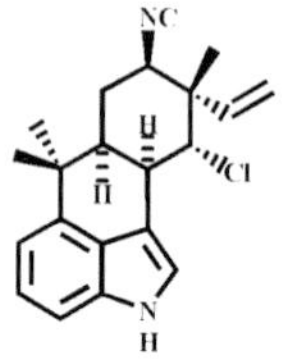

Hapalindole A

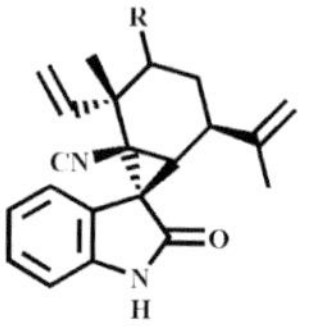

Hapalindolinone A: R=Cl
Hapalindolinone B: R=H

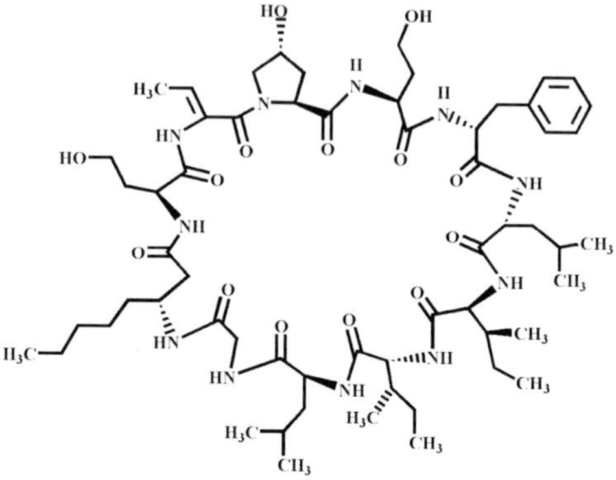

Hormothamnin A

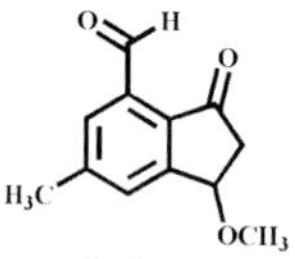

Indanone

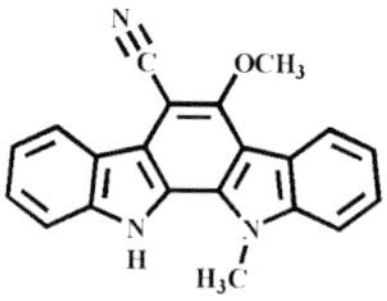

Indolocarbazole

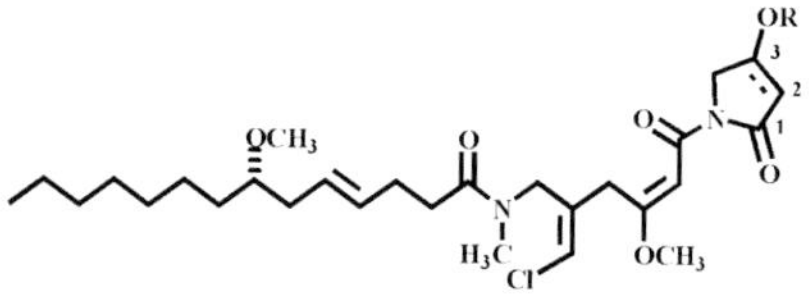

Isomalyngamide A: R=CH₃ (C2-C3 unsaturated)
Isomalyngamide B: R=H (C2-C3 saturated)

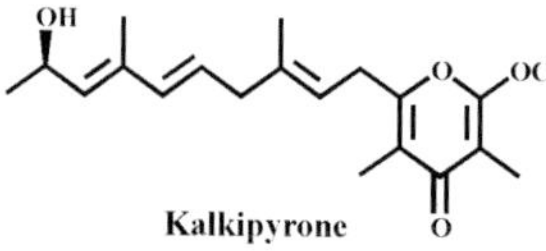

Kalkipyrone

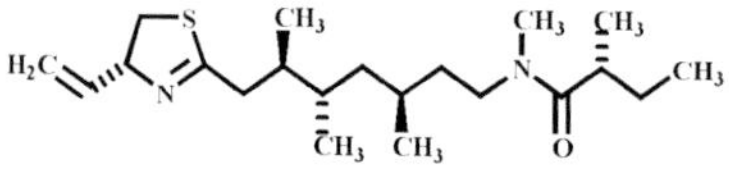

Kalkitoxin

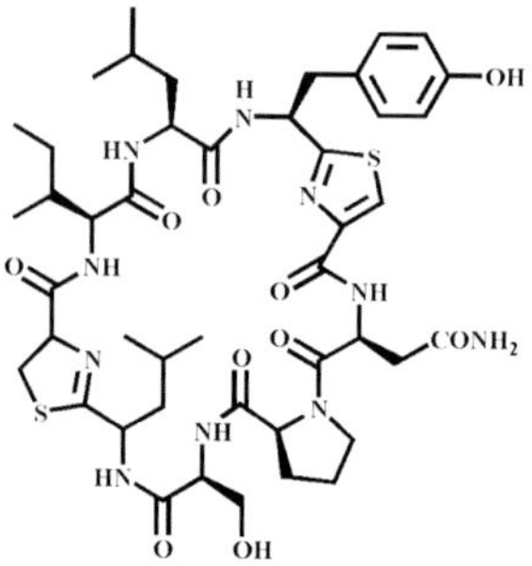

Kororamide

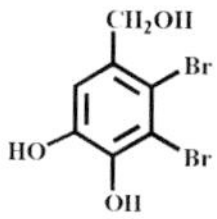

Lanosol

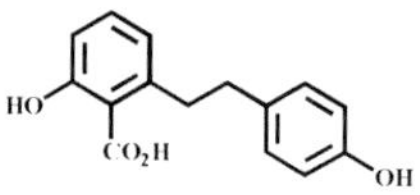

Lunularic Acid

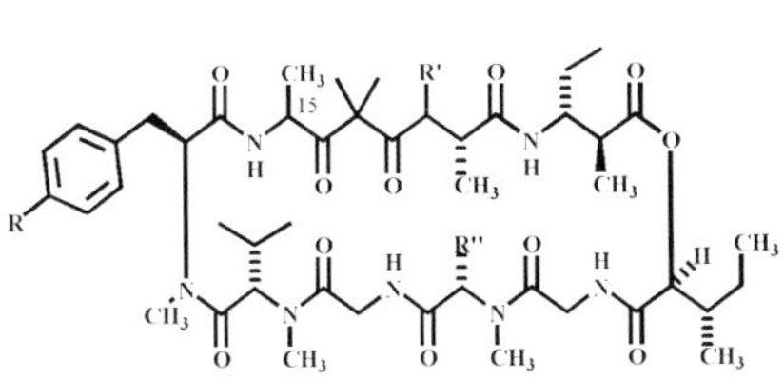

Lynbyastatin 1: R=OCH₃ R'=CH₃ R''=(CH₃)₂CHCH₂-
Dolastatin 12: R=H R'=CH₃ R''=(CH₃)₂CHCH₂-
Majusculamide C: R=OCH₃ R'=H R''=CH₃CH₂(CH₃)CH- (*S* at C-15)

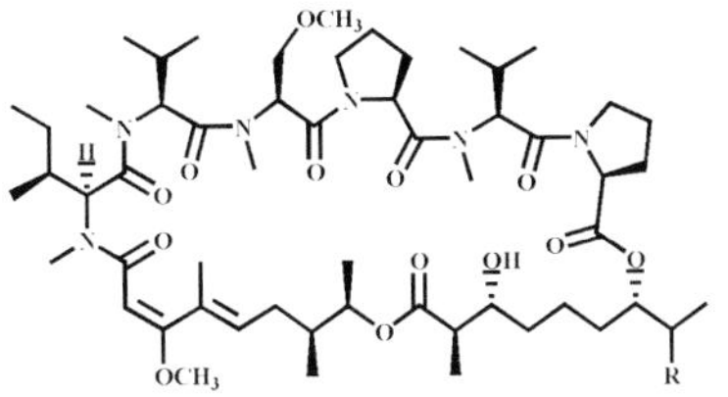

Lynbyastatin 2: R=OCH₃
Norlynbyastatin 2: R=H

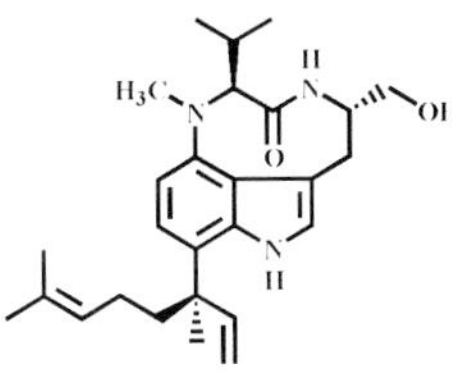

Lyngbyatoxin A

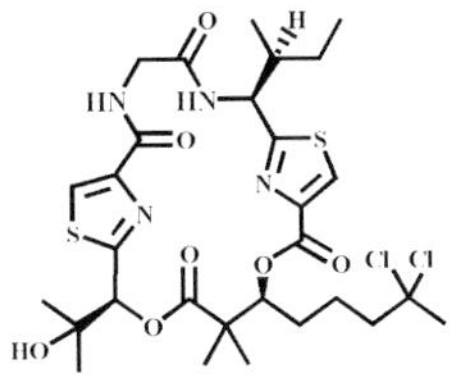

Lyngbyabellin A

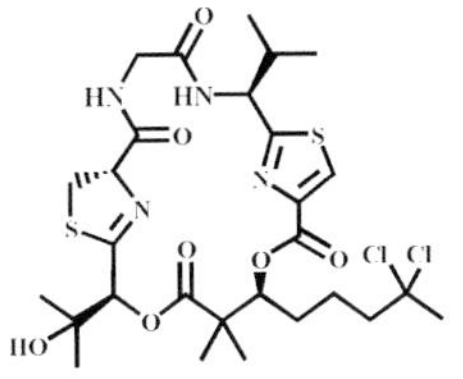

Lyngbyabellin B

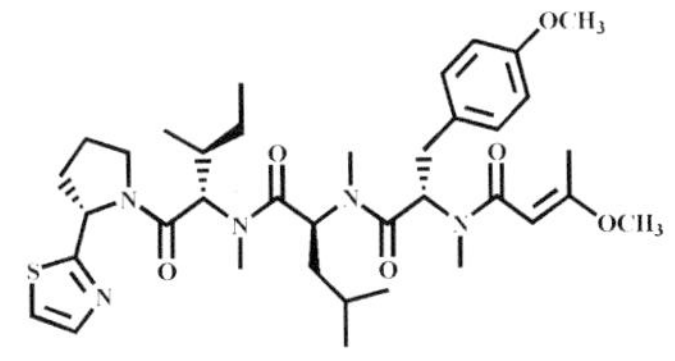

Lyngbyapeptin A

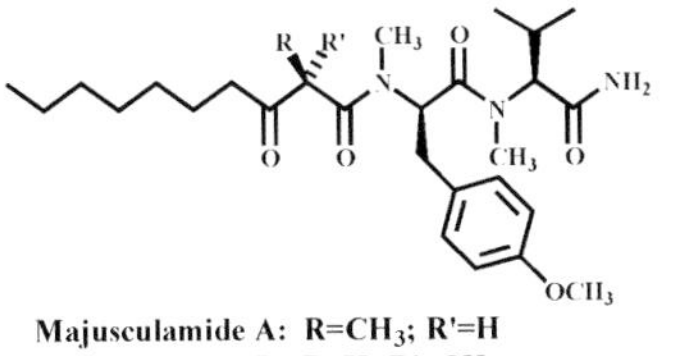

Majusculamide A: R=CH₃; R'=H
Majusculamide B: R=H; R'=CH₃

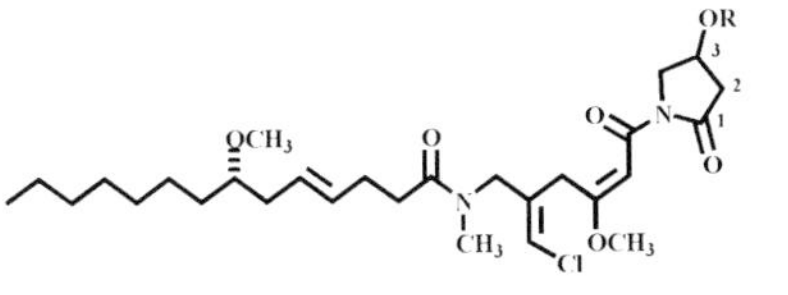

Malyngamide A: R=CH₃ (C2-C3 unsaturated)
Malyngamide B: R=H (C2-C3 saturated)

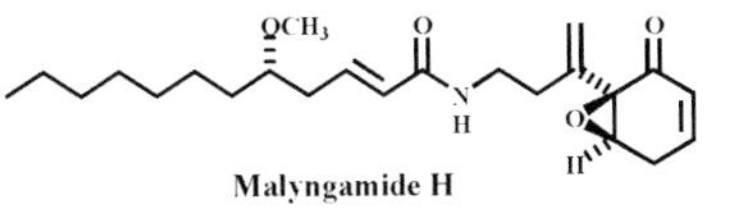

Malyngamide H

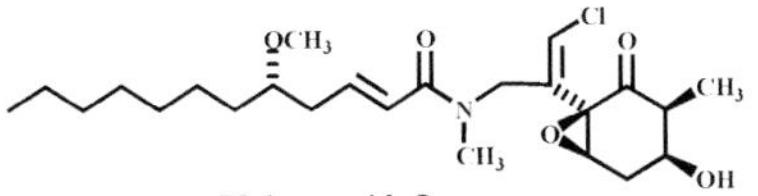

Malyngamide I

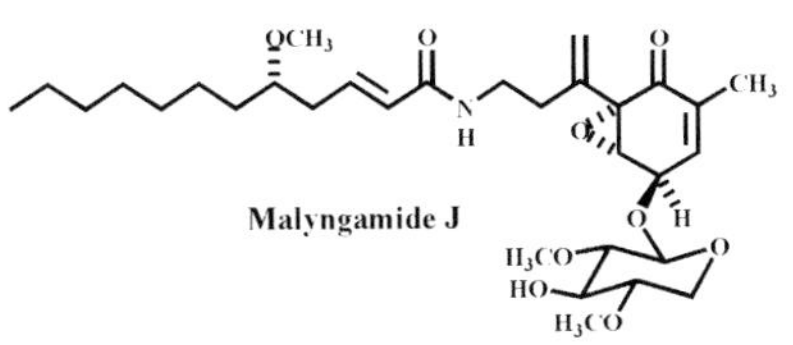

Malyngamide J

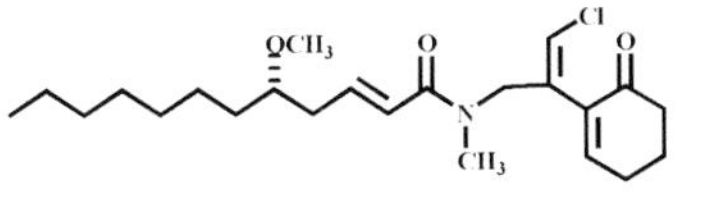

Malyngamide K

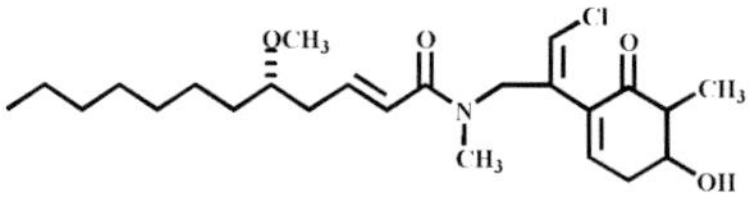

Malyngamide L

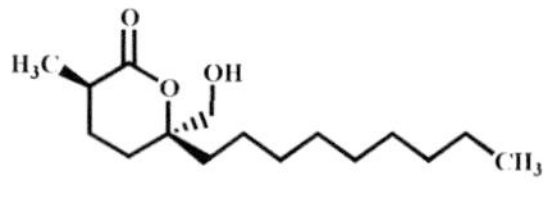

Malyngolide

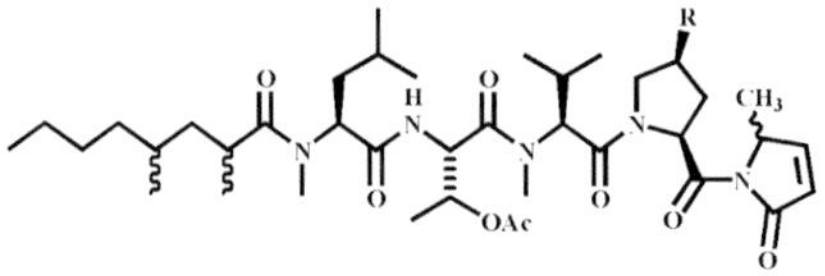

Microcolin A: R=OH
Microcolin B: R=H

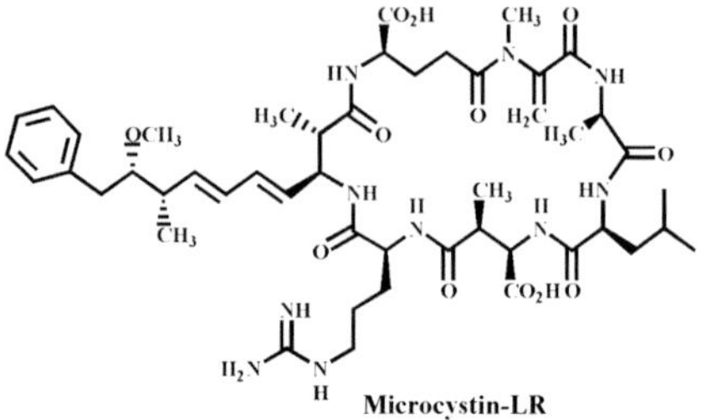

Microcystin-LR

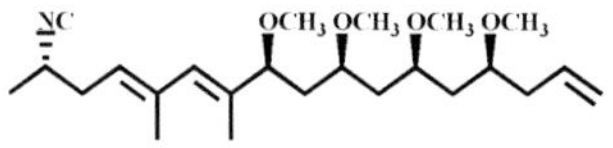

Mirabilene-A isonitrile

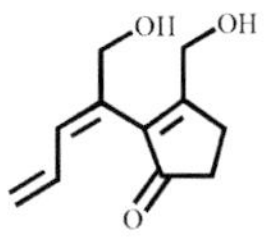

Nakienone A

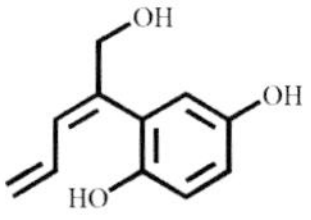

Nakitriol

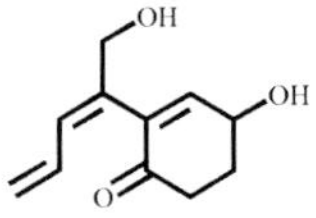

Nakienone B

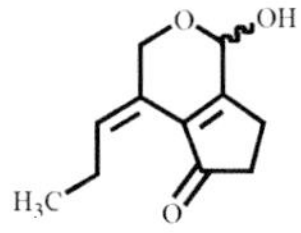

Nakienone C

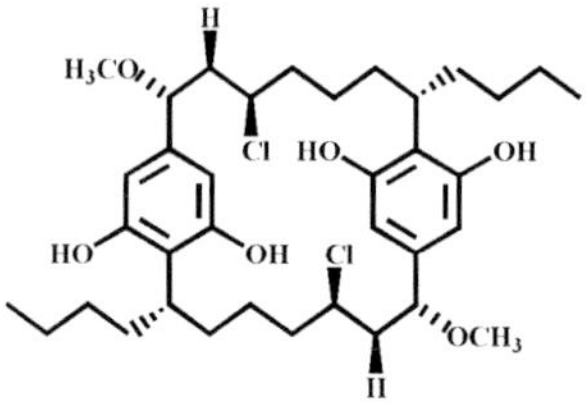

Nostocyclophane

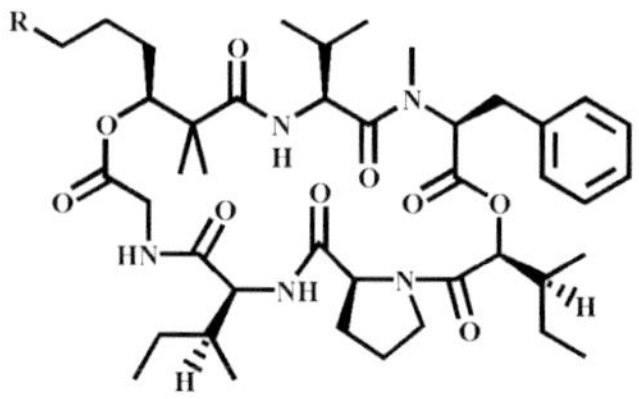

Pitipeptolide A: R= ⌇———≡

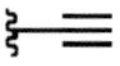

Pitipeptolide B: R= ⌇——╲

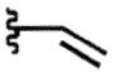

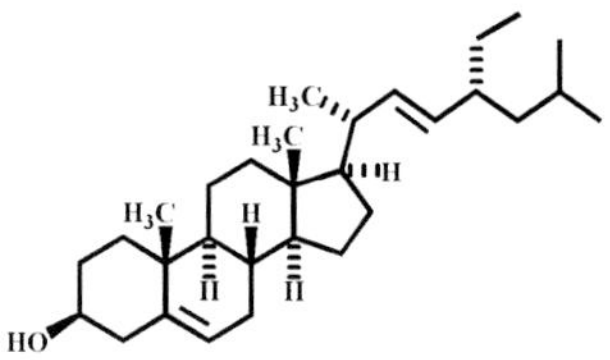

Poriferasterol

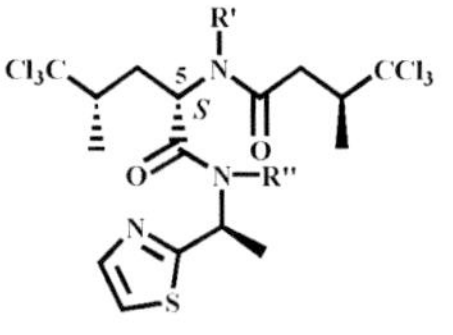

Pseudodysidenin: R'=H   R''=CH₃
Nordysidenin: R'=H   R''=H
Dysidenin: R'=CH₃ R''=H
Isodysidenin: R'=CH₃ R''=H   C-5 (R)

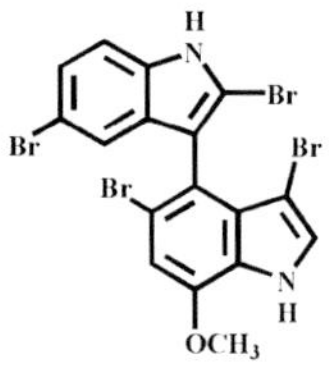

Rivularin D₃

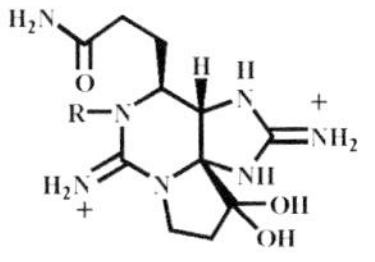

Saxitoxin: R=H
Neosaxitoxin: R=OH

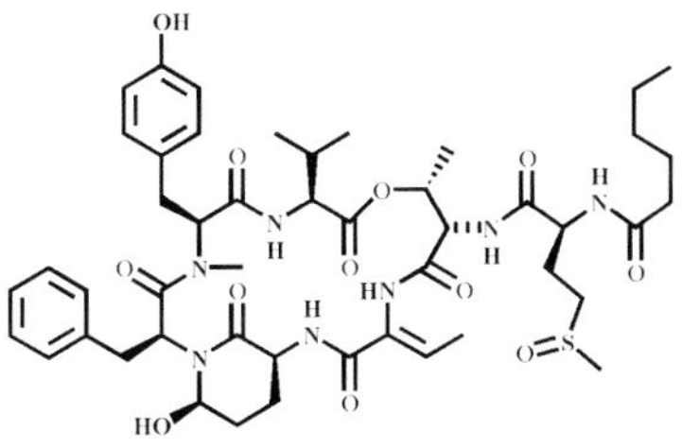

Somamide A

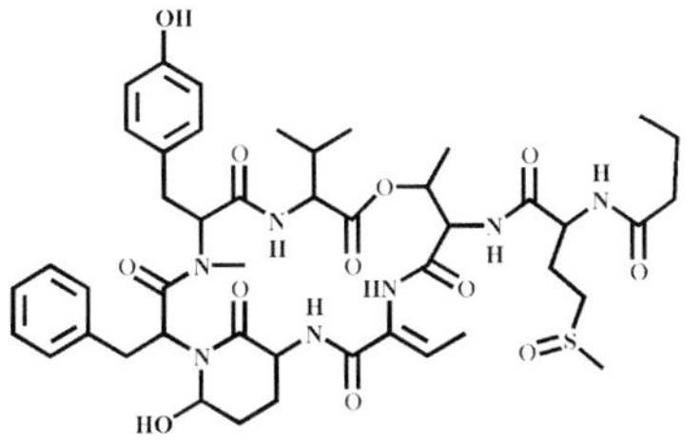

Somamide B

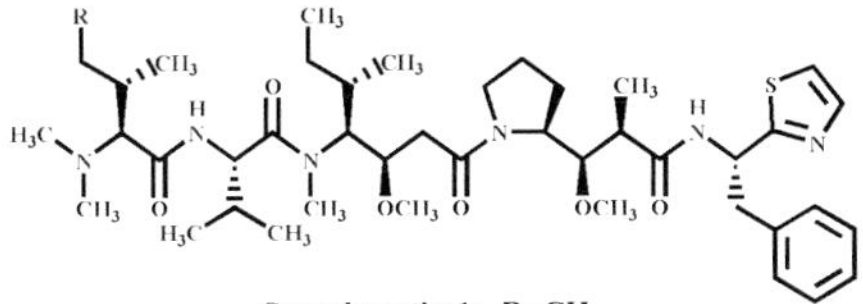

Symplostatin 1: R=CH₃
Dolastatin 10: R=H

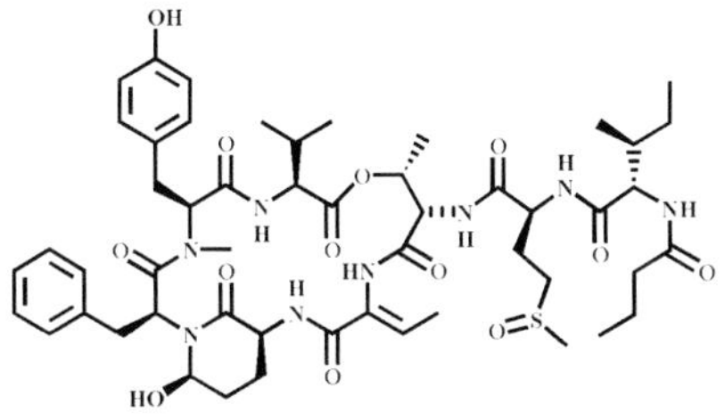

Symplostatin 2

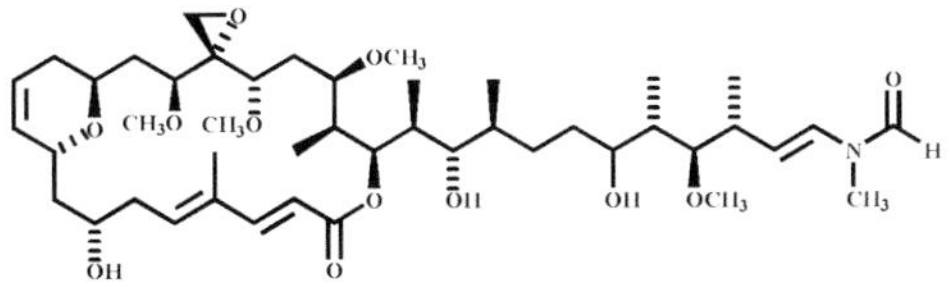

Scytophycin A

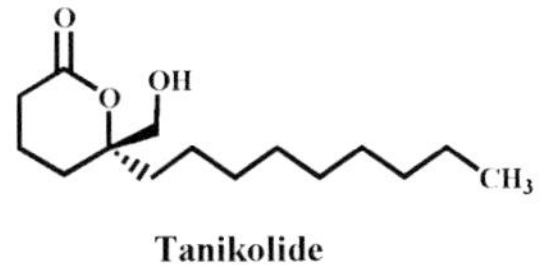

Tanikolide

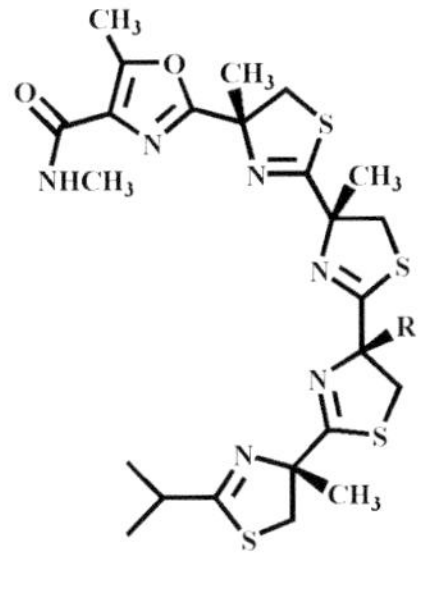

Tantazole A: R=H
Tantazole B: R=CH₃

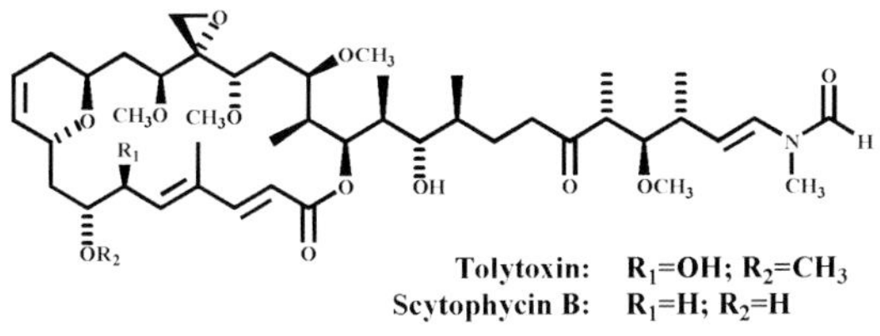

Tolytoxin: R₁=OH; R₂=CH₃
Scytophycin B: R₁=H; R₂=H

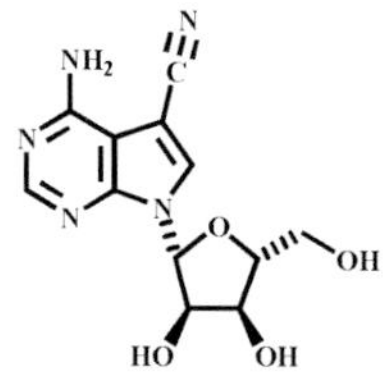

Toyocamycin

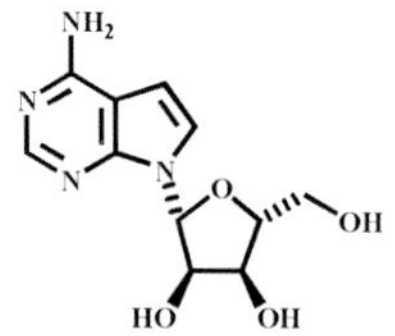

Tubercidin

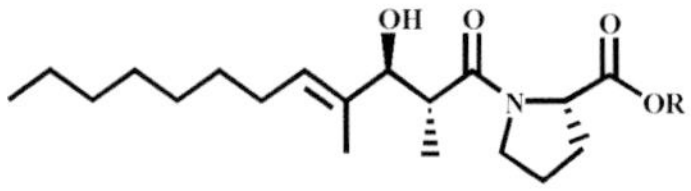

Tumonic Acid A: R=H
Methyl Tumonate A: R=CH$_3$

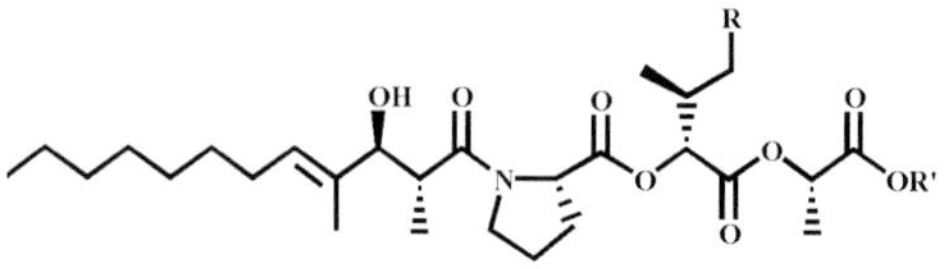

Tumonic Acid B: R=CH$_3$ R'=H
Methyl Tumonate B: R=CH$_3$ R'=CH$_3$
Tumonic Acid C: R=H   R'=H

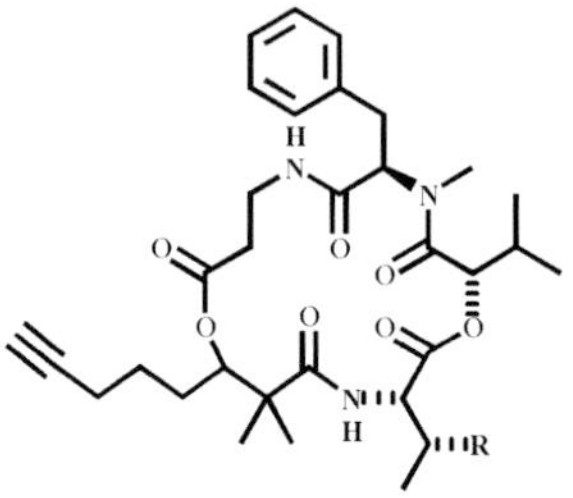

Yanucamide A: R=CH$_3$
Yanucamide B: R=CH$_2$CH$_3$

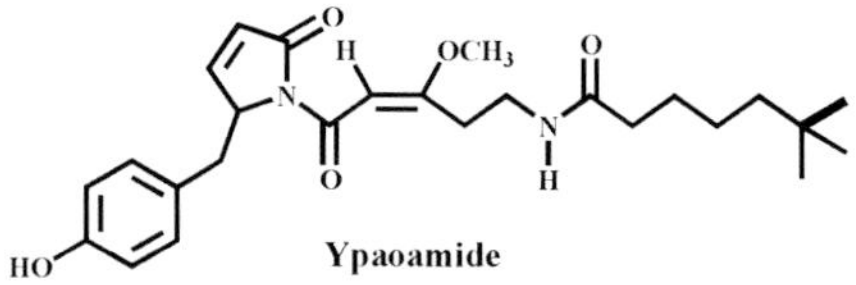

Ypaoamide

References

1 Carmichael WW (1982) Chemical and toxicological studies of the toxic freshwater cyanobacteria *Microcystis aeruginosa*, *Anabaena flos-aquae* and *Aphanizomenon flos-aquae*. *S Afr J Sci* 78: 367–372

2 Carmichael WW (1986) Algal toxins. *In*: JA Callow (ed): *Advances in botanical research.* Academic Press, London, 47–101

3 Carmichael WW (1988) Toxins of freshwater algae. *In*: AT Tu (ed): *Handbook of natural toxins.* Marcel Dekker, New York, 121–147

4 Carmichael WW (1989) Freshwater cyanobacteria (blue-green algae) toxins. *In*: CL Ownby and GV Odell (eds): *Natural toxins: characterization, pharmacology and therapeutics.* Pergamon Press, New York, 3–16

5 Carmichael WW (1992) Cyanobacteria secondary metabolites – the cyanotoxins. *J Appl Bacteriol* 72: 445–459

6 Carmichael WW (1994) The toxins of cyanobacteria. *Scient Amer* 270: 64–70

7 Carmichael WW, Falconer IR (1993) Diseases related to freshwater blue-green algal toxins, and control measures. *In*: IR Falconer (ed): *Algal toxins in seafood and drinking water.* Academic Press, New York, 187–209

8 Carmichael WW, Gorham PR (1977) Factors influencing the toxicity and animal susceptibility of *Anabaena flos-aquae* (Cyanophyta) blooms. *J Phycol* 13: 97–101

9 Carmichael WW, Gorham PR (1978) Anatoxins from clones of *Anabaena flos-aquae* isolated from lakes of Western Canada. *Mitt Int Vern Limnol* 21: 285–295

10 Chapman VJ (1979) Seaweeds in pharmaceuticals and medicine: a review. *In*: HA Hoppe, T Levring, Y Tanaka (eds): *Marine algae in pharmaceutical science.* Walter de Gruyter, Berlin, 139–147

11 Codd GA, Poon GK (1988) Cyanobacteria toxins. *In*: LJ Rogers, JG Gallon (eds): *Biochemistry of the algae and cyanobacteria*. Oxford Science Publishers, Oxford, 283–296

12 Glombitza KW, Koch M (1989) Secondary metabolites of pharmaceutical potential. *In*: RC Cresswell, TAV Rees, N Shah (eds): *Algal and cyanobacterial biotechnology*. Longman Scientific and Technical, New York, 161–238

13 Gorham PR, Carmichael WW (1988) Hazards of freshwater blue-greens (cyanobacteria). *In*: CA Lembi, JR Waaland (eds): *Algae and human affairs*. Cambridge University Press, Cambridge, 403–431

14 Gerwick WH, Roberts MA, Proteau PJ, Chen J-L (1994) Screening cultured marine microalgae for anticancer-type activity. *J Appl Phycol* 6: 143–149

15 Smith CD, Zhang X, Mooberry SL, Moore RE (1994) Crytophycin: a new antimicrotubule agent active against drug-resistant cells. *Cancer Res* 54: 3779–3784

16 Harrigan GG, Luesch H, Yoshida WY, Moore RE, Nagle DG, Paul VJ, Mooberry SL, Corbett TH, Valeriote FA (1998) Symplostatin 1: a dolastatin 10 analog from the marine cyanobacterium, *Symploca hydnoides*. *J Nat Prod* 61: 1075–1077

17 Limtiaco S (1994) Slimy, gooey mess turns out to be algae. *Pac Daily News*, May 7, 25: 1

18 Knuckles W (1995) Something is rotten in Ipan Tolofofo. *Pac Daily News*, March 3, 26: 3

19 Pedersen M, DaSilva EJ (1973) Simple brominated phenols in the blue-green alga *Calothrix brevissima* West. *Planta* 115: 83–86

20 Pryce RJ (1972) The occurrence of lunularic and abscisic acids in plants. *Phytochemistry* 11: 1759–1761

21 Moore RE, Cheuk C, Yang XG, Patterson GML (1987) Haplinodoles, antibacterial and antimycotic alkaloids from the cyanophyte *Hapalosiphon fontinalis*. *J Org Chem* 52: 1036–1043

22 Moore RE, Yang XG, Patterson GML (1987) Fontonamide and anhydrohapaloxindole A, two new alkaloids from the blue-green alga *Hapalosiphon fontinalis*. *J Org Chem* 52: 3773–3777

23 Moore RE, Yang XG, Patterson GML, Bonjouklian R, Smitka TA (1989) Hapalonamides and other oxidized hapalindoles from *Hapalosiphon fontinalis*. *Phytochemistry* 28: 1565–1567

24 Patterson GML, Baldwin CL, Bolis CM, Chaplin IR, Karuso H, Larsen LK, Levine IA, Moore RE, Nelson CS, Tschappat KD et al (1991) Antineoplastic activity of cultured blue-green algae (Cyanophyta). *J Phycol* 27: 530–536

25 Barchi J Jr, Moore RE, Patterson GML (1984) Acutiphycin and 20,21-didehydroacutiphycin, new antineoplastic agents from the cyanophyte *Oscillatoria acutissima*. *J Am Chem Soc* 106: 8193–8197

26 Gleason FK, Case DE (1986) Activity of the natural algicide, cyanobacterin, on angiosperms. *Plant Physiol* 80: 834–837

27 Mason CP, Edwards KR, Carlson RE, Pignatello J, Gleason FK, Wood JM (1982) Isolation of chlorine-containing antibiotic from the freshwater cyanobacterium *Scytonema hofmanni*. *Science* 215: 400–402

28 Pignatello JJ, Porwoll J, Carlson RE, Xavier A, Gleason FK, Wood JM (1983) Structure of the antibiotic cyanobacterin, a chlorine-containing γ-lactone from the freshwater cyanobacterium *Scytonema hofmanni*. *J Org Chem* 48: 4035–4038

29 Schwartz RE, Hirsch CF, Springer JP, Pettibone DJ, Zinke DL (1987) Unusal cyclopropane-containing hapalindolinones from a cultured cyanobacterium. *J Org Chem* 52: 3704–3706

30 Norton RS, Wells RJ (1982) A series of chiral polybrominated biindoles from the marine blue-green alga *Rivularia firma*: application of ^{13}C NMR spin-lattice relaxation data and ^{13}C-^{1}H coupling constants to structure elucidation. *J Am Chem Soc* 104: 3628–3635

31 Moore RE (1981) Constituents of blue-green algae. *In*: PJ Scheuer (ed): *Marine Natural Products, Vol. 4*. Academic Press, New York, 1–52

32 Moore RE (1982) Toxins, anticancer agents, and tumor promoters from marine prokaryotes. *Pure Appl Chem* 54: 1919–1934

33 Moore RE (1996) Cyclic peptides and depsipeptides from cyanobacteria: a review. *J Ind Microbiol* 16: 134–143

34 Cardellina JH II, Moore RE, Arnold EV, Clardy J (1979) Structure and absolute configuration of malyngolide, an antibiotic from the marine blue-green alga *Lyngbya majuscula* Gomont. *J Org Chem* 44: 4039–4042

35 Singh IP, Milligan KE, Gerwick WH (1999) Tanikolide, a toxic and antifungal lactone from the marine cyanobacterium *Lyngbya majuscula*. *J Nat Prod* 62: 1333–1335

36 Cardellina JH II, Marner F-J, Moore RE (1979) Seaweed dermatitis: structure of lyngbyatoxin A.

Science 204: 193–195

37 Sims JK, Zandee Van Rilland RD (1981) Escharotic stomatitis caused by the "stinging seaweed" *Microcoleus lyngbyaceus* (formerly *Lyngbya majuscula*): a case report and review of literature. *Hawaii Med J* 40: 243–248

38 Fujiki H, Suganuma M, Hakii H, Bartolini G, Moore RE, Takayama S, Sugimura T (1984) A two-stage mouse skin carcinogenesis study of lyngbyatoxin A. *J Cancer Res Clin Oncol* 108: 174–176

39 Mynderse JS, Moore RE, Kashiwagi M, Norton TR (1977) Antileukemia activity in the Oscillatoriaceae: isolation of debromoaplysiatoxin from *Lyngbya. Science* 196: 538–539

40 Kato Y, Scheuer PJ (1974) Aplysiatoxin and debromoaplysiatoxin, constituents of the marine mollusk *Stylocheilus longicauda* (Quoy and Gaimard, 1824). *J Am Chem Soc* 96: 2245–2246

41 Kato Y, Scheuer PJ (1975) The aplysiatoxins. *Pure Appl Chem* 41: 1–14

42 Hanne M, Matsubayashi H, Vogt R, Wakida C, Hau S, Nagai H, Hokama Y, Solorzano L (1995) Outbreak of gastrointestinal illness associated with consumption of seaweed – Hawaii, 1994. *Morbid Mortal Week Rep* 44: 724–727

43 Nagai H, Yasumoto T, Hokama Y (1996) Aplysiatoxin and debromoaplysiatoxin as the causative agents of a red alga *Gracilaria coronopifolia* poisoning in Hawaii. *Toxicon* 37: 753–761

44 Nagai H, Yasumoto T, Hokama Y (1997) Manauealides, some of the causative agents of a red alga *Gracilaria coronopifolia* poisoning in Hawaii. *J Nat Prod* 60: 925–928

45 Orjala J, Nagle DG, Gerwick WH (1995) Malyngamide H, an ichthyotoxic amide possessing a new carbon skeleton from the Caribbean cyanobacterium *Lyngbya majuscula. J Nat Prod* 58: 764–768

46 Todd JT, Gerwick WH (1995) Malyngamide I from the tropical marine cyanobacterium *Lyngbya majuscula* and the probable structure revision of Stylocheilamide. *Tetrahedron Lett* 36: 7837–7840

47 Wu M, Okino T, Nogle LM, Marquez BL, Williamson RT, Sitachitta N, Berman FW, Murray TF, McGough K, Jacobs R et al (2000) Structure, synthesis, and biological properties of kalkitoxin, a novel neurotoxin from the marine cyanobacterium *Lyngbya majuscula. J Am Chem Soc* 122: 12,041–12,042

48 Orjala J, Nagle DG, Hsu VL, Gerwick WH (1995) Antillatoxin: an exceptionally ichthyotoxic cyclic lipopeptide from the tropical marine cyanobacterium *Lyngbya majuscula. J Am Chem Soc* 117: 8281–8282

49 Nogle LM, Okino T, Gerwick WH (2001) Antillatoxin B, a neurotoxic lipopeptide from the marine cyanobacterium *Lyngbya majuscula. J Nat Prod* 64: 983–985

50 Wu M, Milligan KE, Gerwick WH (1997) Three new malyngamides from the marine cyanobacterium *Lyngbya majuscula. Tetrahedron* 53: 15,983–15,990

51 Tan LT, Okino T, Gerwick WH (2000) Hermitamides A and B, toxic malyngamide-type natural products from the marine cyanobacterium *Lyngbya majuscula. J Nat Prod* 63: 952–955

52 Graber MA, Gerwick WH (1998) Kalkipyrone, a toxic g-pyrone from an assemblage of the marine cyanobacteria *Lyngbya majuscula* and *Tolypothrix* sp. *J Nat Prod* 61: 677–680

53 Anderson JE, Goetz CM, McLaughlin JL, Suffness M (1991) A blind comparison of simple bench-top bioassays and human tumor cell cytotoxicities as antitumor prescreens. *Phytochem Anal* 2: 107–111

54 Solis PN, Wright CW, Anderson MM, Gupta MP, Phillipson JD (1993) A microwell cytotoxicity assay using *Artemia salina* (brine shrimp). *Planta Med* 59: 250–252

55 Kan Y, Sakamoto B, Fujita T, Nagai H (2000) New malyngamides from the Hawaiian cyanobacteria *Lyngbya majuscula. J Nat Prod* 63: 1599–1602

56 Luesch H, Yoshida WY, Moore RE, Paul VJ, Mooberry SL (2000) Isolation, structure determination, and biological activity of lyngbyabellin A from a marine cyanobacterium *Lyngbya majuscula. J Nat Prod* 63: 611–615

57 Luesch H, Yoshida WY, Moore RE, Paul VJ (2000) Isolation and structure of the cytotoxin lyngbyabellin B and absolute configuration of lyngbyapeptin A from the marine cyanobacterium *Lyngbya majuscula. J Nat Prod* 63: 1437–1439

58 Milligan KE, Marquez BL, Williamson RT, Gerwick WH (2000) Lyngbyabellin B, a toxic and antifungal secondary metabolite from the marine cyanobacterium *Lyngbya majuscula. J Nat Prod* 63: 1440–1443

59 Gerwick WH, Proteau PJ, Nagle DG, Hamel E, Blokhin A, Slate DL (1994) Structure of curacin A, a novel antimitotic, antiproliferative, and brine shrimp toxic natural product from the marine cyanobacterium *Lyngbya majuscula. J Org Chem* 59: 1243–1245

60 Blokhin AV, Yoo H-D, Geralds RS, Nagle DG, Gerwick WH, Hamel E (1995) Characterization of the interaction of the marine cyanobacterial natural product curacin A with the colchicine site of tubulin and initial structure-activity studies with analogs. *Mol Pharmacol* 48: 523–531

61 Verdier-Pinard P, Lai J-Y, Yoo H-D, Yu J, Marquez B, Nagle DG, Nambu M, White JD, Falck JR, Gerwick WH et al (1998) Structure-activity analysis of the interaction of curacin A, the potent colchicine site antimitotic agent, with tubulin and effects of analogs on the growth of MCF-7 breast cancer cells. *Mol Pharmcol* 53: 62–76

62 Sitachitta N, Gerwick WH (1998) Grenadadiene and grenadamide, cyclopropyl-containing fatty acid metabolites from the marine cyanobacterium *Lyngbya majuscula*. *J Nat Prod* 61: 681–684

63 Luesch H, Yoshida WY, Moore RE, Paul VJ, Corbett TH (2001) Total structure determination of apratoxin A, a potent novel cytotoxin from the marine cyanobacterium *Lyngbya majuscula*. *J Am Chem Soc* 123: 5418–5423

64 Luesch H, Pangilinan R, Yoshida WY, Moore RE, Paul VJ (2001) Pitipeptolides A and B, new cycodepsipeptides from the marine cyanobacterium *Lyngbya majuscula*. *J Nat Prod* 64: 304–307

65 Jimémez JI, Scheuer PJ (2001) New lipopeptides from the Caribbean cyanobacterium *Lyngbya majuscula*. *J Nat Prod* 64: 200–203

66 Orjala J, Gerwick WH (1996) Barbamide, a chlorinated metabolite with molluscicidal activity from the Caribbean cyanobacterium *Lyngbya majuscula*. *J Nat Prod* 59: 427–430

67 Hooper GJ, Orjala J, Schatzman RC, Gerwick WH (1998) Carmabins A and B, new lipopeptides from the caribbean cyanobacterium *Lyngbya majuscula*. *J Nat Prod* 61: 529–533

68 Koehn FE, Longley RE, Reed JK (1992) Microcolins A and B, new immunosuppressive peptides from the blue-green alga *Lyngbya majuscula*. *J Nat Prod* 55: 613–619

69 Luesch H, Yoshida WY, Moore RE, Paul VJ (2000) Apramides A-G, novel lipopeptides from the marine cyanobacterium *Lyngbya majuscula*. *J Nat Prod* 63: 1106–1112

70 Harrigan GG, Luesch H, Yoshida W, Moore RE, Nagle DG, Biggs J, Park PU, Paul VJ (1999) Tomonoic acids: novel metabolites from a cyanobacterial assemblage of *Lyngbya majuscula* and *Schizothrix calcicola*. *J Nat Prod* 62: 464–467

71 Nagle DG, Zhou Y-D, Park PU, Paul VJ, Rajbhandari I, Duncan CJG, Pasco DS (2000) A new indanone from the marine cyanobacterium *Lyngbya majuscula* that inhibits hypoxia-induced activation of the VEGF promoter in Hep3B cells. *J Nat Prod* 63: 1431–3143

72 Carter DC, Moore RE, Mynderse JS, Niemezura WP, Todd JS (1984) Structure of majusculamide C, a cyclic depsipeptide from *Lyngbya majuscula*. *J Org Chem* 49: 236–241

73 Moore RE, Mynderse JS (1982) Majusculamide C. U.S. Patent 4342751, Aug

74 Luesch H, Moore RE, Paul VJ, Mooberry SL, Corbett TH (2001) Isolation of dolastatin 10 from the marine cyanobacterium *Symploca* species VP642 and total stereochemistry and biological evaluation of its analogue symplostatin 1. *J Nat Prod* 641: 907–910

75 Luesch H, Yoshida WY, Moore RE, Paul VJ (1999) Lyngbyastatin 2 and norlyngbyastatin 2, analogues of dolastatin G and nordolastatin G from the marine cyanobacterium *Lyngbya majuscula*. *J Nat Prod* 62: 1702–1706

76 Mitchell SS, Faulkner DJ, Rubins K, Bushman FD (2000) Dolastatin 3 and two novel cyclic peptides from a Palauan collection of *Lyngbya majuscula*. *J Nat Prod* 63: 279–282

77 Harrigan GG, Yoshida WY, Moore RE, Nagle DG, Park PU, Biggs J, Paul VJ, Mooberry SL, Corbett TH, Valeriote FA (1998) Isolation, structure determination, and biological activity of dolastatin 12 and lyngbyastatin 1 from *Lyngbya majuscula/Schizothrix calcicola* cyanobacterial assemblages. *J Nat Prod* 61: 1221–1225

78 Nogle LM, Williamson RT, Gerwick WH (2001) Somamides A and B, two new depsipeptide analogues of dolastatin 13 from a Fijian cyanobacterial assemblage of *Lyngbya majuscula* and *Schizothrix* species. *J Nat Prod* 64: 716–719

79 Harrigan GG, Luesch H, Yoshida W, Moore RE, Nagle DG, Paul VJ (1999) Symplostatin 2: a dolastatin 13 analogue from the marine cyanobacterium *Symploca hydnoides*. *J Nat Prod* 62: 655–658

80 Persinos G (1998) Update: promising new compounds. *Wash Insight* 11: 6

81 Pettit GR, Kamano Y, Herald CL, Tuinman AA, Boettner FE, Kizu H, Schmidt JM, Baczynskyj L, Tomer KB, Bontems RJ (1987) The isolation and structure of a remarkable marine animal antineoplastic constituent: dolastatin 10. *J Am Chem Soc* 109: 6883–6885

82 Gerwick WH, Jiang ZD, Agarwal SK, Farmer BT (1992) Total structure of hormothamnin A, a toxic cyclic undecapeptide from the tropical marine cyanobacterium *Hormothamnion enteromorphoides*. *Tetrahedron* 48: 2313–2324

83 Nagle DG, Gerwick WH (1995) Nakienones A-C and nakitriol, new cytotoxic cyclic C_{11} metabolites from an Okinawan microalgal (*Synechocystis* sp.) overgrowth of coral. *Tetrahedron Lett* 36: 849–852

84 Barrow RA, Hemscheidt T, Liang J, Paik S, Moore RE, Tius MA (1995) Total synthesis of cryptophycins: revision of the structures of cryptophycins A and C. *J Am Chem Soc* 117: 2479–2490

85 Panda D, Himes RH, Moore RE, Wilson L, Jordan MA (1997) Mechanism of action of the unusually potent microtublue inhibitor cryptophycin 1. *Biochemistry* 36: 12,948–12,953

86 Trimurtulu G, Ohtani I, Patterson GML, Moore RE, Corbett TH, Valeriote FA, Demchik L (1994) Total structures of cryptophycins, potent antitumor depsipeptides from the blue-green alga *Nostoc* sp. strain GSV 224. *J Am Chem Soc* 116: 4729–4737

87 Carmichael WW, Mahmood NA, Hyde EG (1990) Natural toxins from cyanobacteria (blue-greens) algae. *In*: S Hall, G Strichartz (eds): *Marine toxins: origin, structure and molecular pharmacology*. American Chemical Society, Washington, 87–106

88 Carmichael WW, Biggs DF, Gorham PR (1975) Toxicological and pharmacological action of *Anabaena flos-aquae* toxins. *Science* 187: 542–544

89 Watanabe MM, Kaya K, Takamura N (1992) Fate of the toxic cyclic heptapeptides, the microcystins, from blooms of *Microcystis* (Cyanobacteria) in a hypertrophic lake. *J Phycol* 28: 761–767

90 Azevedo SMFO, Evans WR, Carmichael WW, Namikoshi M (1994) First report of microcystins from a Brazilian isolate of the cyanobacterium *Microcystis aeruginosa*. *J Appl Phycol* 6: 261–266

91 Moore RE, Blackman AJ, Cheuk CE, Mynderse JS, Matsumoto GK, Clardy J, Woodard RW, Craig JC (1984) Absolute sterochemistries of the aplysatoxins and oscillatoxin A. *J Org Chem* 49: 2484–2489

92 Bagchi SN, Chauhan VS, Marwah JB (1993) Effect of an antibiotic from *Oscillatoria late-virens* on growth, photosynthesis, and toxicity of *Microcystis aeruginosa*. *Curr Microbiol* 26: 223–228

93 Chauhan VS, Marwah JB, Bagchi SN (1992) Effect of an antibiotic from *Oscillatoria* sp. on photoplankters, higher plants and mice. *New Phytol* 120: 251–257

94 Gross EM, Wolk CP, Juttner F (1991) Fischerellin, a new allelochemical from the freshwater cyanobacteria *Fischerella muscicola*. *J Phycol* 27: 686–692

95 Kohlhase M, Pohl P (1988) Saturated and unsaturated sterols of nitrogen-fixing blue-green algae (cyanobacteria). *Phytochemistry* 27: 1735–1740

96 Ikawa M, Sasner JJ, Haney JF (1994) Lipids of cyanobacterium *Aphanizomenon flos-aquae* and inhibition of *Chlorella* growth. *J Chem Ecol* 20: 2429–2436

97 Ireland CM, Roll DM, Molinski TF, Mckee TC, Zabriskie TM, Swersey JC (1998) Uniqueness of the marine chemical environment: categories of marine natural products from invertebrates. *In*: Fautin DG (ed): *Biomedical importance of marine organisms*. California Academy of Sciences, San Francisco, 41–57

98 Hughes T, Szmant AM, Steneck R, Carpenter R, Miller S (1999) Algal blooms on coral reefs: what are the causes? comments. *Limnol Oceanogr* 44: 1583–1586

99 Hay ME (1991) Fish-seaweed interactions on coral reefs: effects of herbivorous fishes and adaptations of their prey. *In*: PF Sale (ed): *The ecology of fishes on coral reefs*. Academic Press, New York, 96–119

100 Thacker RW, Nagle DG, Paul VJ (1997) Effects of repeated exposures to marine cyanobacterial secondary metabolites on feeding by juvenile rabbitfish and parrotfish. *Mar Ecol Prog Ser* 167: 21–29

101 Paul VJ, Pennings SC (1991) Diet derived chemical defenses in the sea hare *Stylocheilus longicauda* (Quoy et Gaimard 1824). *J Exp Mar Biol Ecol* 151: 227–243

102 Pennings SC, Wiess AM, Paul VJ (1996) Secondary metabolites of the cyanobacterium *Microcoleus lyngbyaceus* and the sea hare *Stylocheilus longicauda*: palatability and toxicity. *Mar Biol* 126: 735–743

103 Nagle DG, Paul VJ, Roberts MA (1996) Ypaoamide, a new broadly acting feeding deterrent from the marine cyanobacterium *Lyngbya majuscula*. *Tetrahedron Lett* 37: 6263–6266

104 Nagle DG, Paul VJ (1998) Chemical defense of a marine cyanobacterial bloom. *J Exp Mar Biol Ecol* 225: 29–38

105 Sitachitta N, Williamson RT, Gerwick WH (2000) Yanucamides A and B, two new depsipeptides from an assemblage of the marine cyanobacteria *Lyngbya majuscula* and *Schizothrix* species. *J Nat Prod* 63: 197–200

106 Pennings SC, Paul VJ (1993) Secondary chemistry does not limit dietary range of the specialist sea hare *Stylocheilus longicauda* (Quoy et Gaimard 1824). *J Exp Mar Biol Ecol* 174: 97–113

107 Pennings SC, Pablo SR, Paul VJ (1997) Chemical defenses of the tropical benthic marine cyanobacterium *Hormothamnion enteromorphoides*: diverse consumers and synergisms. *Limnol Oceanogr* 42: 911–917

108 Nagle DG, Camacho FT, Paul VJ (1998) Dietary preferences of the opisthobranch mollusc *Stylocheilus longicauda* for secondary metabolites produced by the tropical cyanobacterium *Lyngbya majuscula*. *Mar Biol* 132: 267–273

109 Nagle DG, Paul VJ (1999) Production of secondary metabolites by filamentous tropical marine cyanobacteria: ecological functions of the compounds. *J Phycol* 35 Suppl. 607: 1412–1421

110 Nagle DG, Park PU, Paul VJ (1997) Pitiamide, a new chlorinated lipid from a mixed marine cyanobacterial assemblage. *Tetrahedron Lett* 38: 6969–6972

111 Pennings SC, Paul VJ (1993) Sequestration of dietary metabolites by three species of sea hares: location, specificity, and dynamics. *Mar Biol* 117: 535–546

112 Tsuda RT, Bryan PG (1973) Food preference of the juvenile *Siganus rostratus* and *S. Spinus* in Guam. *Copeia* 3: 604–606

113 Bryan PG (1975) Food habits, functional digestive morphology, and assimilation efficiency of the rabbitfish *Siganus spinus* (Pisces, Siganidae) on Guam. *Pac Sci* 29: 269–277

114 Serdula M, Bartolini G, Moore RE, Gooch J, Wiebenga N (1982) Seaweed itch on windward Oahu. *Hawaii Med J* 41: 200–201

115 Harder R (1917) Ernährungsphysiologische Untersuchungen an Cyanophyceen, hauptsächlich dem endophytischen *Nostoc punctiformae*. *Z Bot* 9: 154–242

116 Rice EL (1984) *Allelopathy*. Academic Press, Orlando

117 Akehurst SC (1931) Observations on pond life with special reference to the possible causation of swarming of phytoplankton. *R Microsc Soc J* 51: 237–265

118 Inderjit, Dakshini KMM (1994) Algal allelopathy. *Bot Rev* 60: 182–196

119 Lefevre M, Jakob H, Nisbet M (1950) Sur la secretion par certaines Cyanophytes, de substances algostatiques dans leur collections d'eau naturelles. *C R Acad Sci* 230: 2226–2227

120 Keating KI (1977) Allelopathic influence on blue green bloom in a eutrophic lake. *Science* 196: 885–887

121 Keating KI (1978) Blue green algal inhibition of diatom growth: transition from mesotrophic to eutrophic community structure. *Science* 199: 971–973

122 Schlegel I, Doan NT, de Chazal N, Smith GD (1999) Antibiotic activity of new cyanobacteria isolates from Australia, Asia against green algae and cyanobacteria. *J Appl Phycol* 10: 471–479

123 Gross E (1999) Allelopathy in benthic and littoral areas: case studies on allelochemicals of benthic cyanobacteria and submerged macrophytes. *In*: Inderjit, KMM Dakshini, CL Foy (eds): *Principles and practices in plant ecology: allelochemical interactions*. CRC Press, Boca Raton, 179–199

124 Papke U, Gross EM, Francke W (1997) Isolation, identification and determination of the absolute configuration of fischerellin B: a new algicide from the freshwater cyanobacterium *Fischerella muscicola* (Thuret.). *Tetrahedron Lett* 38: 379–382

125 Srivastava A, Juttner F, Strasser RJ (1998) Action of the allelochemical, Fischerellin A on photosystem II. *Biochim Biophys Acta* 1364: 326–336

126 Pushparaj B, Pelosi E, Jüttner F (1999) Toxicological analysis of the marine cyanobacterium, *Nodularia harveyana*. *J Appl Phycol* 10: 527–530

127 Flores E, Wolk CP (1986) Production, by filamentous, nitrogen-fixing cyanobacteria, of a bacteriocin and of other antibiotics that kill related strains. *Arch Microbiol* 145: 215–219

128 Todorova AK, Jüttner F, Linden A, Plüss T, von Philipsborn W (1995) Nostocyclamide: a new macrocyclic, thiazole-containing allelochemical from *Nostoc* sp. 31 (cyanobacteria). *J Org Chem* 60: 7891–7895

129 Jüttner F, Todorova AK, Walch N, von Philipsborn W (2001) Nostocyclamide M: a cyanobacterial cyclic peptide with allelopathic activity from *Nostoc* 31. *Phytochemistry* 57: 613–619

130 Smith GD, Doan NT (1999) Cyanobacterial metabolites with bioactivity against photosynthesis in cyanobacteria, algae and higher plants. *J Appl Phycol* 11: 337–344

131 Doan NT, Richards RW, Rothschild JM, Smith GD (2000) Allelopathic action of the alkaloid 12-*epi*-hapalindole E-isonitrile and calothrixin A from cyanobacterium of the genera *Fischerella* and *Calothrix*. *J Appl Phycol* 12: 409–416

132 Utkilen H, Gjølme N (1992) Toxin production by *Microcystis aeruginosa* as a function of light

in continuous cultures and its ecological significance. *Appl Environ Microbiol* 58: 1321–1325

133 Van der Westhuizen AJ, Eloff JN (1985) Effect of temperature and light on the toxicity and growth of the blue-green alga *Microcystis aeruginosa*. *Z Pflanzenphysiol* 110: 157–163

134 Watanabe MF, Oishi S (1985) Effect of environmental factors as toxicity of a cyanobacteria (*Microcystis aeruginosa*) under culture conditions. *Appl Environ Microbiol* 49: 1342–1344

135 Gross EM, von Elert E, Jüttner F (1994) Production of allelochemicals in *Fischerella muscicola* under different environmental conditions. *Verh Int Verein Limnol* 25: 2231–2233

136 Skulberg OM (2000) Microalgae as a source of bioactive molecules – experience from cyanophyte research. *J Appl Phycol* 12: 341–348

137 Inderjit, Dakshini KMM (1997) Effects of cyanobacterial inoculum on soil characteristics and cereal growth. *Can J Bot* 75: 1267–1272

138 Inderjit, Weiner J (2001) Plant allelochemical interference or soil chemical ecology? *Perspect Plant Ecol Evol Syst* 4: 4–12

139 Keating KI (1999) Allelochemistry in plankton communities. *In*: Inderjit, KMM Dakshini, CL Foy (eds): *Principles and practices in plant ecology: allelochemical interactions.* CRC Press, Boca Raton, 165–178

140 Pflugmacher S, Wiegand C, Beattie KA, Krause E, Steinberg EW, Codd GA (1998) Uptake, effects, and metabolism of cyanobacterial toxins in the emergent reed plant *Phragmites australis* (Cav.) Trin. Ex Steud. *Environ Toxicol Chem* 20: 846–852

141 Pflugmacher S, Wiegand C, Beattie KA, Krause E, Codd GA, Steinberg EW (1998) Uptake of the cyanobacterial hepatotoxin microcystin-LR by aquatic macrophytes. *J Appl Bot* 72: 228–232

Phytotoxic compounds from *Typha domingensis* P.

Maria T. Gallardo-Williams[1] and Dean F. Martin[2]

[1] *Laboratory of Molecular Toxicology, National Institute of Environmental Health Sciences, P.O. Box 1233, Mail drop E1-06, Research Triangle Park, NC 27709, USA*
[2] *Institute for Environmental Studies, Department of Chemistry –SCA 400, University of South Florida, 4202 East Fowler Avenue, Tampa, FL 33620-5202, USA*

Introduction

Cattails make up the genus *Typha* of the family Typhaceae, and are probably the most familiar of all wetland plants in the world. Their brown flower clusters can be seen at the edges of ponds, rivers, lakes, or just about any place where there is shallow, standing water for at least part of the year. Cattails are tall, erect plants that may grow to 6 to 8 feet tall. They usually grow along the shoreline but may also grow in water 3 to 4 feet deep. Creeping rootstalks and seeds spread cattails. Different species of cattail occur commonly in wet soil, marshes, swamps, and shallow fresh and brackish waters throughout the world. The genus consists of 10 species that have significant differences in physical appearance, allowing for identification. The broad-leafed cattail, *T. domingensis*, is widely distributed in southern Florida. The species has long pistillate, separated from the upper stamen, and the blades are broad and less convex than other species occurring in the area, such as *T. latifolia* [1].

Cattail seeds are small and tightly packed in characteristic pods. The pods first appear in early summer and are green colored. As they mature in the fall and following spring, the pods turn brown and release the tiny seeds that can be dispersed by wind or transported by water or water fowl [2]. Germination of *Typha* seeds, which can remain viable for up to two years, requires uninfested and unshaded areas [2]. Seedlings are reported to be rare in undisturbed stands [3–6]. Because of the small size of seeds, however, it is unclear whether the lack of regeneration from seed is the result of an inhibition of germination or because of seedling mortality.

Utility

Cattails have several useful characteristics. The plants and their flowers are edible to humans and animals during certain stages of growth [7]; cattail stands are preferred by muskrats and beavers for feeding and nesting [8]. To the wetland community, cattails offer a resting place for nesting and protective cover [2]. Of all wild plants, cattails have been called the most useful emergency food source and traditionally have been important foods for native peoples. The root-like rhizomes contain a firm core that can be eaten raw, roasted, or boiled. It is pungent, fibrous, and rich in edible starch, which can be separated from the fiber as a flour. Even the yellow pollen is edible [7]. In times of food scarcity, several nations have considered cattail rhizomes as a source of starch and the seeds as a source of edible oil and animal feed. Aside from many food products and medicines, Native Americans used cattails for basket weaving and even as a roofing material [9]. Cattails are also used in traditional Chinese medicinal formulations as a hemostatic agent, which is useful in the treatment of stagnant blood and bloody stools [10].

Nuisance characteristics

As useful as cattails can be, when cattail populations spread in an uncontrolled manner, they become noxious weeds. An aquatic weed is defined as "an aquatic plant which, when growing in abundance, is not desired by the manager of its place of occurrence" [7]. Cattails can become weeds that present serious problems. Fast growing, they invade new waterways, rice fields, other irrigated agricultural lands, farm ponds, lakes, and canals and must be rigorously controlled [9].

The ability of cattails to expand over areas that were previously occupied by other species has been the subject of many studies [11]. One species can be troublesome in wetland environments of South Florida. *T. domingensis* is believed to be a natural component of the Everglades ecosystem, occurring largely in scattered, diffuse stands [12]. The problem is that in many cases, it has become the dominant marsh species, out-competing native plants such as sawgrass (*Cladium jamaicense*) [13]. Its uncontrolled growth can lead to the formation of dense, monotypic stands and even to the clogging of waterways. Cattails are particularly prone to uncontrollable spread in recently disturbed sites; cattail invasions often follow natural or anthropogenic disturbances [14, 15].

Mechanisms of cattail dominance

General characteristics and allelopathy

In Florida public waters, cattails are the most dominant of the emergent species of aquatic plants, according to Schardt [16]. There are several factors that enable the cattail to accomplish its opportunistic expansion, including its size, growth habits, adaptability to changes in the surroundings, and the release of compounds that can prevent the growth of other species. This chemical-signaling mechanism is what we know as allelopathy: "any indirect or direct harmful effect by one plant on another through production of chemical compounds that are released into the environment" [17].

It is possible that the reason for the excessive and undesirable expansion of cattails is a combination of the above-mentioned factors. To date, some studies have looked at different aspects of the problem. Several of the mechanisms that have been proposed to explain the widespread occurrence of cattail stands should be reviewed.

Nutrient factors

Nutrient dominance has been proposed as a possible cause of cattail infestations [13]. Accordingly, shallow waters tend to have low cattail populations, while deeper waters have denser growth. In shallow areas, because of the inability of cattails to store adequate nutrient reserves for the winter, cattail expansion is restricted. In deeper, nutrient-rich waters, cattails have optimal growth conditions, and this environment provides for domination by the opportunistic *Typha*, inhibiting the growth of other marsh plants [13]. Once established, these cattail communities can replace native plants, depriving native animals of their natural habitats and food supplies. However, in experimental mixed cultures, and in the absence of nutrient limitations, cattails are able to outperform other weeds [18, 19], perhaps indicating the contribution of other factors.

Evidence for chemical dominance and autotoxicity

Even though cattails produce thousands of very small, wind-dispersed seeds with high viability, cattail seedlings are uncommon in undisturbed stands [3, 4]. This observation has prompted some studies designed to investigate the possibility of autotoxicity of cattail to the germination of its own seeds. McNaughton [3] pioneered such studies and found complete inhibition of the germination of viable cattail seeds (*T. latifolia*) when exposed to aqueous extracts of cattail leaves. Slight inhibition was observed with water from cattail marshes and high inhibition, with water squeezed from soil where cattails were growing. Moreover, he showed that upon treatment with an adsorbent material, the autotoxic effect could be eliminated [3]. Van der Valk and Davis [4] also found positive evidence for autotoxic behavior in *T. glauca*. But Sharma and Gopel [20], working with *T. angustata* and *T. elephantina*, and Grace [21], who looked at *T. latifolia*, found no evidence of such autotoxic effect under similar (but not identical) conditions.

The work of Szczepanska [18, 19] and her collaboration with Szczepanski [22] have been very important contributions to the understanding of the mechanisms of chemical dominance among the aquatic plants. By carefully planting mixed cultures of *T. latifolia* and other aquatic weeds (specially *Phragmites australis*) under controlled conditions to compensate for shade and nutrient competition, the interaction between the plants was determined on the basis of growth and production characteristics. Results, described in the literature, provide ample evidence of a chemical (allelopathic) interaction between the cattail and the other wetland species, but with some variability as indicated by these examples:

- First, despite Szczepanska's predictions, Bonasera et al. [23] found only very low inhibitory effects when testing weak aqueous extracts of *T. latifolia* against lettuce, radish, cucumber, and tomato.
- Second, Elakovich and Wooten [24] found similar results against lettuce and duckweed.
- Third, Prindle and Martin [25] found that aqueous extracts from different portions of *T. domingensis* could inhibit the growth of lettuce and radish seeds; the same extracts were also able to inhibit the oxygen production of *Lyngbya majuscula* [25]. The observed phytotoxic properties were conserved even upon autoclaving at 120°C for one hour.

There are many sources of variability that could explain the contradictory accounts found in the literature. It is known that the phytotoxins produced by aquatic macrophytes are neither highly specific nor very potent. The phytotoxic effects observed in aquatic ecosystems are most likely due to the effect of several compounds that are, individually considered, nontoxic in nature [26]. Seasonal variations [27], phytochemical induction [28], and other factors must be considered as sources of variability that can influence the observed phytotoxic effects.

Factors affecting the production of allelopathic materials

As mentioned above, many different factors affect the release of phytotoxins at a given time. The main natural abiotic factors affecting allelopathic interactions are moisture, heat, light levels, nutrient deficiencies, and organic content of the soil. Other abiotic sources of variability include seasonal variations in phytotoxin levels (a combined heat-light-moisture effect). Seasonal variation was investigated by Lohdi [29] and subsequently by Dolling et al. [27] and Nilsson et al. [30]. Lohdi [29] found that toxin levels in forest soils decrease in the fall. Although each of these works finds that toxin levels peak at different times, in every case, there is a seasonality associated with the release of the allelopathic compounds.

Recently, we have found inhibitory effects of cattail on the growth and oxygen production of *Salvinia minima* [31, 32] as well as detrimental effects of cattail on the germination of other wetland species [33]. In our own experience, there are many sources of variability that could account for the contradictory literature. The phytotoxins produced by aquatic macrophytes are neither highly specific nor very potent [36]. Seasonal variations [27], phytochemical induction, and uptake of compounds of anthropogenic origin [34] are some of the major sources of variability, together with inconsistencies in the extraction procedure.

Bioassays

We have examined the effect of concentration and decomposition time on the phytotoxic properties of aqueous extracts of *T. domingensis* [35]. Cattail tissues contained water-soluble materials that were toxic to germination and growth of lettuce and cattail seeds. A strong correlation was found between extract concentration and toxicity. At very low concentrations, cattail aqueous extracts stimulated germination and root elongation, but as the concentration increased, the phytotoxic effects became apparent. The phytotoxic effects observed included germination inhibition, inversion of the seedlings, and inhibition of growth. The extracts became generally more toxic after a week of decomposition at 23°C. Longer decomposition time, however, made the extracts less active. Addition of increasing amounts of activated charcoal removed the phytotoxic properties; there was a direct correlation between the amount of adsorbent used (0 to 1.5 g) and the decrease in phytotoxicity under the conditions of the bioassay.

For all bioassays, standard statistical comparisons were made with mean values and standard deviations or standard errors. We used Student's t-tests to test for the statistical validity of differences between test and control runs at the $P = 0.05$ level (or less).

Modeling bioassay responses

When the cattail extracts were bioassayed against lettuce seeds, it was found that the phytotoxic effects followed the mathematical models suggested by An et al. [36]. At very low phytotoxin concentrations, the extracts were stimulatory to germination and growth; as the concentrations increased, the extracts became inhibitory (Fig. 1). In this case, graphs can be constructed that approximate the proposed allelopathic response. The model implies that as concentration changes, the relative domination of stimulation and inhibition by the allelochemical mixture is altered and that the biological response of the test organism (the biological property of the allelochemical) will comprise responses to both stimuli. The response is also highly organism-specific [37]. This is particularly clear if we consider root elongation as a diagnostic parameter [37]. Even at concentrations that reduce the percent of germinated seedlings, the net effect on root elongation typically is stimulatory.

Practical implications

In order to assess the practical ecological significance of these findings, bioassays were performed using cattail seeds as test organisms. In general, the cattail seeds followed the same percent germination and dry weight trends as the lettuce seeds, which seemed to indicate that the observed effects correlate well

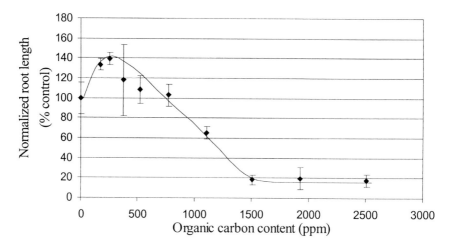

Figure 1. Effect of increasing concentrations of cattail extracts on the root length of lettuce seedlings. Mean values ± standard error.

for the two species [33]. However, in the case of the cattail bioassay, the standard deviations of the measurements were higher than for the lettuce, indicating the non-domesticated nature of the seeds used. Larger variations between individuals in the wild population were responsible for more scattered data. Cattail seedlings are also very fragile, and for this reason, root elongation was not determined for the cattail bioassay.

In practical terms, these findings serve to reconcile the conflicting literature. Small differences in concentration can mean a different response; different test organisms and the parameters used to evaluate the effect also become important [38]. Inconsistencies in the literature can be attributed to differences in sampling times, extract preparation, sources of plant material, and extract concentration. It is important to standardize the actual concentration of the extracts under study in order to make a meaningful correlation between different sets of data.

The phytotoxic effect on root elongation increased with decomposition time in the first week of the study (Fig. 2) and decreased afterwards. However, the effect of the extracts on germination inhibition was less than the effects of the fresh extract. This observation accounts for higher dry weights, since more seedlings were germinating, but has less viability because those seedlings that germinated had short, stunted roots. Without the root elongation data, it would appear as if the 7-day decomposition of the extract removed the phytotoxic properties, but upon inclusion of these data, it becomes apparent that the microbial decomposition actually serves to increase the phytotoxicity of the extracts. By propitiating a fitness-reducing event [39] that would eventually lead to increased mortality, the extracts are potentially useful as weed-management agents.

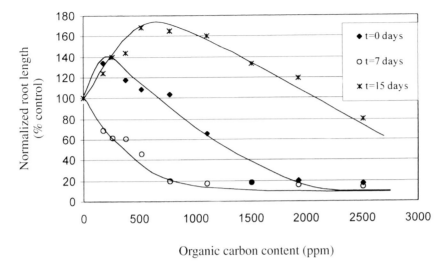

Figure 2. Effect of decomposition time of cattail extracts of varying concentrations on the root length of lettuce seedlings.

Almost all inhibitory activity toward root elongation was lost after decomposition proceeded for 15 days (except for the most concentrated extract); germination inhibition was lower than for the fresh or 7-day extracts, and dry weights showed little variation [35]. This trend—initial increase in toxicity followed by decline probably due to microbial activity—is in general agreement with the literature [37, 40].

We believe that microbial decomposition plays an important role in the dynamics of the phytotoxic interaction of cattail with other species. In natural systems, continuous release of fresh toxins ensures the persistence of the phytotoxic effects, but microbial decomposition enhances the overall inhibition and broadens the spectrum of species that can be affected.

Phytotoxins from cattails

There is ample evidence regarding the presence of a number of biologically active substances in cattail tissue (see Fig. 3). Eleven kinds of phenolic compounds have been detected in female flowers of *T. latifolia* [41]. The compound present in the greatest concentration was 5-*trans*-caffeoylshikimic acid (**1**), an enzymatic browning substrate that was considered to be involved in the characteristic browning of the cattail flowers. Other phenolic substances isolated and identified were epicatechin (**2**) and afzelechin (**3**), but their relative amounts were very low. The rest of the materials could not be fully identified at the time.

(1)

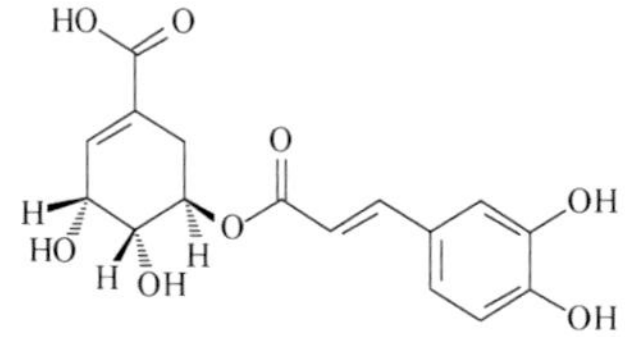

(2)

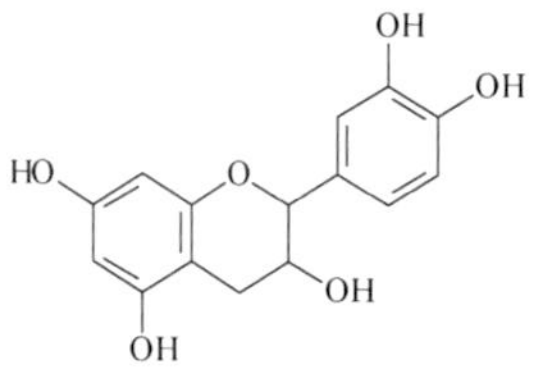

(3)

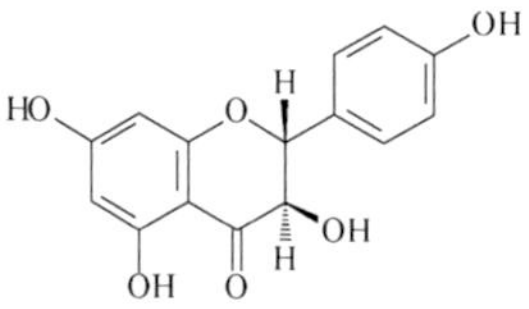

(4)

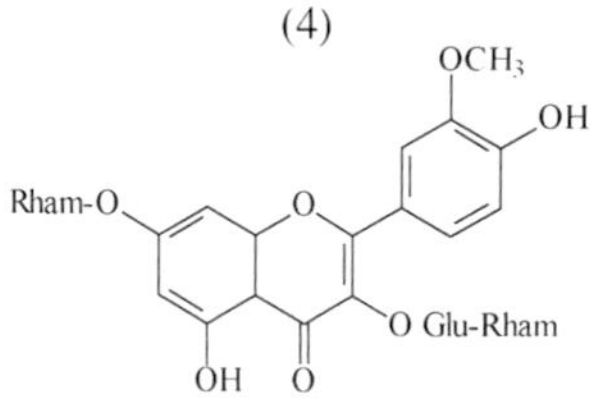

(5)

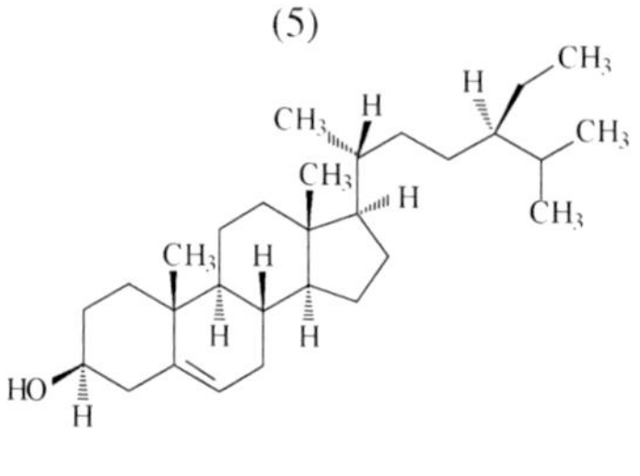

(6)

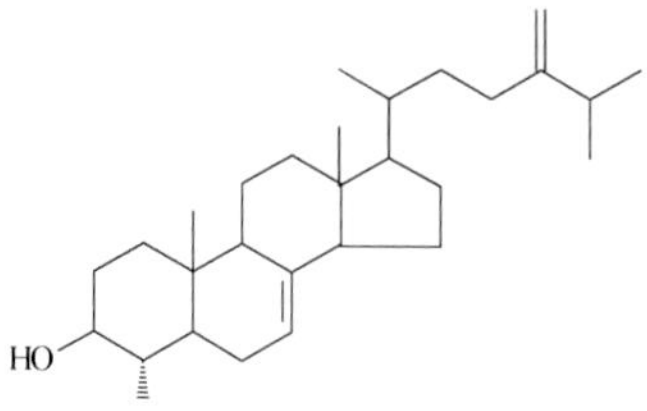

(7)

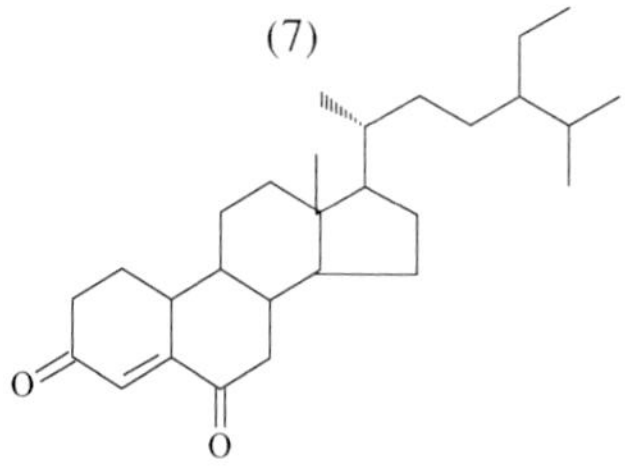

(8)

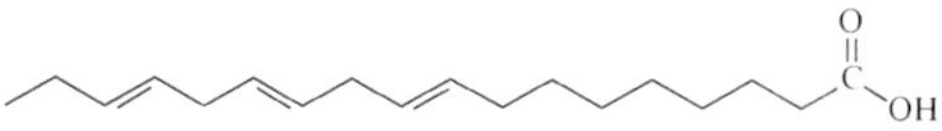

(9)

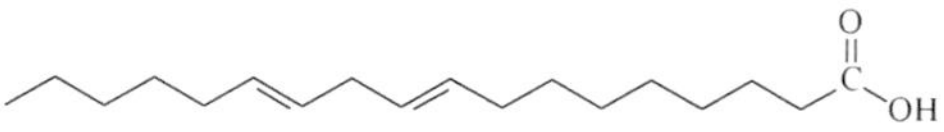

(10)

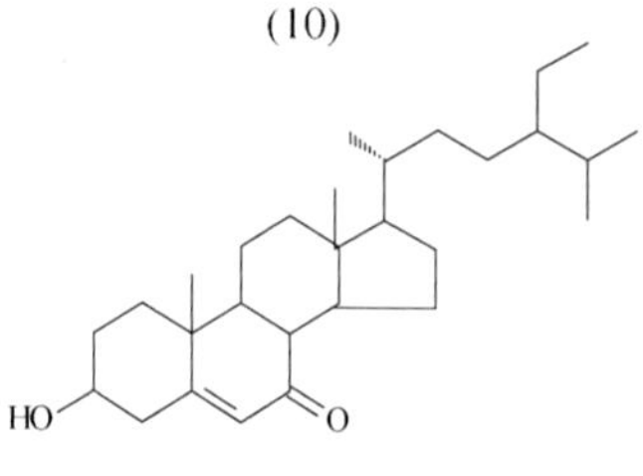

Figure 3. Compounds with potential biological activity isolated from cattails. (Continued on page 66.)

Extensive work by several different groups has concentrated on the isolation and characterization of new bioactive materials from *Typha*.

• By following the antihemorrhagic properties of the pollen of *T. latifolia*, Ishida et al. [10] isolated a new flavonol glucoside (isorhamnetin 3-rutinoside-7-rhamnoside) (**4**) from concentrated extracts obtained by refluxing the pollen with hot water. The compound was tested and found to have hemostatic action in mice [10].

• Aliotta et al. [42] successfully isolated three steroids (β-sitosterol (**5**), (20*S*)24-methylenlophenol (**6**), and stigmast-4-ene-3,6-dione (**7**) and three fatty acids (α-linolenic (**8**), linoleic (**9**) and an unidentified $C_{18:2}$ fatty acid) from organic extracts of dried *T. latifolia*. The inhibitory action of the isolated compounds was tested against several algal strains. The compound with the highest inhibitory potential was α-linolenic acid (**8**), which had activity similar to $CuSO_4$ (a commercial algaecide) and was active against 10 different algal strains at low concentration (0.5 mg crude *versus* 0.5 and 1 μmole $CuSO_4$).

• Several free and acylglucosilated stigmasterols (7-oxositosterol (**10**), 7β-hydroxysitosterol (**11**), and, 7α-hydroxysitosterol (**12**), as well as a newly reported enone (**13**), dione (**14**), and enedione (**15**); (see Fig. 1) were also found in *T. latifolia*, from ether extracts of the dry tissue [43]. The oxidation pattern of the stigmasterols corresponds to that obtained by microbial oxidation of cholesterol [44] and suggests the presence in the cattail of an enzymatic system able to transform β-sitosterol into the sterols reported. However, it is also possible that these compounds are being produced through an autooxidation process of β-sitosterol and stigmast-4-ene-3,6-dione.

• The structure of (20*S*)-4α-metylenecholest-7-en-3β-ol (**16**), another sterol isolated from *T. latifolia*, was determined by means of spectroscopic studies [45]. This compound (2 μmol) was tested in a paper disk bioassay against a cyanobacterium, *Anabaena flos-aquae,* and a green alga, *Chlorella vulgaris.* The compound was found to be a growth inhibitor for these two species.

• Two carotenoid-like compounds, Blumenol A (**17**) and the novel (3*R*,5*R*,6*S*,9ϵ)-5,6-epoxy-3-hydroxy-β-ionol, also have been isolated from methanolic extracts of *T. latifolia* [46]. No phytotoxic properties have been reported for these compounds yet.

• A different set of compounds has been isolated from ether extracts of dry *T. latifolia*: 5β,8β-epidioxyergosta-6,22-dien-3β-ol (**19**), 5α,8α-epidioxyergosta-6,9(11), 22-trien-3β-ol (**20**), and 5α,8α-epidioxyergosta-6,22-trien-3β-ol (**21**) [47]. In the cattail tissue, the three compounds were found in a 1:2:18 ratio, as determined by high performance liquid chromatography (HPLC). The ergosterol epidioxide (**19**) has been isolated from other natural sources and it is considered a true metabolite by some [48, 49].

• Research dealing with aqueous extracts of *T. domingensis* has noted its allelopathic properties [25]. It was found to inhibit the germination of lettuce and radish seeds and also to reduce the oxygen production rates of the

(11)

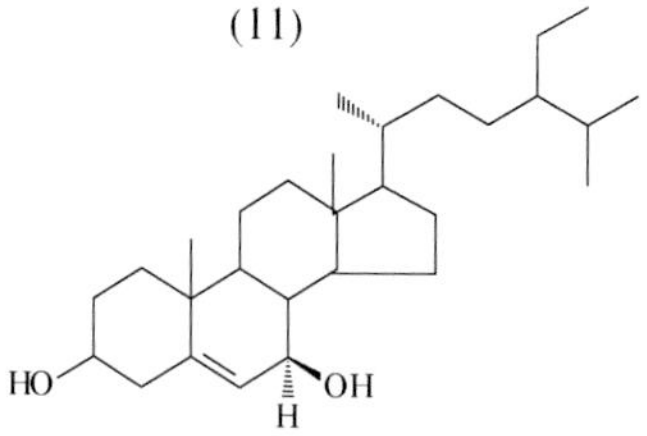

(12)

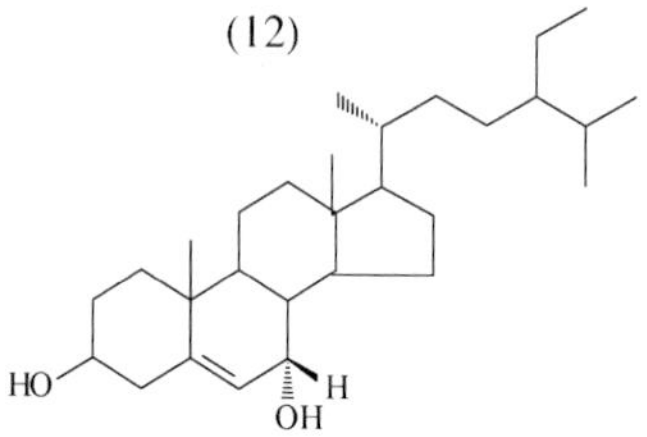

(13)

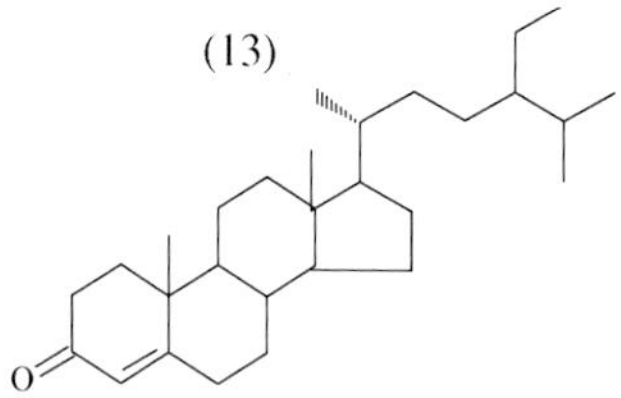

(14)

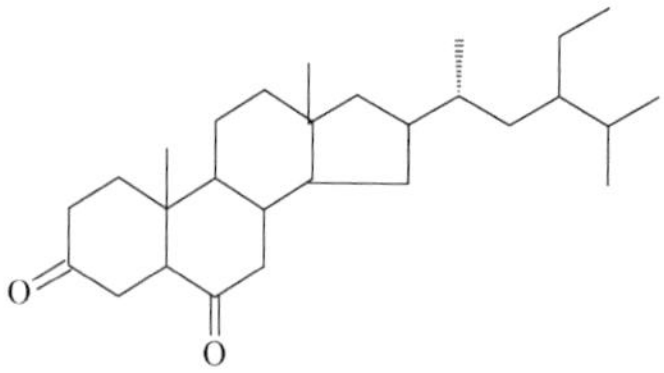

(15)

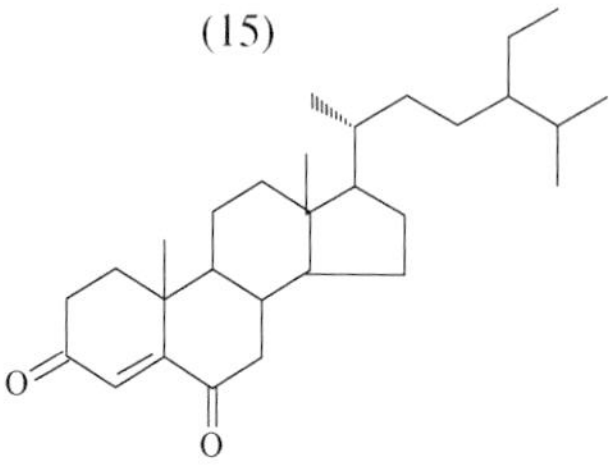

(16)

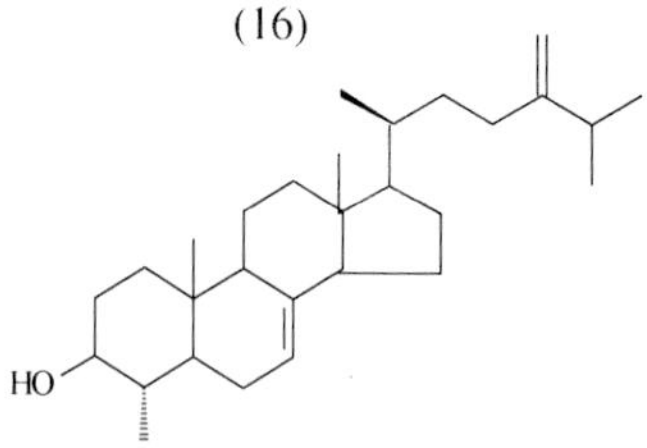

(17)

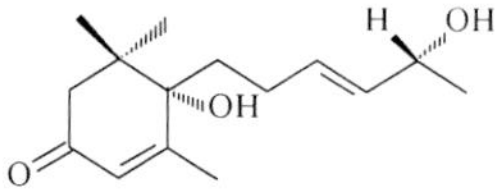

(18)

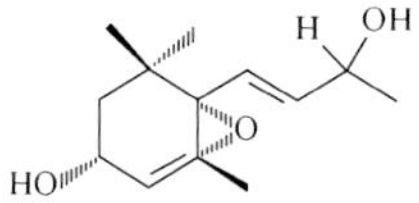

(19)

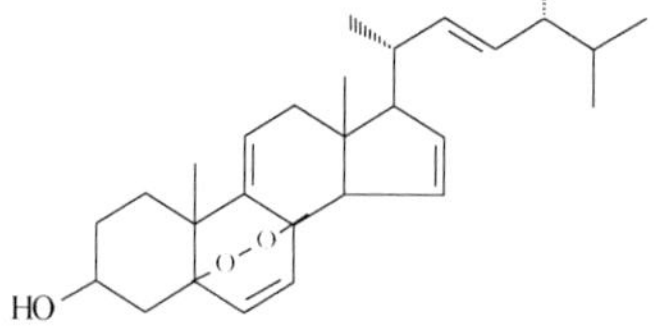

(20)

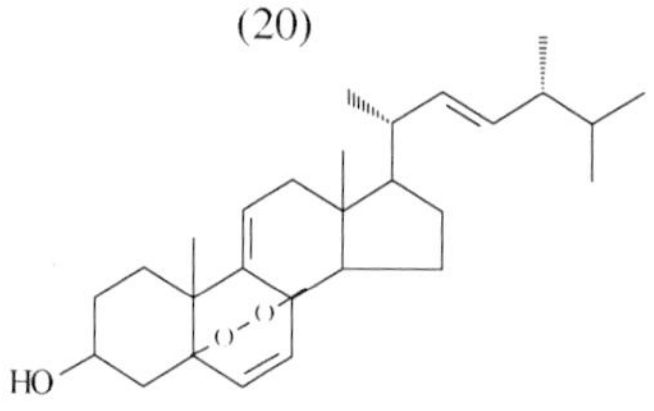

Figure 3 (Continued).

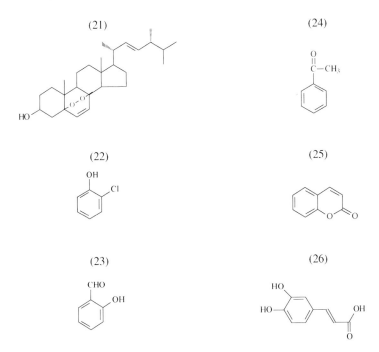

(21)   (24)

(22)   (25)

(23)   (26)

Figure 3 (Continued).

filamentous alga *Lyngbya majuscula*; the extracts were found to be active even after autoclaving. When studied using gas chromatography/mass spectrometry (GC/MS), two of the major components of the extracts were found to be 2-chlorophenol (**22**) and salicylaldehyde (**23**) [50]. The relative concentrations of these compounds in the extracts were unknown at the time. These compounds have been found to be present in sediments in the *T. domingensis* growth front but not in sediments 2 m away from the cattail infestation [51].

- In order to determine whether the compounds previously isolated from *T. domingensis* were true available metabolites, cattails were grown in a controlled environment (Environmental Growth Chamber) for a period of eight weeks prior to the preparation of aqueous extracts. The conditions were constant temperature (26°C), 12-hour photoperiod with a light intensity of 190 $\mu Esm^{-2}sec^{-1}$ (as measured by a LI-COR model LI-185A photometer) at ground level and a relative humidity of 80%. A bioassay-driven separation procedure was used to isolate the fractions of potential activity from the non-active materials [33]. In addition to the linoleic and α-linolenic acids previously reported in *T. latifolia* [42], acetophenone (**24**), coumarin (**25**), and caffeic acid (**26**) were isolated and characterized by GC/MS. All five compounds came from biologically active fractions, all were identified by GC/MS, and all are known to be biologically active [54].

- Cattail leachates also have been investigated [33]. The motivation for studying cattail leachates was to obtain an accurate representation of the bioavailability of secondary metabolites from cattails in the aquatic ecosystem. The concentration of bioactive materials in the leachates was very low, and extraction of very large volumes with activated charcoal was used as a preconcentration step.

The diethyl ether extract of the adsorbent material contained several fractions that possessed distinct phytotoxic activity against lettuce seeds. Upon analysis, the following compounds were identified: linoleic, α-linolenic, caffeic, gallic, and p-coumaric acids.

The phenolic acids found in the cattail leachates had not been previously described in cattail tissue, but they are known to be common in aquatic macrophytes and are responsible for many allelopathic interactions [52]. These compounds, which are only mildly toxic [53], were isolated from the last fractions eluted from the chromatography column. The two fatty acids were the major components of the cattail leachates, comprising more than 80% of the total isolated material.

Cattail mulches

Once it became clear that cattail extracts had phytotoxic properties that could be used for the management of undesirable species, we realized that there was a need to develop a suitable vehicle for the dispersal of the phytotoxins. Aqueous extracts, although environmentally friendly and very potent when freshly prepared, seemed an improbable alternative because of the magnitude and extension of the problem. Aquatic nuisance species are not confined to small ponds; usually the challenge is to manage species that cover large areas.

The use of dried materials, processed to a small but manageable particle size, offered an alternative to the use of liquid preparations. Mulches were thus prepared under different conditions. Once dried, it was found that the mulches retain their phytotoxic properties [33].

In the experiment, the dried material was placed underneath a piece of filter paper and saturated with water. Black Seeded Simpson lettuce seeds were used as a target and were placed on top of the filter paper (Fig. 4). This was done in order to assess the phytotoxic potential of the different mulches while excluding the effect of shading. However, in a realistic field situation, the seeds or plants to be inhibited would be under the mulch. In such a situation, the effect of shading combined with the phytotoxicity would make the mulch more effective.

The mulch prepared using cattail roots cut into small (~1–2 cm) pieces was found to be very effective as a germination inhibitor in this study, reducing the germination of lettuce by about 80%. Mulches have the potential to become suitable vehicles for the dispersal of phytotoxins. Once in the soil, the phytotoxic effects are retained for a limited time (in our experience, mulches were

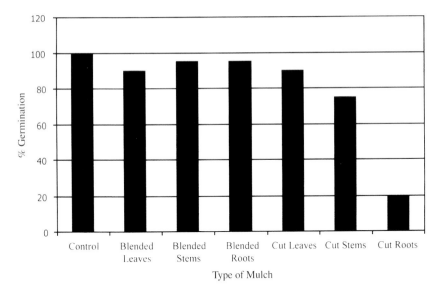

Figure 4. Germination of lettuce seeds in the presence of *Typha domingensis* mulches. Samples dried in Environmental Growth Chamber.

not inhibitory after being stored for three weeks). After that time, phytotoxic properties are lost as the mulch is decomposed by soil microorganisms and is incorporated into the soil as organic matter.

Implications

A better understanding of the chemical basis for the ecological processes in which cattails (*Typha domingensis*) participate in the wetland environment is essential for the formulation of control strategies for this very invasive plant. At the same time, such knowledge is important as a building block in the elucidation of the large-scale events that take place in the aquatic ecosystem.

Cattails are currently one of the most invasive emergent species in the world, and millions of dollars are spent each year in efforts to manage its undesirable expansion with conventional herbicides. However, the cattail populations keep gaining territory, particularly in ecosystems that have undergone recent disturbances. Perhaps herbicide spraying is not the solution to this problem; rather, the solution to cattail expansion may be linked to its use as a crop and in the development of useful products.

Research has shown that cattail phytotoxins are available in cattail extracts and leachates. Also, we know that the type of compounds isolated from cattail aqueous by-products (fatty acids and phenolic acids) are inhibitory towards the germination and growth of a number of ecologically relevant targets. The mode of action for the cattail extracts is in accordance with the models proposed for allelopathic materials and is affected by microbial decomposition.

Cattail mulches could be used as natural ground covers in restoration projects. Since the mulches are phytotoxic to seed germination and development, their use could mitigate recolonization of recently restored environments by undesirable species present in the seed bank, as well as by invasive species such as cattail itself. Even though the mulches may not be useful as a standalone tactic and may require some intervention in order to achieve total weed control, the growth suppression provided by the cattail phytotoxins could reduce post-emergence herbicide rates and the frequency of cultivation.

Acknowledgments
We thank Mrs. Barbara B. Martin for editorial assistance.

References

1 Long RW, Lakela O (1976) *A flora of tropical Florida.* Banyan Books, Miami
2 Pieterse A, Murphy K (eds) (1993) *Aquatic weed: the ecology and management of nuisance aquatic vegetation.* Oxford University Press, New York
3 McNaughton SJ (1968) Autotoxic feedback in relation to germination and seedling growth in *Typha latifolia. Ecology* 49: 367–369
4 van der Valk A, Davis CB (1978) The role of seed banks in the vegetation dynamics of prairie glacial marshes. *Ecology* 59: 322–335
5 Grace JB, Wetzel RG (1981) Habitat partitioning and competitive displacement in cattails (*Typha*): experimental field studies. *Am Nat* 118: 4474
6 Grace JB, Wetzel RG (1981) Phenotypic and genotypic components of growth and reproduction in *Typha latifolia*: experimental studies in marshes of differing successional maturity. *Ecology* 62: 789–801
7 Peterson LA (1977) *A field guide to edible wild plants. Eastern and Central North America.* Houghton Mifflin Co., Boston
8 Fink DF (1994) *A guide to aquatic plants.* Minnesota Dept. of Natural Res. Minneapolis, MN
9 National Academy Of Sciences (1976) *Making aquatic plants useful: some perspectives for developing countries.* National Academy of Sciences, Washington
10 Ishida H, Umino T, Tsuji K, Kosuge T (1988) Studies on the antihemorrhagic substances in herbs clasified as hemostatics in Chinese medicine IX. On the antihemorrhagic principles in *Typha latifolia* L. *Chem Pharm Bull* 36: 4414–4420
11 Gallardo MT, Martin BB, Martin DF (1998) An annotated bibliography of allelopathic properties of cattails, *Typha* spp. *Florida Scient* 61: 52–58
12 Davis SM (1994) Phosphorus inputs and vegetation sensitivity in the Everglades. *In:* SM Davis, JC Odgen (eds): *Everglades: the ecosystem and its restoration.* St. Lucie Press, Delray Beach, 357–378
13 Toth LA (1988) Effects of hydrologic regimes on lifetime production and nutrient dynamics of cattail. Technical publication 88-6, South Florida Water Management District, West Palm Beach, Florida
14 Grace JB (1987) The impact of preemption on the zonation of two *Typha* species along lakeshores. *Ecol Monogr* 57: 283–303
15 Hellsten S, Dieme C, Mbengue M, Janauer GA, Den Hollander N, Pieterse AH (1999) *Typha* control efficiency of a weed-cutting boat in the Lac de Guiers in Senegal: a preliminary study on mowing speed and re-growth capacity. *Hydrobiologia* 415: 249–255
16 Schardt JD (1997) 1994 *Florida aquatic plant survey. Dominant plants in public waters.* Tech. Report # 972-CGA. Bureau of Aquatic Plant Management, Florida Department of Natural Resources, Tallahassee, Florida

17 Rice EL (1984) *Allelopathy* (2nd ed.) Academic Press, Orlando

18 Szczepanska W (1971) Allelopathy among the aquatic plants. *Pol Arch Hydrobiol* 18: 17–30

19 Szczepanska W (1987) Allelopathy in halophytes. *Arch Hydrobiol Beih* 27: 173–179

20 Sharma KP, Gopel B (1978) Seed germination and occurrence of *Typha* species in nature. *Aquat Bot* 4: 353–358

21 Grace JB (1983) Autotoxic inhibition of seed germination by *Typha latifolia*: an evaluation. *Oecologia* 59: 366–369

22 Szczepanska W, Szczepanski A (1982) Interactions between *Phragmites Australis* (Cav.) Trin. Ex Steud. and *Typha latifolia* L. *Ekolog Pol* 30: 165–186

23 Bonasera J, Lynch J, Leck MA (1979) Comparison of the allelopathic potential of four marsh species. *Bull Torr Bot Club* 106: 217–222

24 Elakovich SD, Wooten JW (1989) Allelopathic potential of sixteen aquatic and wetland plants. *J Aquat Plant Manage* 27: 78–84

25 Prindle V, Martin DF (1996) Allelopathic properties of cattails, *Typha domingensis*, in Hillsborough County, Florida. *Florida Scient* 59: 155–162

26 Muller CH, Chou CH (1972) Phytotoxins: an ecological phase of phytochemistry. *In*: JB Harborne (ed): *Phytochemical ecology*. Academic Press, New York, 201–216

27 Dolling A, Zacrisson O, Nilsson MC (1994) Seasonal variation in phytotoxicity of bracken (*Pteridium aquilinum* L. Kuhn). *J Chem Ecol* 20: 3163–3172

28 Einhellig A (1999) An integrated view of allelochemicals amid multiple stresses. *In*: Inderjit, KMM Dakshini, CL Foy (eds): *Principles and practices in plant ecology: allelochemical interactions*. CRC Press, Boca Raton, 479–494

29 Lohdi MAK (1975) Soil-plant phytotoxicity and its possible significance in patterning of herbaceous vegetation in a bottomland forest. *Am J Bot* 62: 618–622

30 Nilsson M, Gallet C, Wallstedt A (1998) Temporal variability of phenolics and batatasin-III in *Empetrum hermaphroditum* leaves over an 8-year period. interpretations of ecological function. *Oikos* 81: 6–16

31 Gallardo MT, Martin BB, Martin DF (1988) Inhibition of water fern *Salvinia minima* by cattail (*Typha domingensis*) extracts and by 2-chlorophenol and salicylaldehyde. *J Chem Ecol* 24: 1483–1490

32 Gallardo MT, Ascher JR, Collier MJ, Martin BB, Martin DF (2000) Effect of cattail (*Typha domingensis*) extracts, leachates, and selected phenolic compounds on rates of oxygen production by salvinia (*Salvinia minima*). *J Aquat Plant Manage* 37: 80–82

33 Gallardo M (1999) Chemical basis for ecological processes of cattails (*Typha domingensis*). Ph.D. Dissertation. University of South Florida, Tampa, Florida

34 Gallardo MT, Sawyers WG, Martin DF (1999) Concentrations of two phytotoxic materials in cattail extracts: 2-chlorophenol and salicylaldehyde. *Florida Scient* 62: 164–171

35 Gallardo-Williams MT, Meadows DG, Mendoza-Galvez LB, Papachristou MD,. Martin DF (2001) Effect of concentration and decomposition time on the phytotoxicity of cattail (*Typha domingensis*) extracts. *Florida Scient* 64: 44–55

36 An M, Johnson IR, Lovett JV (1993) Mathematical modeling of allelopathy: biological response to allelochemicals and its interpretation. *J Chem Ecol* 19: 2379–2388

37 An M, Pratley JE, Haig T (1997) Phytotoxicity of vulpia residues: I. Investigation of aqueous extracts. *J Chem Ecol* 23: 1979–1995

38 Inderjit, Dakshini KMM (1995) Quercetin and quercitrin from *Pluchea lanceolata* and their effect on growth of asparagus bean. *In*: Inderjit, KMM Dakshini, FA Einhellig (eds): *Allelopathy: organisms, processes and applications*. ACS Sym. Series 582, American Chemical Society, Washington, DC, 86–95

39 Williams DH, Stone MJ, Hauck PR, Rahman SH (1989) Why are secondary metabolites (natural products) biosynthesized? *J Nat Prod* 52: 1189–1208

40 Mersie W, Singh M (1987) Allelopathic effect of parthenium (*Parthenium hysterophorus* L.) extracts and residue in some agronomic crops and weeds. *J Chem Ecol* 13: 1739–1747

41 OzawaT, Imagawa H (1988) Polyphenolic compounds from female flowers of *Typha latifolia* L. *Agric Biol Chem* 52: 595–597

42 Aliotta G, Della Greca M, Monaco P, Pinto G, Pollio A, Previtera L (1990) *In vitro* algal growth inhibition by phytotoxins of *Typha latifolia* L. *J Chem Ecol* 16: 2637–2646

43 Della Greca M, Monaco P, Previtera L (1990) Stigmasterols from *Typha latifolia*. *J Nat Prod* 53: 1430–1435

44 Bridgeman JE, Cherry PC, Clegg AS, Evans JM, Jones ERH, Kasal A, Kumar V, Meakins GD, Morisawa Y, Richards EE et al (1970) Microbiological hydroxylation of steroids. Part I. Proton magnetic resonance spectra of ketones, alcohols, and acetates in the androstane, pregnane, and oestrane series. *J Chem Soc* 250–257

45 Della Greca M, Mangoni L, Molinaro A, Monaco P, Previtera L (1990) (20*S*)-4α-Methyl-24-methylenecholest-7-en-3β-ol, an allelopatic sterol from *Typha latifolia*. *Phytochemistry* 29: 1797–1798

46 Della Greca M, Monaco P, Previtera L, Aliotta G, Pinto G (1990) Carotenoid-like compounds from *Typha latifolia*. *J Nat Prod* 53: 972–974

47 Della Greca M, Mangoni L, Molinaro A, Monaco P, Previtera L (1990) 5β,8β-epidioxyergosta-6,22-dien-3β-ol from *Typha latifolia*. *Gazz Chim Ital* 120: 391–392

48 Sheikh YM, Djerassi C (1974) Steroids from sponges. *Tetrahedron* 30: 4095–4103

49 Kokke WCMC, Fenical W, CDjerassi C (1981) Sterols with unusual nuclear unsaturation from three cultured marine dinoflagellates. *Phytochemistry* 20: 127–134

50 Prindle V, Sawyers WG, Martin BB, Martin DF (1996) Identification of allelopathic substances from cattails, *Typha domingensis*. *Florida Scient* 60 (SP 1): 24–25

51 Albalat Domènech J, Pérez E, Gallardo MT, Martin DF (1997) Identification of allelopathic chemicals in a pond of cattails. *Florida Scient* 60: 202–203

52 Walstad D (1995) Allelopathy in aquatic plants, part 2. The subtle nature of aquatic plant allelopathy. *Aquat Gardener* 8: 148–156

53 Aliotta G, Molinaro A, Monaco P, Pinto G, Previtera L (1992) Three biologically active phenylpropanoids glucosides from *Myriophyllum verticillatum*. *Phytochemistry* 31: 109–111

54 Harborne JB, Baxter H (1993) *Phytochemical dictionary. a handbook of bioactive compounds from plants*. Taylor and Francis, Washington

Chemical Ecology of Plants: Allelopathy in Aquatic and Terrestrial Ecosystems
ed. by Inderjit and Azim U. Mallik
© 2002 Birkhäuser Verlag/Switzerland

Allelochemicals from sunflowers: chemistry, bioactivity and applications

Francisco A. Macías, Rosa M. Varela, Ascensión Torres, José L.G. Galindo, and José M.G. Molinillo

Departamento de Química Orgánica, Facultad de Ciencias, Universidad de Cádiz, , Apdo. 40, 11510-Puerto Real, Cádiz, Spain

Introduction

Significant progress in the field of natural product chemistry could be due to advances in technology, new molecules of substantial interest, and changing ethical principles for organism collection [1]. During the last two decades, there have been many changes in natural product chemistry, such as plant selection and collection, isolation techniques, structure elucidation, biological evaluation, semi-synthesis, and biosynthesis. The aim of this chapter is to describe these changes by using sunflower as an example.

Approaches generally taken to select the plant materials for study are (1) locally random, (2) taxonomic, and (3) ethnobotanical and use (4) information available in the literature about its chemistry. The goal of several traditional studies was to isolate new compounds. Although novel compounds may have biological and pharmacological activities, and may be of taxonomic value, there should be a basis to judge the quality of the discovery. The search for new metabolites was the principal goal for phytochemists until the 1980s. Plants were extracted using standard methods and then fractionated. A large number of new compounds were isolated from plants and other organisms. Some of the novel compounds were tested for their biological and pharmacologic activities.

Cultivated sunflower (*Helianthus annuus*) was considered a pharmacognosic challenge because of its complex chemistry and readily available plant material. A number of sesquiterpene lactones, mainly germacranolides, were isolated from sunflower. Melek et al. [2] identified argophyllin A as the major component, and another epoxide, argophyllin B, along with niveusin B and 4,5-dihydroniveusin A, as minor components. Spring et al. [3] isolated 15-hydroxy-3-dehydrodesoxytifruticin; its 3-hemiketal, annuithrin; 1-*O*-methyl-4,5-niveusin A, its oxo form; and 15-hydroxy-3-dehydro-4,5-dihydro-desoxytifruticin (Fig. 1). Their study focused on a taxonomic goal that identified sesquiterpenes in order to make taxonomic relationships between a large number of species using high performance liquid chromatography (HPLC) and

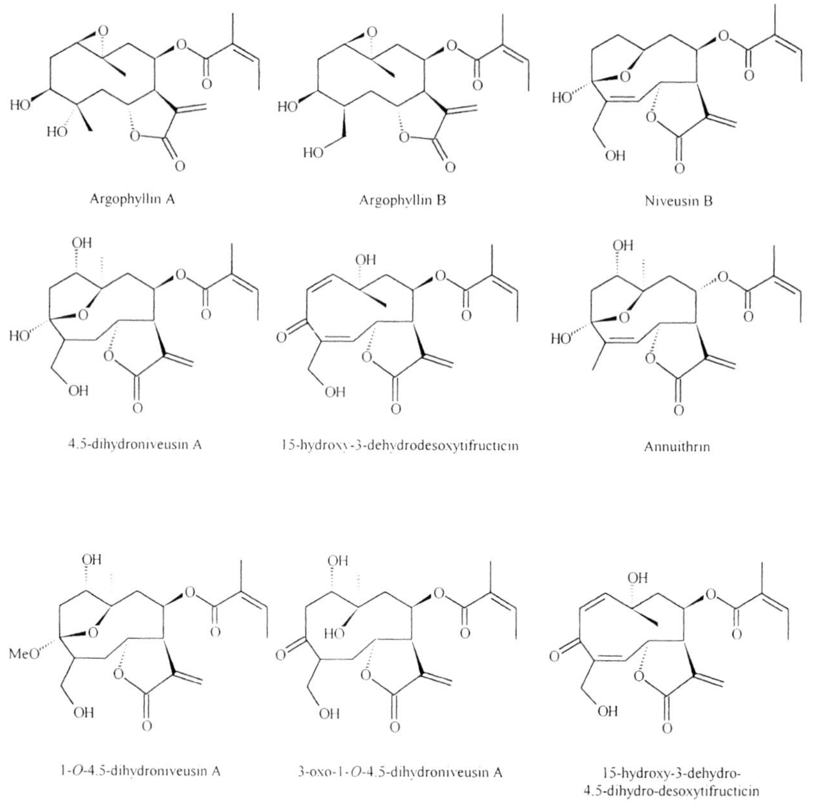

Argophyllin A

Argophyllin B

Niveusin B

4.5-dihydroniveusin A

15-hydroxy-3-dehydrodesoxytifructicin

Annuithrin

1-O-4.5-dihydroniveusin A

3-oxo-1-O-4.5-dihydroniveusin A

15-hydroxy-3-dehydro-
4.5-dihydro-desoxytifructicin

Figure 1.

samples taken from capitate glandular trichomes. Using HPLC, the study of Spring et al. [3] generated important taxonomic data. Spring et al. [4], however, obtained three bisabolene-type compounds from noncapitate glandular trichomes.

Other constituents from sunflower extracts contained 33 diterpenoids [5–8], including 14 seed gibberellins and 8 epicuticular flavonoids. In 1992, Alfatafta

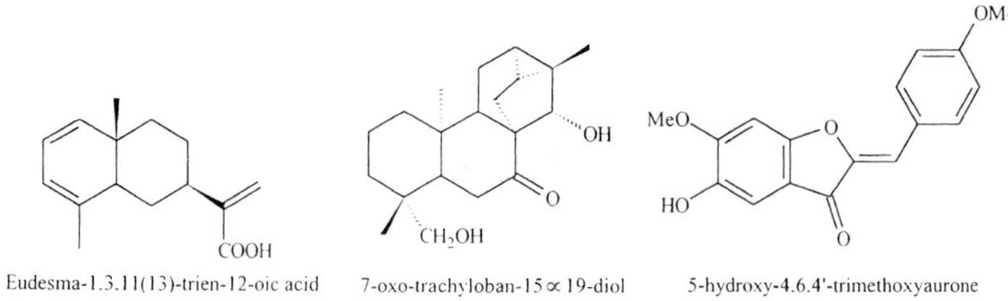

Eudesma-1.3.11(13)-trien-12-oic acid

7-oxo-trachyloban-15∝19-diol

5-hydroxy-4.6.4'-trimethoxyaurone

Figure 2.

and Mullin [9] identified the germacranolides 3-*O*-methylniveusine A and 1,10-*O*-dimethyl-3-dehydroargophyllin B diol, eudesmanoic acid eudesma-1,3,11(13)-trien-12-oic acid, diterpene 7-oxo-trachyloban-15α, 19-diol, and 5-hydroxy-4,6,4'-trimethoxyaurone (Fig. 2). They found strong antifeedant activity for sesquiterpene compounds against adult western corn rootworm (*Diabrotica virgifera virgifera*).

Sunflower: some reasons for rediscovering its chemistry

One of the main changes in natural product chemistry is a new approach to plant selection. New reasons for studying sunflower appeared. One of these reasons was the agronomic interest in sunflower. Cultivation of sunflower was introduced in Europe in the early 1960s and today plays an important role in southern parts of Spain and France. More than a million ha are dedicated to this crop in Spain, where it is grown primarily to produce oil. Sunflower has become more and more important as an oilseed crop and ranks as the second most important source of vegetable oil in the world. Some reasons for the rapid expansion of sunflower research are (1) it is a good crop for dry lands, (2) its mechanization is easy and economic, and (3) new hybrid varieties with high oil content adapt to a wide diversity of soils and climates. These varieties present good resistance to plant diseases and infections and acceptable incomes and easy commercialization.

The economic importance of sunflower has provoked many agronomic and genetic studies to improve the yield and quality of this crop. More recently, biochemical studies have become important. It is well-known that sunflower is allelopathic to weeds and shows biological activity [10–13] against some troublesome weeds such as morning glory (*Ipomoea hederacea*), velvetleaf (*Abutilon theophrasti*), pigweed (*Amaranthus* sp.), jimson weed (*Datura stramonium*), and wild mustard (*Brassica* sp.). On the other hand, sunflower produces a variety of secondary metabolites, most commonly terpenoids and phenolics possessing a wide variety of biological properties, such as (1) nectarguide flowers, (2) larval-growth inhibitor, (3) cytotoxic agents, (4) phytoalexins, (5) effects on the oxidative properties of intact mitochondria, and (6) effects on the enzymatic regulation of the levels of indole-3-acetoic acid (IAA) [14, 15].

Technical advances

There are two important fields where technical advances have influenced natural product chemistry in general and sunflower in particular. These aspects are analytical techniques and structure-elucidation techniques. The development of new analytical techniques allows the isolation of new compounds in very small amounts from more and more complex matrices. The natural product chemist can use a wide range of methods for separation: droplet counter-current chromatography (DCCC), rotation locular counter-current chromatography (RLCC), centrifugal partition chromatography (CPC), and, of course, HPLC. These new techniques have changed our laboratories and the routines of our daily work. Many different possibilities now exist: the solvent, gradients, and the stationary phase, in all combinations. One example of this could be the previously cited works of Spring et al. [3], where the application of the HPLC led him to study complex mixtures of very small amounts of extract from capitate glandular trichomes. On the other hand, the isolation of small amounts of minor compounds would be useless if we do not have proper structure-elucidation techniques available. Therefore, the development of these techniques has been impressive, especially in the field of nuclear magnetic resonance (NMR). The range of NMR techniques that are available is almost bewildering. The sensitivity of these techniques with new 900 MHz equipment and the list of acronyms that describe the techniques (COSY, DQF-COSY, HOHAHA, NOESY, ROESY, HETCOR, HMQC, HMBC, FLOCK, COLOC, INADEQUATE, etc.) allows natural product chemists the almost unambiguous elucidation of very complex structures with no destructive methods.

Search for new bioactive compounds

For a long time the search for new compounds was one of many goals of natural product chemists. Chemists looked for the most rare and novel structures that they could find. Although these products were sometime tested for the different types of activities mentioned above, most of them were tested for similar types of activity. Indeed, the biological test was mostly an excuse than the real target.

Because of the wide available range of bioassays, the main target for the study of metabolites has become the isolation of those compounds that are responsible for the specific activities.

Now it is possible to assay for different activities by using low amounts of compounds. The search for compounds of interest is efficient these days because of the large number of active compounds that are available nowadays. Compounds with better characteristics than those already available are needed; therefore, there is a need to look for specific, low-concentration biological activities. There are numerous studies on isolation of compounds from sunflower. We will highlight relevant recent ones that indicate the high levels of biological activity.

Phytoalexins

The production and excretion of coumarin phytoalexins in different plant organs of sunflower by using an abiotic elicitation system has been shown [16, 17]. Coumarins studied were scopoletin, scopolin, and ayapin (Fig. 3).

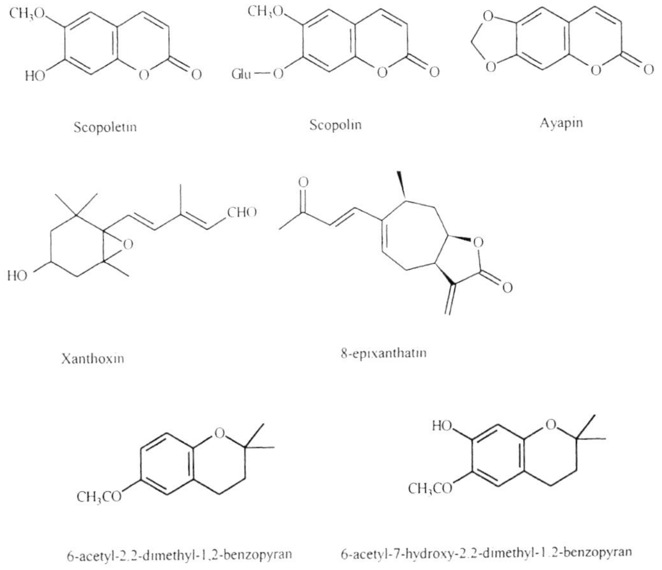

Scopoletin   Scopolin   Ayapin

Xanthoxin   8-epixanthatin

6-acetyl-2,2-dimethyl-1,2-benzopyran   6-acetyl-7-hydroxy-2,2-dimethyl-1,2-benzopyran

Figure 3.

Coumarins can be defined as stress metabolites. Thus, in sunflower these compounds are not present or are at a very low concentration in tissue extracts from healthy plants, but they are induced in response to sufficient biotic and abiotic stress; nutritional deficiencies; and plant hormones, metabolites, and xenobiotics, and they are tissue-specific.

Light-induced auxin-inhibiting substance

Phototropism is one of the common growth responses of plants to unilateral irradiation. It seems that this is produced by the inhibition of growth in the parts of the plant where it is irradiated, so the growth in the other plant parts produces growth towards the light. Spring et al. [18, 19] found that two sesquiterpene lactones isolated from sunflower are auxin-inhibiting substances. Furthermore, xantoxin and caprolactam were identified from sunflower as light-induced plant-growth inhibitors [20]. Recently, 8-epixantatin has been isolated from sunflower seedlings [21]. This compound inhibits auxin-induced growth of sunflower hypocotyls. Because the light-grown seedlings were observed to contain a higher level of 8-epixantatin than did the dark control, and because the difference in its level was noticeable before the growth inhibition by light appeared, this compound may play an important role in light-induced growth-inhibition of sunflower.

Antimicrobial compounds

Satoh et al. [22], investigated antimicrobial substances in sunflower and isolated two antifungal benzopyran derivatives, 6-acetyl-2,2-dimethyl-1,2-ben-
zopyran and 6-acetyl-7-hydroxy-2,2-dimethyl-1,2-benzopyran (Fig. 3).
Additionally, an antifungal protein (AFP) has been isolated from sunflower
flowers. It inhibits the germination of fungal spores [23]. This protein causes
the complete inhibition of *Sclerotinia sclerotiorum* ascospores germination at
a concentration of 5 µg/mL and a clear reduction of mycelial growth at lower
concentrations. *Sclerotinia sclerotiorum* is a natural pathogen of sunflower
that is responsible for severe yield losses in crops. These results suggest a
potential role of this protein in the innate defense of sunflower by a direct inhibition of fungal growth.

Anti-inflammatory triterpenes

Akihisa et al. [24, 25] isolated 3,4-seco-triterpene alcohol, helianol, together
with 7 known triterpene alcohols. The antiinflammatory effects of these compounds were evaluated and compared with the effects provoked by other triterpene alcohols isolated from flowers of Compositae and those of the commercially available antiinflammatory drugs indomethacin and hydrocortisone.
Helianol exhibited the strongest antiinflammatory effect.

Search for new herbicides

Despite of modern methods of weed management, weeds are responsible for
declines in crop yield. Additionally, the use of herbicides has provoked an
increasing incidence of resistance to herbicides in weeds, and new, more efficient and specific herbicides are therefore required [26]. Indeed, the agrochemical industries have been producing new herbicides every year without
the help of natural product chemistry. Newly commercialized herbicides have
principally the same sites of action that were discovered before 1985 [27].

Another important factor that should be considered is sustainability. New
compounds must be environmentally friendly [28]. But the term "sustainable
technology" includes different points of view: agrochemical, environmental,
and economical. There is a need to discover safe compounds with new sites of
action; however, this is not enough. Novel compounds should have a wide rage
of activities but be targeted at one particular problem.

Plants have their own defense mechanisms, and certain allelochemicals may
act as natural herbicides. The knowledge of chemical relationships between
plants may allow the development of new herbicides [29]. If we mimic nature,
we could get new compounds with different sites of action. Sunflower is a good
candidate to provide this type of compound. Reports of its allelopathic proper-

ties date back to 1931, when Cooper and Stoesz [30] observed the atypical decrease in the number of plants, size, and inflorescences of *Helianthus scubberimus* around wild sunflower *Helianthus rigidus*. This phenomenon was observed for the common sunflower in a first step of weed succession in abandoned plots [31]. Chlorogenic and *iso*-chlorogenic acids were the first compounds that were proposed as being responsible for above-discussed activity. They were not present in root exudates or leaf leachates, so it was suggested that they came into the environment *via* the decomposition of plant residues.

Thereafter, several phenolic and fatty acids isolated from roots and leaves of *H. annuus* and *H. tuberosus* were correlated with the allelopathic growth suppression of nearby species [32]. Neither the levels of phenolic acids in the plant and the surrounding soil nor the concentration of these compounds can explain the allelopathic properties of *Helianthus* spp. In 1990, our group began a systematic study of the allelopathic activity of the *H. annuus* cultivar with the following objectives:

1. Study of the allelopathic potential of sunflower cultivars of the Andalusia region
2. Selection of sunflower varieties with higher levels of biological activities and the development period where the activity is enhanced.
3. Bioassay-directed biochemical study
4. Establishment of the activity profiles of phytotoxicity of the isolated compounds
5. Chemical transformation and partial synthesis of allelochemicals to study the structure-activity relationships
6. Study of their mechanisms of action
7. Greenhouse bioassays in order to design new formulations of natural agrochemicals.

The above-stated objectives involved following changes in the traditional methodology of natural product chemistry [33]. The first step of any chemical analysis is sampling. In this case, as in the following steps of the process, bioassay was the guide to the selection of the sample. Moreover, sampling and the bioassay itself must mimic field situations, so extraction could not be made using traditional extraction techniques but, rather, in similar conditions that occur in nature. We selected water as our solvent, simulating lixiviation produced by rain or dew. Another important difference was the use of fresh plant material. In most phytochemical analyses, plant material is dry in order to avoid water in the extract, to allow the storage of plant material for longer time, and to allow for easier handling. This introduced the additional technical problem of making quick extraction in order to protect the sample from microorganism and enzyme degradative attack.

Cultivars of sunflowers were collected at four different development stages: (1) 15–20 cm tall plants, (2) 50 cm tall plants, (3) 1 m tall plants (one month before harvest), and (4) 1–1.5 m tall plants (close to the harvest). The third developmental stage (i.e., 1 m tall plants) showed better profiles of activity and provided information on the appropriate period to collect plant material [34].

After the selection of three varieties in the third plant development stage, we began bioactivity-directed isolation of the active compounds. This methodology consists of the evaluation of the allelopathic activity of every fraction of the chromatographic process. This approach needed a reproducible bioassay that would allow comparison of bioactivity data. We proposed a standard bioassay with selected standard target species that belonged to monocotyledonous (e.g., onion, *Allium cepa;* barley; *Hordeum vulgare;* wheat, *Triticum aestivum;* and corn, *Zea mays*) and dicotyledonous (e.g., tomato, *Lycopersicom esculentum;* lettuce, *Lactuca sativum;* carrot, *Daucus carota;* and cress, *Lepidium sativum*) species [35]. This bioassay also included a commercial herbicide (Logran®) with known levels of activity. The use of this internal standard serves to compare the levels of phytotoxicity shown by the compounds with the levels of a commercial formulation in the same conditions. The second important advantage is that it allows comparison of results of bioassays carried out at different places and/or times.

Sunflower extracts had more active compounds than we expected. We could isolate more than 50 compounds from active fractions. They belonged to different compound class skeletons, being mainly terpenoids. We could not isolate monoterpenes, but some related compounds were isolated from sunflower cultivar: three new bioactive ionone-type bisnorsesquiterpenes, annuionones A–C, and the new norbisabolene helinorbisabone [36] (Fig. 4). According to allelopathic bioassays performed, the most relevant effects on dicotyledonous species (*Lactuca sativa* and *Lepidium sativum*) are those shown by helinorbisabone, which inhibited the germination of *L. sativa* in all tested concentrations with an average of –50%. Sesquiterpenes are the most abundant compounds present in sunflower extracts. The number and structural variability was very important, and we discovered two new skeletons of this type of compound: heliannuols [37–40] (Fig. 5) and heliespirones [41] (Fig. 6).

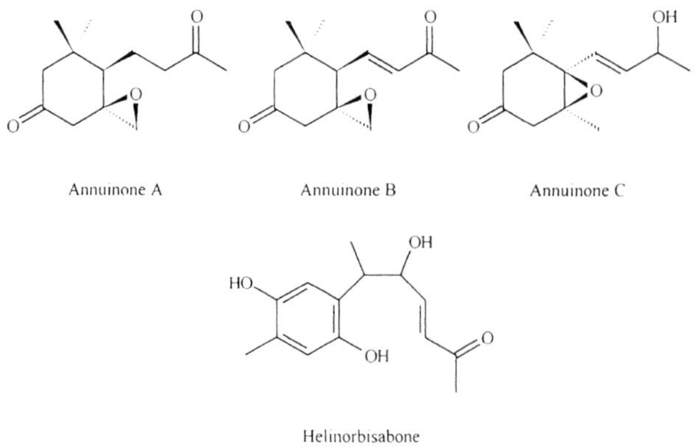

Annuinone A   Annuinone B   Annuinone C

Helinorbisabone

Figure 4.

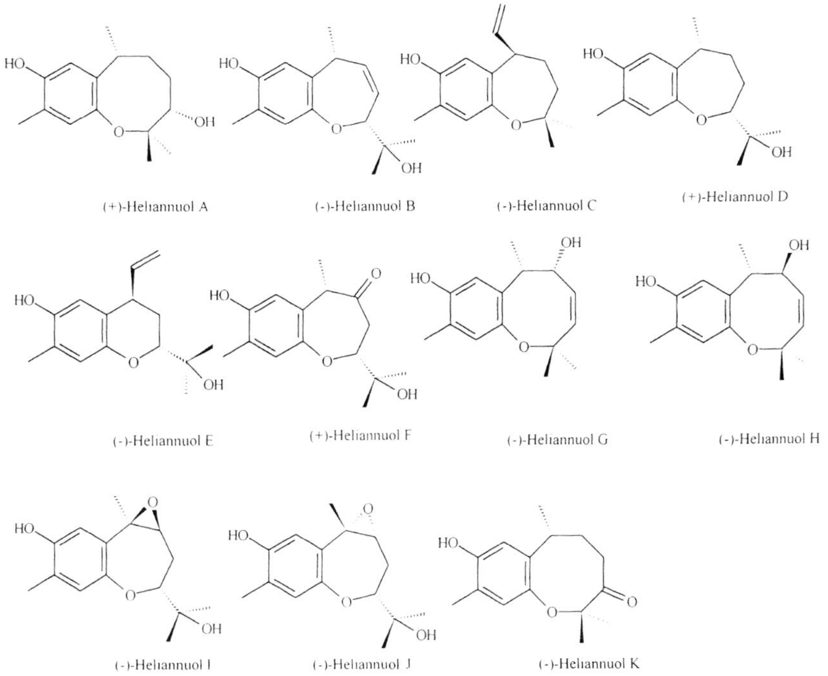

(+)-Heliannuol A   (-)-Heliannuol B   (-)-Heliannuol C   (+)-Heliannuol D

(-)-Heliannuol E   (+)-Heliannuol F   (-)-Heliannuol G   (-)-Heliannuol H

(-)-Heliannuol I   (-)-Heliannuol J   (-)-Heliannuol K

Figure 5.

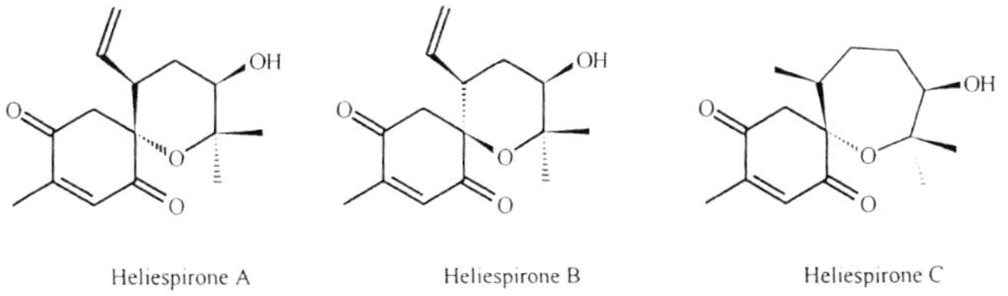

Heliespirone A   Heliespirone B   Heliespirone C

Figure 6.

A number of compounds from the novel sesquiterpene family heliannuol have been isolated from cultivar sunflower. To evaluate their potential allelopathic activity and to obtain information about the specific requirements needed for their bioactivity, the effects of aqueous solutions from 10^{-4} to 10^{-9} M of these compounds were evaluated on root and shoot lengths of lettuce, barley, wheat, cress, tomato and onion seedlings. Comparison of active heliannuols [42] with the commercial herbicide Logran® showed that the most important effects were those caused by heliannuols A, C, H, I, and K inhibiting germination of lettuce and by heliannuols C, G, H, I, and K stimulating root growth of barley.

We have isolated 18 sesquiterpene lactones from sunflower [43–45]. They have different skeletons: guaianolides, germacranolides, heliangolide, *cis,cis-*

germacranolide, and melampolide. In the case of guaianolides, those with lower functionalization (lower number of hydroxyl groups) strongly inhibited the germination of lettuce seeds but were not good growth inhibitors. One interesting aspect that has been widely accepted is that the presence of the moiety α-methylene-γ-lactone is a requirement for the biological activity. In our case, this is not a clear requirement, and sometimes a decrease of the activity can be observed, i.e., annuolide B has good levels of activity on the germination of lettuce, while the corresponding annuolide A that has this moiety has no activity. Again, configuration seems to play an important role in the activity; thus, the persistence of the activity with dilution is higher in annuolide E, compared to its epimer annuolide D (Fig. 7). The different observed profiles of activity of compounds that contain an ester at C-8, can be attributed to the steric hindrance on the β side of the molecule and, consequently, less accessibility of the α-methylene-γ-lactone moiety.

Germacranolides have more flexibility in their skeleton and different possibilities of conformation. Observed activities are less intense, but we can deduce from results that those compounds that possess a double bond with Z geometry between C-4 and C-5 are more active on root and shoot length of dicotyledonous species. The effects of conformational changes are important because of greater flexibility of the molecule. This factor will influence germacranolides more strongly than guaianolides.

Extraction of fresh sunflower (var. VYP) leaf aqueous extract with methylene dichloride afforded, from low-polar fractions, six kaurenes and one linear natural diterpene [46]. In general, inhibitory effects were observed on germination and shoot length of lettuce, cress, and onion. In contrast, root length was

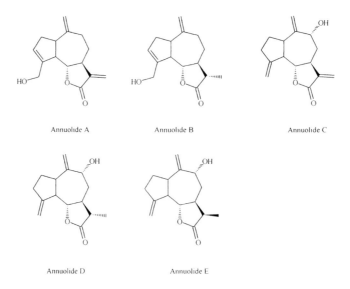

Figure 7.

increased in all three species. The most active compounds affecting germination and growth of lettuce and onion were (+)-*Trans*-fitol and (−)-kaur-16-en-19-oic acid. The compound 16β-hydroxykaurene (Fig. 8) had stimulatory effects on the radical length of cress.

From medium polar-active fractions we have isolated five flavonoids [47]. Despite the large number and wide distribution of these compounds, only a few have been implicated in allelopathy. Inderjit and Dakshini [48, 49] have related ononin, hesperidin, and taxifolin 3-arabinoside with the allelopathic activity of *Pluchea lanceolata*. Some interesting examples have been described, i.e., ceratiolin (Fig. 9) is an inactive dihydrochalcone but has been degraded to hydrocinnamic acid and then to acetophenone. Both byproducts inhibited germination and growth of Florida grasses and pines [50]. The flavonol tambulin did not show any effect against germination and radical

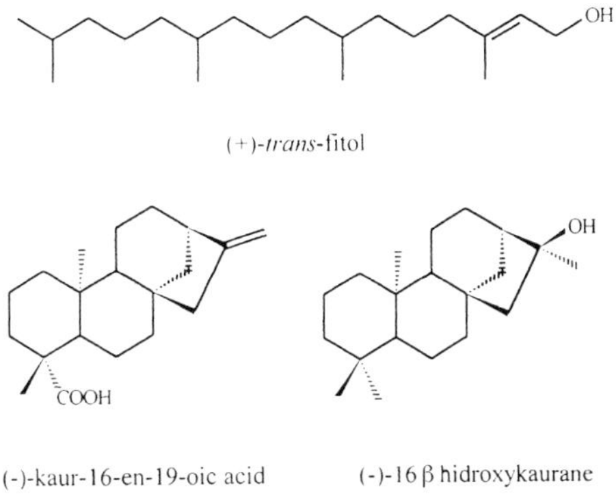

(+)-*trans*-fitol

(-)-kaur-16-en-19-oic acid   (-)-16 β hidroxykaurane

Figure 8.

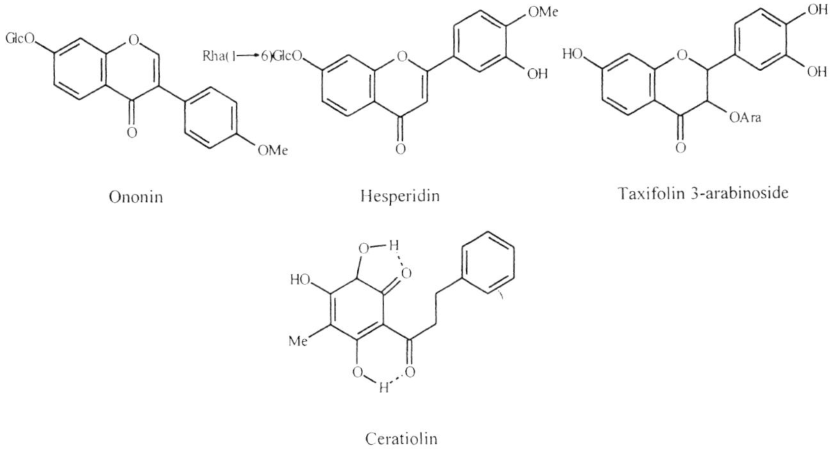

Ononin   Hesperidin   Taxifolin 3-arabinoside

Ceratiolin

Figure 9.

length of tomato and barley, but it inhibited shoot growth of tomato (average −25%) and barley (−22%, 10^{-5} M).

The chalcones kukulkanin B and heliannone A; have very close structures. They differ in the presence of an additional methyl group in heliannone A; however, their activities are quite different. Heliannone A inhibited the germination of tomato and barley and slightly affected barley shoot growth with an inhibitory profile. On the other hand, kukulkanin B only showed inhibitory activity on the shoot growth of tomato.

The flavanones heliannones B and C showed similar activity profile, although the activities produced by heliannone B solutions are, in general, higher than those produced by heliannone C (Fig. 10). Both compounds showed inhibitory effects on the shoot and radical length of tomato. These flavonoids influence, principally, the shoot growth of seedlings, but germination and root growth can be affected by chalcones.

Ohno et al. have studied the exudates of sunflower seeds during germination [51]. They have isolated a stereoisomer of diversifolide, sundiversifolide (Fig. 11), that inhibits shoot and root growth of cat's-eyes. This compound has been tested on lettuce, cockscomb (*Celosia argentea*), tomato, crabgrass

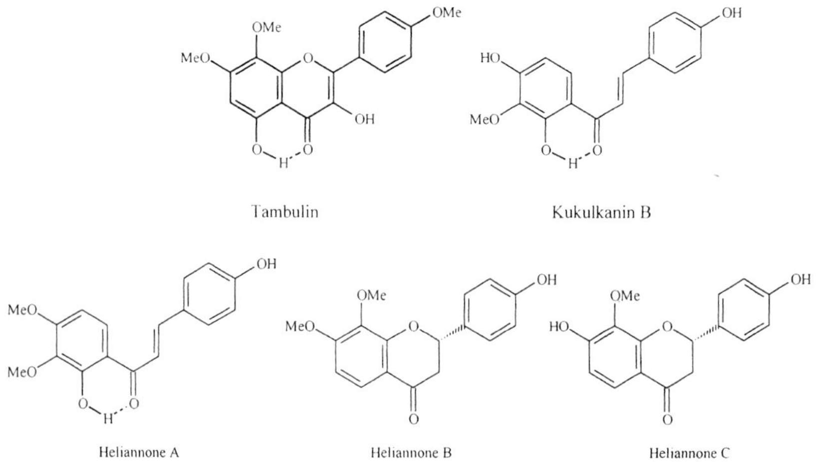

Figure 10.

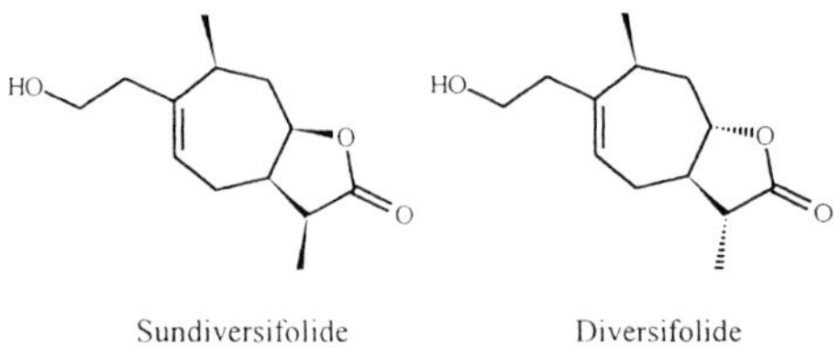

Figure 11.

(*Digitaria singuinalis*), and barnyard grass (*Echinochloa crus-galli*) and shows selective biological activity.

Despite the experience of the past 14 years of allelopathic studies on sunflower, there are still many questions to be addressed, including quantification of the original water extract or in the artificial leaches and explanations of the variability, in terms of chemistry, from one cultivar to another. Discovery of the modes of action of these new molecules and the possibility to find which genera are responsible for the production of one particular allelochemical, among other issues, suggest reasons to continue with these fascinating studies.

Acknowledgments
This research has been supported by the Dirección General de Investigación Científica y Técnica, Spain (DGICYT, PB98-0575).

References

1 Cordell GA (1995) Changing strategies in natural products chemistry. *Phytochemistry* 40: 1585–1612
2 Melek FR, Cage DA, Gersherson J, Mabry TJ (1985) Sesquiterpene lactone and diterpene constituents of *Helianthus annuus*. *Phytochemistry* 24: 1537–1539
3 Spring O, Benz T, Ilg M (1989) Sesquiterpene lactones of the capitate glandular trichomes of *Heliantus annuus*. *Phytochemistry* 28: 745–749
4 Spring O, Rondon U, Macías FA (1992) Sesquiterpene from noncapitate glandular trichomes of *Helianthus annuus*. *Phytochemistry* 31: 1541–1544
5 Pyrek JSt (1979) New pentacyclic diterpene acid; Thrachyloban-19-oic acid form sunflowers. *Tetrahedron* 26: 5029–5032
6 Panizo FM, Rodríguez B (1979) Some diterpenic constituents of the sunflower (*Helianthus annuus* L.). *Anal Quim* 75: 428–430
7 Fergunson G, McCrindle R, Murphy ST, Parvez M, Mitscher LA, Rao GSR, Veysoglu T, Drake S, Haas T (1983) Isolation and identification of trachyloban-19-oic and (–)-kaur-16-en-19-oic acids as antimicrobial agents from the prairie sunflower, *Helianthus annuus*. *J Nat Prod* 46: 745–746
8 Pyrek JS (1984) Neutral diterpenoids of *Helianthus annuus*. *J Nat Prod* 47: 822–827
9 Alfatafta AA, Mullin CA (1992) Epicuticular terpenoids and an aurone from flowers of *Helianthus annuus*. *Phytochemistry* 31: 4109–4113
10 Irons SM, Brunside OC (1982) Competitive and allelopathic effects of sunflower (*Helianthus annuus*). *Weed Sci* 30: 372–377
11 Leather GR (1982) Weed controls using allelopathic crop plants. *J Chem Ecol* 9: 983–989
12 Leather GR (1983) Sunflower (*Helianthus annuus*) are allelopathic to weeds. *Weed Sci* 31: 37–42
13 Leather GR (1987) Weed control using allelopathic sunflower and herbicide. *Plant Soil* 98: 17–23
14 Macias FA, Molinillo JMG, Varela RM, Torres A, Galindo JCG (1999) Bioactive compounds from the genus *Helianthus*. *In*: FA Macias, JCG Galindo, JMG Molinillo, HG Cutler (eds): *Recent advances in allelopathy. Volume I: a science for the future.* Servicio de publicaciones de la Universidad de Cádiz, Puerto Real, Cádiz, 121–148
15 Macias FA, Molinillo JMG, Galindo JCG, Varela RM, Torres A, Simonet AM (1999) Terpenoids with potential use as natural herbicide templates. *In*: HG Cutler, SJ Cutler (eds): *Biologically active natural products: agrochemicals.* CRC Press. Boca Raton, 15–31
16 Gutiérrez-Mellado MC, Edwards R, Tena M, Cabello F, Serghini K, Jorrin J (1996) The production of coumarin phytoalexins in different plant organs of sunflower (*Helianthus annuus* L.). *J Plant Physiol* 149: 261–266
17 Jorrín J, Prats E (1999) Allelochemicals, phytoalexins and insect-feeding deterrents: different definitions for 7-hydroxylated coumarins. *In*: FA Macias, JCG Galindo, JMG Molinillo, HG Cutler (eds): *Recent advances in allelopathy. Volume I: a science for the future.* Servicio de publicaciones

de la Universidad de Cádiz, Puerto Real, Cádiz, 179–192

18 Spring O, Hager A (1982) Inhibition of elongation growth by two sesquiterpene lactones isolated from *Helianthus annuus* L. Possible molecular mechanism. *Planta* 156: 433–440

19 Spring O, Prieter T, Hager A (1986) Light induced accumulation of sesquiterpene lactones in sunflower seedlings. *J Plant Physiol* 123: 79–89

20 Hagegawa K, Knegt E, Bruinsma J (1983) Caprolactam, a light-promoted growth inhibitor in sunflower seedling. *Phytochemistry* 22: 2611–2612

21 Yokotani-Tomita K, Kato J, Kosemura S, Yamamura S, Kushima M, Kakuta H, Hasegawa K (1997) Light-induced auxin-inhibiting substance from sunflower seedlings. *Phytochemistry* 46: 503–506

22 Satoh A, Utamura H, Ishizuka M, Endo N, Tsuji M, Nishimura H (1996) Antimicrobial benzopyrans from the receptacle of sunflower. *Biosci Biotechnol Biochem* 60: 664–665

23 Giudici AM, Regente MC, de la Canal L (2000) A potent antifungal protein from *Helianthus annuus* flowers is a trypsin inhibitor. *Plant Physiol Biochem* 38: 881–888

24 Akihisa T, Oinuma H, Yasukawa K, Kasahara Y, Kimura Y, Takase SI, Yamanouchi S, Takido M, Kumaki K, Tamura T (1996) Helianol [3, 4-seco-19(10→9)abeo-8alpha, 9beta, 10alpha-eupha-4, 24-dien-3-ol], a novel triterpene alcohol from the tabular flowers of *Helianthus annuus* L. *Chem Pharm Bull* 44: 1255–1257

25 Akihisa T, Yasukawa K, Oinuma H, Kasahara Y, Yamanouchi S, Takido M, Kimura Y, Tamura T (1996) Triterpene alcohols from the flowers of compositae and their anti-inflammatory effects. *Phytochemistry* 43: 1255–1260

26 Petsko GA, Ringe D, Hogan J (1999) A new paradigm for the structure-guide pesticide design using combinatorial chemistry. *In*: GT Brooks, TR Roberts (eds): *Pesticide chemistry and bioscience: the food-environment challenge*. The Royal Society of Chemistry, Cambridge, UK, 66–70

27 Hay JV (1999) Herbicide discovery in the 21st century-a look into the crystal ball. *In*: GT Brooks, TR Roberts (eds): *Pesticide chemistry and bioscience: the food-environment challenge*. The Royal Society of Chemistry, Cambridge, UK, 55–65

28 Evans DA (1999) How can technology feed the world safely and sustainably?. *In*: GT Brooks, TR Roberts (eds): *Pesticide chemistry and bioscience: the food-environment challenge*. The Royal Society of Chemistry, Cambridge, UK, 3–24

29 Macías FA, Molinillo JMG, Galindo JCG, Varela RM, Simonet AM, Castellano D (2001) The use of allelopathic studies in the search for natural herbicides. *J Crop Prod* 4: 237–255

30 Cooper WS, Stoesz AD (1931) The submediterranean organs of *Helianthus scubberimus*. *Bull Torr Bot Club* 58: 67–72

31 Wilson RE, Rice EL (1968) Allelopathy as expressed by *Helianthus annuus* and its role in old field succession. *Bull Torr Bot Club* 95: 432–448

32 Leather GR, Florrence LE (1979) Allelopathic potential in thirteen varieties of sunflower. *Abstracts of 1979 Meeting of the Weed Science Society of America*. Frederick (Md., USA). U.S. Department of Agriculture

33 Macias FA (1995) Allelopathy in the search for natural herbicide models. *In*: Inderjit, Dakshini KMM, Einhellig FA (eds): *Allelopathy: organisms, processes, and application*. ACS Symposium Series 582. American Chemical Society, Washington, DC, 310–329

34 Macias FA, Varela RM, Torres A, Molinillo JMG (1999) Potential of cultivar sunflower (*Helianthus annuus* L.) as a source of natural herbicide templates. *In*: Inderjit, Dakshini KMM, Foy CL (eds): *Principles and practices in plant ecology: allelochemical interactions*. CRC Press, Boca Raton, 531–550

35 Macias FA, Castellano D, Molinillo JMG (2000) In the search for a standard phytotoxic bioassay for allelochemicals: selection of standard target species (STS). *J Agric Food Chem* 48: 2512–2521

36 Macias FA, Varela RM, Torres A, Oliva RM, Molinillo JMG (1998) Bioactive norsesquiterpenes from *Helianthus annuus*: potential allelopathic activity. *Phytochemistry* 48: 631–636

37 Macias FA, Varela RM, Torres A, Molinillo JMG, Fronczek FR (1993) Novel sesquiterpene from bioactive fractions of cultivar sunflower. *Tetrahedron Lett* 34: 1999–2002

38 Macias FA, Varela RM, Torres A, G Molinillo Fronczek FR (1994) Structural elucidation and chemistry of a novel family of bioactive sesquiterpenes, Heliannuols. *J Org Chem* 59: 8261–8266

39 Macias FA, Varela RM, Torres A, Molinillo JMG (1999) Heliannuol E: a novel bioactive sesquiterpene of the heliannane family. *Tetrahedron Lett* 40: 4725–4728

40 Macias FA, Varela RM, Torres A, Molinillo JMG (1999) New bioactive plants heliannuols from

cultivar sunflower leaves. *J Nat Prod* 62: 1636–1639

41　Macias FA, Varela RM, Torres A, Molinillo JMG (1998) Heliaespirone A: the first member of a novel family of bioactive sesquiterpenes. *Tetrahedron Lett* 39: 427–430

42　Macias FA, Varela RM, Torres A, Molinillo JMG (2000) Potential allelopathic activity of natural plant heliannanes: a proposal of absolute configuration and nomenclatura. *J Chem Ecol* 26: 2173–2186

43　Macias FA, Varela RM, Torres A, Molinillo JMG (1993) Potential allelopathic guaianolides from cultivar sunflower leaves, var. SH-222. *Phytochemistry* 34: 669–674

44　Macias FA, Torres A, Varela RM, Molinillo JMG, Varela RM, Castellano D (1996) Potential allelopathic sesquiterpene lactones from cultivar sunflower leaves. *Phytochemistry* 43: 1205–1215

45　Macias FA, Oliva RM, Varela RM, Torres A, Molinillo JMG (1999) Allelochemicals from sunflower leaves cv. Peredovick. *Phytochemistry* 52: 613–621

46　Macias FA, Molinillo JMG, Torres A, Varela RM, Chinchilla D (2002) Natural diterpenes from cultivar sunflower leaves with potential use as natural herbicide models. *Phytochemistry; in press*

47　Macias FA, Molinillo JMG, Torres A, Varela RM, Castellano D (1997) Bioactive flavonoids from *Helianthus annuus* cultivars. *Phytochemistry* 45: 683–687

48　Inderjit, Dakshini KMM (1991) Hesperetin 7-rutinoside (hesperidin) and taxifolin 3-arabinoside as germination and growth inhibitors in soils associated with the weed, *Pluchea lanceolata* (DC) C.B. Clarke (Asteraceae). *J Chem Ecol* 17: 1585–1591

49　Inderjit, Dakshini KMM (1992) Formononetin 7-O-glucoside (onorin) and additional growth inhibitor in soils associated with the weed *Pluchea lanceolata* (DC) C.B. Clarke (Asteraceae). *J Chem Ecol* 18: 713–718

50　Fischer NH, Williamson B, Weidenhamer JD, Richardson DR (1994) In search of allelopathy in the Florida scrub: the role of terpenoids. *J Chem Ecol* 20: 1355–1380

51　Ohno S, Yokotani KT, Kosemura S, Node M, Suzuki T, Amano M, Yasui K, Goto T, Yamamura S, Hasegawa K (2001) A species-selective allelopathic substance from germinating sunflower (*Helianthus annuus* L.) seeds. *Phytochemistry* 56: 577–581

Chemical Ecology of Plants: Allelopathy in Aquatic and Terrestrial Ecosystems
ed. by Inderjit and Azim U. Mallik
© 2002 Birkhäuser Verlag/Switzerland

Feedback mechanism in the chemical ecology of plants: role of soil microorganisms

X. Carlos Souto[1] and François Pellissier[2]

[1] *Departamento de Ingeniería de los Recursos Naturales y Medio Ambiente, E.U.E.T. Forestal, Universidade de Vigo, 36005 Pontevedra, Spain*
[2] *Dynamics of Altitude Ecosystems Laboratory, University of Savoie, F-73 376 Le Bourget-du-Lac, France*

Introduction

As primary producers in any ecosystem, plants regulate the functioning of other components; thus, processes that regulate plant communities are of fundamental importance to ecosystem function [1–3]. Various ecosystem components often are related closely to each other, and factors that alter one component will have flow-through effects on other components [1, 4]. Most net primary production enters into the soil system as plant leaves and roots detritus [5], which is then decomposed or transformed by soil microflora. This decomposition process is determined to a large extent by litter quality, i.e., secondary compounds [6–8]. Indeed, vegetation type is known to have a direct influence on the size and composition of the soil microbial community [1, 9]. This direct plant influence on microbes can induce important feedback effects on nutrient cycling, thereby affecting plant growth and vegetation community structure [1, 10]. Therefore, allelopathic interactions, as a result of plant secondary compounds, can have consequences at the ecosystem level. These occur not only as direct plant-plant relationships but also *via* soil microorganisms. A focus on the latter in studies of allelopathy will help us to better understand feedback processes of chemical interactions.

Schmidt and Ley [11] suggested that most putative allelochemicals could not build up to phytotoxic levels under natural conditions because of (1) the greater uptake potential of microorganisms compared with plant roots, (2) the distribution of microbes in soil, (3) the motility and chemotactic abilities of many soil bacteria, and (4) the potential of microbial degradation. Schmidt and Ley [11] proposed several strategies of allelochemical release that may overcome these hurdles, involving the release of compounds that are recalcitrant to microbial attack and variation in space and time of allelochemicals release. In the second case, in order to avoid microbial destruction, chemicals would be released only occasionally but in large doses, swamping the demand of the existing microbial populations and causing harm to surrounding plants before microbial populations could grow. In the first case, allelochemicals would be compounds that are resistant to microbial attack. These authors suggested that microbes do evolve rapidly and can adapt to degrade even complex compounds that are regularly released into the soil, resulting in the loss of effectiveness of that chemical as a phytotoxin. In any case, if chemicals are rapidly degraded, as proposed by Schmidt and Ley [11], we should reconsider the perspective given by Wardle et al. [1], where in allelopathic effects must be evaluated at the ecosystem level; chemicals released by plants and used by soil microbes as a source of nutrients cannot be considered phytotoxins, but they probably enhance microbial growth, which implies more nutrient uptake and less nutrient availability to plant roots. The end result is that feedback processes influence vegetation structure and dynamics through the release of chemical compounds. A compound, therefore, may not have a direct role in plant-growth inhibition, but it might interfere with plant growth *via* its influence over soil microbial ecology [12].

Soil – a "black box"

Soil is generally considered a "black box" [13–16] because of the difficulties in understanding its biochemistry and the complexity of related biological processes. However, soil is the place in which most allelopathic interactions occur in terrestrial systems, and microorganisms play an important role in these processes [17].

Allelochemicals produced by plants are subject to the action of microbes and various physical and chemical degradation processes in soil [17–20]. Allelopathic interactions are strongly related to microbial activity in soil. Microorganisms both produce and degrade chemicals, and in turn, they can be directly affected by plant secondary metabolites [17, 21, 22]. Many changes can occur once allelochemicals reach the soil, from inactivating their phytotoxicity [23] to enhancing the toxicity of otherwise inactive secondary metabolites [24] by various mechanisms including microbial breakdown, surface adsorption, polymerization, pH change, altered oxygen concentration, etc. [14, 25–28].

As a source of carbon, plants (roots and plant residues) provide a food source to the microflora that contribute to soil structure formation and stabilization and provide physical protection of the soil surface against structure-altering processes [29]. At the same time, however, plants act as a source of potential allelochemicals that can affect microbial community structure and activity. For example, most phenolic substances in decomposing leaf litter leach into the soil, where they are partially or totally degraded, are subject to condensation reactions, or are surface adsorbed onto organic matter, thereby contributing to humic substances formation [15, 30, 31].

Positive feedback describes favorable conditions during plant succession [15, 32, 33]. Foremost among these feedbacks is the input of organic matter, which profoundly influences soil as a medium for plant growth and nutrient cycling [34]. The quality of plant litter is critical in regulating plant growth and community composition through biotic interactions in the decomposer sub-system [1]. Many cases have been reported in which phenolics and condensed tannins act as toxic chemicals, inhibiting (at least partially) plant and microorganism metabolism [35–38].

Competition between plant species and the resulting species composition of the plant community are strongly affected by changes in nutrient supply [39–41]. As a positive feedback, some authors have suggested that species located in high nutrient sites should produce high-quality litter that fosters high rates of nutrient recycling, and species from low nutrient sites should produce recalcitrant litter that retards nutrient recycling [16, 42, 43]. Toxic metabolites contained in leaf litter may alter soil conditions to the disadvantage of competing species if they are released in an amount high enough to induce a biological disorder [15]. Plants tend to produce more allelochemicals under stress conditions [14, 44, 45], and the biological effects of allelochemicals on target plants increase under stress conditions, depending on the level of stress and on the system studied [14, 46, 47].

Allelopathy—a chemical interface between microorganisms and plant populations dynamics

Phenolic compounds have been reported to be an important carbon source for microbial populations in numerous ecosystems [18, 25, 35, 48, 49]. Wardle and Nilsson [50], however, attributed the low microbial biomass levels [51] and scarcity of other species of vascular plants under *Empetrum hermaphroditum* to the high levels of total phenolics in the humus layer.

In some cases, allelopathy could explain the scarcity of understory species and their cover in forest ecosystems. We have shown that decomposition of aerial parts from *Eucalyptus globulus* and *Acacia melanoxylon* releases phytotoxic compounds that inhibit the germination and growth of understory species [23]. One of the most interesting results was the importance of soil microorganisms in detoxifying these inhibitory compounds, particularly in this forest ecosystem where biodiversity of the understory was high and successional stage was close to the climax. In the same stands, the seasonal population dynamics of nitrogen cycle microorganisms was recorded. The lowest number of microbes was found in spring, but populations came back to normal values (meaning values not statistically different from the climax forest stand) within a few weeks following decomposition of leaf allelochemicals [22]. However, the short period in which inhibition occurred coincided with germination and early growth for most of the autochthonous understory species and with a period of increased microbial activity in the soil. Thus, short-term effects can be important in population dynamics because of the extremely sensitive period of the year during which they occur. Small changes in soil properties derived from release of chemical substances and from changes in the microbial community may have important effects on plant population dynamics.

In addition to the beneficial effects of rhizosphere organisms on plants, Westover et al. [52] described how rhizosphere bacteria and fungi are influenced by plants, demonstrating that patterns of community structure in rhizosphere populations of free-living bacteria and fungi may be determined in part by plant species composition. Bever et al. [53] provided a framework for the interrelations between the composition of plant and soil communities in which a feedback process is involved. The model proposes two possibilities: (1) a positive feedback leading to the loss of species diversity at a local scale or (2) a negative feedback leading to its maintenance. In the first case, the relative rate of population growth of a plant species in association with its local soil community increases over time. In this sense, allelopathic interactions should be added to the model as a competitive mechanism that may affect the plant population dynamics [54, 55].

Allelochemicals, microorganisms, and nitrogen cycling

There are several possible mechanisms for plants to influence nitrogen cycling [56]. The first is through "controlling" litter quality [57, 50], thereby controlling microbes through the quality of carbon susbtrate. An alternate mechanism is that trees may produce and release compounds (substrates, toxins, ligands) that directly alter microbial activity [56]. In both instances, nitrogen cycling and soil processes may be directly influenced and, thus, feed back to the vegetation.

Soil nitrification rates have been found to be controlled more by the presence or absence of particular tree species than by other factors such as soil pH [15, 58]. Polyphenols seem to act as determining factors, controlling the rate of nitrification in many forest floors [15, 59]. However, this concept is controversial and needs to be more widely studied. Northup et al. [60] proposed that high polyphenol production by the fern *Dicranopteris* may create edaphic conditions unfavorable for competing species, allowing the fern to remain dominant on soils that rapidly accumulate nutrients in organic matter over time. They suggested that there has been natural selection for feedbacks between soil conditions and polyphenol production. Northup et al. [61] reported that polyphenol concentration of decomposing *Pinus muricata* litter controls the proportion of nitrogen released in dissolved organic forms relative to mineral forms. They suggested that this regulation of soil conditions apparently controls the dominant mobilized form of nitrogen, facilitates nitrogen recovery through pine-mycorrhizal associations, minimizes nitrogen availability to competing organisms, and attenuates nitrogen losses from leaching and denitrification.

Stark and Hart [62] found that gross nitrification rates were surprisingly high in all the forests they examined. The soil microbial communities had the capacity to assimilate almost all of the nitrate produced; therefore, net nitrification rates poorly predicted gross nitrification rates. They suggested that current models greatly underestimate the role of the microbial community in preventing nitrate loss.

There is clearly feedback between soil conditions and polyphenols production. Polyphenols have multiple effects on plant-litter-soil interactions, including the suppression of detritivore activity [15], formation of mor-type humus, retardation of nitrogen mineralization, and sequestration of nutrients into a very slowly mineralizing pool of organic matter. These interactions are considered examples of negative feedback.

Allelochemicals as regulators of mycorrhizal symbiosis

Melin [63] first reported inhibition of mycorrhizal fungi growth by root exudates. In contrast, Handley [64] reported ectomycorrhizal fungi growth to be stimulated by humus extract from beneath *Calluna vulgaris*. Since then, many studies have confirmed different response of fungi to leaf extract, humic solution, or allelochemicals, according to the fungal species [65–68]. At the cellular level, allelochemicals may alter fungal enzymatic activity [69] or electron-transfer pathway of mycelium mitochondria [70, 71]. Souto et al. [36] compared the mycelial growth of two mycorhizal fungi in the presence of the same phenolics: *Hymenoscyphus ericae* (symbiotic of *Vaccinium myrtillus*, the main allelochemicals producer plant in sub-alpine spruce forest) and *Hebeloma crustuliniforme* (symbiotic of *Picea abies*, the target plant). This work revealed the metabolism of *H. ericae* to be highly adapted to soil phenolics. This mycorrhizal fungus was able not only to survive in the presence of allelochemicals in its environment but also to utilize them as an additional carbon source, thereby allowing *H. ericae* to better compete with other mycorrhizal fungi. Moreover, the biological performance (improved growth and respiration) of *H. ericae* in this allelopathic context could also benefit *V. myrtillus*, its photosynthetic host, perhaps partially explaining the dominance of this understory plant.

Although these biological effects concerned mycorrhizal fungi, when associated with an autotrophic host, the metabolism of such fungi differs from that of strictly free-living heterotrophs. Most of the studies focusing on mycorrhizas quantify the number of mycorrhizal tips in allelopathic environment. Rose et al. [65] reported inhibition of Douglas colonization by *Rhizopogon* spp. that was due to the litter water-soluble compounds (not identified) from shrubs and coniferous species. In the same way, lichens of the *Cladonia* genus have also been reported to inhibit ectomycorrhiza formation on various tree species [72]. There is no doubt that the status of mycorrhizal fungi in soil plays an important role in population dynamics of higher plants. However, the condition of this status also should be considered in the context of allelopathy. For instance, there is now a need for studies focusing on which fungal species (which strains) are able to protect root systems from soil allelochemicals.

Conclusion

The effects of microbial activity on the pool of chemical compounds must not be underestimated. Any change in soil characteristics, as minor as they may be, could have dramatic effects on the ecosystem. Thus, we must continue to strive for a better knowledge of soil microorganisms and their activities if we are to understand chemical interactions in the ecosystem and feedback mechanisms in the chemical ecology of plants. A review was published recently on allelopathic bacteria and their impact on higher plants [73]. It states that these bacteria are not plant-specific. We suggest that some chemical specificity between plant species and bacteria could occur and may explain the fate of these bacteria in soils. Questions regarding the key role of allelochemicals in ecosystems dynamics should be a focus of future research.

Acknowledgements
We thank Dr. T.M. De Luca, The University of Montana, for his helpful comments and discussion on this manuscript. We are grateful to professor Inderjit for reviewing the manuscript prior to publication.

References

1 Wardle DA, Nilsson M-C, Gallet C, Zackrisson O (1998) An ecosystem-level perspective of allelopathy. *Biol Rev* 73: 305–319
2 Vitousek PM, Walker LR (1989) Biological invasion by *Myrica faya* in Hawaii: plant demography, nitrogen fixation, ecosystem effects. *Ecol Monogr* 59: 247–265
3 Oksanen L (1990) Predation, herbivory, and plant strategies along gradients of primary productivity. *In*: JB Grace, D Tilman (eds): *Perspectives on plant competition*. Academic Press, San Diego, 445–474
4 Abrams P, Menge BA, Mitterlbach GG, Spiller D, Yodzis P (1996) The role of indirect effects in food webs. *In*: GA Polis, KO Winemiller (eds): *Food webs. integration of patterns and dynamics*. Chapman and Hall, New York, 371–395
5 Mc Naughton SJ (1993) Biodiversity and function of grazing systems. *In*: ED Schulze, HA Mooney (eds): *Biodiversity and ecosystem function*. Springer-Verlag, Berlin, 362–383
6 Lavelle P, Blanchart E, Martin A, Martin S, Spain AV, Toutain F, Barois I, Schaefer R (1993) A hierarchical model for decomposition in terrestrial ecosystems: application to soils of the humid tropics. *Biotropica* 25: 130–150
7 Couteaux MM, Bottner P, Berg B (1995) Litter decomposition, climate and litter quality. *Trends Ecol Evol* 10: 63–66
8 Heal OW, Anderson JM, Swift MJ (1997) Plant litter quality and decomposition: an historical overview. *In*: G Cadisch, KE Giller (eds): *Driven by nature. Plant litter quality and decomposition*. CAB International, Wallingford, 3–30
9 Widden P (1986) Microfungal community structure from forest soils in southern Quebec, using discriminant function and factor analysis. *Can J Bot* 64: 1402–1412
10 Wardle DA, Zackrisson O, Hörnberg G, Gallet C (1997) The influence of island area on ecosystem properties. *Science* 277: 1296–1299
11 Schmidt SK, Ley RE (1999) Microbial competition and soil structure limit the expression of allelochemicals in nature. *In*: Inderjit, KMM Dakshini, CL Foy (eds): *Principles and practices in plant ecology: allelochemical interactions*. CRC Press, New York, 339–351
12 Inderjit, Weiner J (2001) Plant allelochemical interference or soil chemical ecology? *Perspect Plant Ecol Evol Syst* 4: 3–12
13 Reigosa MJ, Souto XC, González L (1996) Allelopathic research: methodological, ecological and evolutionary aspects. *In*: SS Narwal, P Tauro (eds): *Allelopathy: field observations and methodology*. Scientific Publishers, Jodhpur, 213–231

14 Reigosa MJ, Sánchez-Moreiras AM, González L (1999) Ecophysiological approach to allelopathy. *Crit Rev Plant Sci* 18: 577–608

15 Northup RR, Dahlgren RA, McColl JG (1998) Polyphenols as regulators of plant-litter-soil interactions in northern California's pygmy forest: a positive feedback. *Biogeochemistry* 42: 189–220

16 Binkley D, Giardina C (1998) Why do tree species affects soils? The warp and woof of tree-soil interactions. *Biogeochemistry* 42: 89–106

17 Pellissier F, Souto XC (1999) Allelopathy in northern temperate and boreal semi-natural woodland. *Crit Rev Plant Sci* 18: 637–652

18 Cheng HH (1992) A conceptual framework for assessing allelochemicals in the soil environment. *In*: SJH Rizvi, V Rizvi (eds): *Allelopathy: basic and applied aspects*. Chapman and Hall, London, 21–29

19 Inderjit, Dakshini KMM (1995) On laboratory bioassays in allelopathy. *Bot Rev* 61: 28–44

20 Dalton BR (1999) The occurrence and behavior of plant phenolic acids in soil environments and their potential involvement in allelochemical interference interactions: methodological limitations in establishing conclusive proof of allelopathy. *In*: Inderjit, KMM Dakshini, CL Foy (eds): *Principles and practices in plant ecology: allelochemical interactions*. CRC Press, New York, 57–74

21 Kaminsky R (1980) The determination and extraction of available soil organic compounds. *Soil Sci* 130: 118–123

22 Souto XC, Bolano JC, González L, Reigosa MJ (2001) Allelopathic effects of tree species on some soil microbial populations and herbaceous plants. *Biol Plant* 44: 269–275

23 Souto XC, González L, Reigosa MJ (1994) Comparative analysis of allelopathic effects produced by four forestry species during decomposition process in their soils in Galicia (NW Spain). *J Chem Ecol* 20: 3005–3015

24 Weidenhamer JD (1996) Distinguishing resource competition and chemical interference: overcoming the methodological impasse. *Agron J* 88: 866–875

25 Blum U (1998) Effects of microbial utilization of phenolic acids and their phenolic acid breakdown products on allelopathic interactions. *J Chem Ecol* 24: 685–708

26 Zackrisson O, Nilsson M-C (1992) Allelopathic effects by *Empetrum hermaphroditum* on seed germination of two boreal tree species. *Can J For Res* 22: 1310–1319

27 Huang PM, Wang MC, Wang MK (1999) Catalytic transformation of phenolic compounds in the soils. *In*: Inderjit, KMM Dakshini, CL Foy (eds): *Principles and practices in plant ecology: allelochemical interactions*. CRC Press, New York, 287–306

28 Inderjit (2001) Soil: environmental effects on allelochemical activity. *Agron J* 93: 79–84

29 Angers DA, Caron J (1998) Plant-induced changes in soil structure: processes and feedbacks. *Biogeochemistry* 42: 55–72

30 Shindo H, Kuwatsuka S (1976) Behavior of phenolic substances in the decaying process of plants – IV. Adsorption and movement of phenolis acids in soils. *Soil Sci Plant Nutr* 22: 23–33

31 Schnitzer M, Barr M, Hartenstein R (1984) Kinetics and characteristics of humic acids produced from simple phenols. *Soil Biol Biochem* 16: 371–376

32 Chapin F (1993) The evolutionary basis of biogeochemical soil development. *Geoderma* 57: 223–227

33 Perry D, Amaranthus M, Borchers J, Borchers S, Brainerd R (1989) Bootstrapping in ecosystems. *Bioscience* 39: 230–237

34 Van Breemen N (1993) Soils as biotic constructs favouring net primary productivity. *Geoderma* 57: 183–211

35 Souto XC, Chiapusio G, Pellissier F (2000) Relationship between phenolics and soil microorganisms in spruce forests: significance for natural regeneration. *J Chem Ecol* 26: 2025–2034

36 Souto XC, Pellissier F, Chiapusio G (2000) Allelopathic effects of humus phenolics on growth and respiration of mycorrhizal fungi. *J Chem Ecol* 26: 2015–2023

37 Blum U, Shafer SR, Lehman ME (1999) Evidence for inhibitory allelopathic interactions involving phenolic acids in field soils: concepts *vs.* an experimental model. *Crit Rev Plant Sci* 18: 673–693

38 Mizutani J (1999) Selected allelochemicals. *Crit Rev Plant Sci* 18: 653–671

39 Berendse F (1983) Interspecific competition and niche differentiation between *Plantago lanceolata* and *Anthoxanthum odoratum* in a natural hayfield. *J Ecol* 71: 379–390

40 Berendse F (1998) Effects of dominant plant species on soils during succession in nutrient-poor ecosystems. *Biogeochemistry* 42: 73–88

41 Tilman D (ed) (1988) *Plant strategies and the structure and dynamics of plant communities.* Monographs in Population Biology, Princeton University Press, New York

42 Hobbie S (1992) Effects of plant species on nutrient cycling. *Trends Res Ecol Evol* 7: 336–339

43 Van Breemen N (1995) Nutrient cycling strategies. *Plant Soil* 168–169: 321–326

44 Einhelling FA (1996) Interactions involving allelopathy in cropping systems. *Agron J* 88: 886–893

45 Tang CH-SH, Cai W-F, Kohl K, Nishimoto RK (1995) Plant Stress and Allelopathy. *In*: Inderjit, KMM Dakshini, A Einhellig (eds): *Allelopathy. Organisms, processes, and applications.* American Chemical Society, Washington, DC, 142–147

46 Inderjit, Mallik AU (1997) Effects of *Ledum groenlandicum* amendments on soil characteristics and black spruce seedlling growth. *Plant Ecol* 133: 29–36

47 Shibuya T, Fujii Y, Askawa Y (1994) Effects of soil factors on manifestation of allelopathy in *Cytisus scoparius. Weed Res* 39: 222–228

48 Blum U, Shafer SR (1988) Microbial populations and phenolic acids in soil. *Soil Biol Biochem* 20: 793–800

49 Sugai SF, Schimel JP (1993) Decomposition and biomass incorporation of ^{14}C-labeled glucose and phenolics in taiga forest floor: effects of substrate quality, successional state, and season. *Soil Biol Biochem* 25: 1379–1389

50 Wardle DA, Nilsson MC (1997) Microbe-plant competition, allelopathy and arctic plants. *Oecologia* 109: 291–293

51 Wardle DA, Lavelle P (1997) Linkages between soil biota, plant litter quality and decomposition. *In*: G Cadisch, KE Giller (eds): *Driven by nature. Plant litter quality and decomposition.* CAB International, Wallingford, 107–124

52 Westover KM, Kennedy AC, Kelley SE (1997) Patterns of rhizosphere microbial community structure associated with co-occurring plant species. *J Ecol* 85: 863–873

53 Bever JD, Westover KM, Antonovics J (1997) Incorporating the soil community into plant population dynamics: the utility of the feedback approach. *J Ecol* 85: 561–573

54 Bever JD, Westover KM, Antonovics J (1998) Reply correspondence. *Trends Ecol Evol* 13: 407–408

55 Pellissier F (1998) The role of soil community in plant population dynamics: is allelopathy a key component? *Trends Ecol Evol* 13: 407

56 Schimel JP, Cates RG, Ruess R (1998) The role of balsam poplar secondary chemicals in controlling soil nutrient dynamics through succession in the Alaskan taiga. *Biogeochemistry* 42: 221–234

57 Aber JD, Melillo JM, McClaugherty CA (1990) Predicting long-term patterns of mass loss, nitrogen dynamics, and soil organic matter formation from initial fine litter chemistry in temperate forest ecosystems. *Can J Bot* 68: 2201–2208

58 Ellis R, Pennington P (1989) Nitrification in soils of secondary vegetational successions from *Eucalyptus* forest and grassland to cool temperate rainforest in Tasmania. *Plant Soil* 115: 59–73

59 Killham K (1990) Nitrification in coniferous forest floors. *Plant Soil* 128: 31–44

60 Northup RR, Dahlgren RA, Aide TM, Zimmerman JK (1999) Effect of plant polyphenols on nutrient cycling and implications for community structure. *In*: Inderjit, KMM Dakshini, CL Foy (eds): *Principles and practices in plant ecology: allelochemical interactions.* CRC Press, New York, 369–380

61 Northup RR, Yu Z, Dahlgren RA, Vogt KA (1995) Polyphenol control of nitrogen release from pine litter. *Nature* 377: 227–229

62 Stark JM, Hart SC (1997) High rates of nitrification and nitrate turnover in undisturbed coniferous forests. *Nature* 385: 61–64

63 Melin E (1925) *Untersuchungen über die Bedeutung der Baummykorrhiza. Eine ökologisch-physiologische Studie.* Fisher, Jena

64 Handley WRC (1963) *Mycorrhizal* associations and *Calluna* heathland afforestation. *Bull For Commun* 36: 1–70

65 Rose SL, Perry DA, Pilz D, Schoeneberger MM (1983) Allelopathic effects of litter on the growth and colonization of mycorrhizal fungi. *J Chem Ecol* 9: 1153–1162

66 Coté JF, Thibault JR (1988) Allelopathic potential of raspberry foliar leachates on growth of ectomycorrhizal fungi associated with black spruce. *Am J Bot* 75: 966–970

67 Mallik AU, Zhu H (1995) Overcoming allelopathic growth inhibition by mycorrhizal inoculation. *In*: Inderjit, KM Dakshini, FA Einhellig (eds): *Allelopathy: organisms, processes and prospects.*

American Chemical Society, Washington, DC. 39–57

68 Pellissier F (1996) Allelopathy from ecosystem to cell in spruce (*Picea abies* (L.) Karst.) forests. *In*: SS Narwal, P Tauro (eds): *Allelopathy: field observations and methodology*. Scientific Publishers, Jodhpur, 65–79

69 Giltrap NJ (1982) Production of polyphenol oxidases by ectomycorrhizal fungi with special reference to *Lactarius* spp. *Trans Br Mycol Soc* 78: 75–81

70 Pellissier F (1993) Allelopathic effect of phenolic acids from humic solutions on two spruce mycorrhizal fungi: *Cenococcum graniforme* and *Laccaria laccata*. *J Chem Ecol* 19: 2105–2114

71 Boufalis A, Pellissier F (1994) Allelopathic effects of phenolic mixtures on respiration of two spruce mycorrhizal fungi. *J Chem Ecol* 20: 2283–2289

72 Brown RT, Mikola P (1974) The influence of fructose soil lichens upon the mycorrhizal and seedling growth of forest trees. *Acta Forest Fenn* 141: 1–22

73 Barazani O, Friedman J (1999) Allelopathic bacteria and their impact on higher plants. *Crit Rev Plant Sci* 18: 741–756

Do allelochemicals operate independent of substratum factors?

Harleen Kaur[1], Inderjit[2], and K.I. Keating[3]

[1] Department of Botany, Panjab University, Chandigarh 160014, India
[2] Department of Botany, University of Delhi, Delhi 110007, India
[3] Department of Environmental Sciences, Rutgers University of New Jersey, Cook Campus, New Brunswick, New Jersey 08903, USA

Introduction

"The plant world is not colored green; it is colored morphine, caffeine, tannin, phenol, terpene, canavanine, latex, phytohaem-agglutinin, oxalic acid, saponin, L-dopa, etc."—Janzen [1]. Plants synthesize a variety of secondary metabolites playing various roles in plant defense, plant interference (allelopathy), nutrient dynamics, waste elimination, mycorrhizae formation, and substratum ecology [2]. The term "plant allelochemical" was coined by Whittaker and Feeny [3]. Although both primary and secondary metabolites are being considered as allelochemicals, we will discuss only secondary metabolites as allelochemicals for the present discussion. Conn [4] suggested that secondary metabolites are compounds that do not have a direct role in the growth and reproduction of an organism. Although we should not use the terms "allelochemical" and "secondary metabolites" as synonyms, secondary substances largely act as allelochemicals [3]. Secondary metabolites can be very close to primary metabolites structurally. While kaurenoic acid and proline are primary metabolites, the closely related compounds abioetic acid and pipecolic acid are considered secondary metabolites [5]. Berenbaum [6] opined that the two terms (allelochemical and secondary metabolites) are not interchangeable. Secondary metabolites, however, may have some role in primary metabolism. For example, some allelochemicals may act as biosynthetic intermediates, growth regulators, or storage molecules for elements in short supply [7]. Secondary metabolites may exist in the dynamic equilibrium and may be recycled to primary metabolism [8]. Antioxidants such as vitamins A, C, and E can reduce phytotoxicity of photosensitizing allelochemicals [9]. Vitamins and the medicines we refer to as "antibiotics" could be considered allelochemicals, labeling the vitamins "pro-metabolites." In their role of protection in the producing organisms, the antioxidants function as would be anticipated with and without the intrusion of allelochemical interpretation. Various forms of B_{12} are needed by different organisms, depending on what "repair" or "completion" might be done by the receiving organism. Humans need the most complex form, mammalian-type, because they seem to be incompetent at assembling even the simplest additions to the basic cyanocobalamine molecule. The *Leischmania*, on the other hand, are quite capable of this. The form of B_{12} that humans use is an allelochemical because the active molecule, microbially produced, is ingested intact. Because the molecule is completed by *Leischmania*, however, the "incomplete" form of B_{12} is a nutrient rather than an allelochemical, since it is used as a substrate to build functional B_{12} in the organism.

Stahl [10] was the first to suggest that secondary metabolites play a role in the interaction of a plant with its environment and with other organisms to produce defenses against infection, predation, and environment stress. He termed secondary compounds *"Schutzexcrete"*. Plants produce great quantities of chemicals that have effects on other organisms. These are often referred to as secondary compounds because their function in primary metabolism is, or was, not fully known. This may be the case in phytoplankters. Unusually high percentages of the photosynthetic product of these cells are "dumped" into surrounding waters. This level of apparent waste must have a value to the producer, otherwise the producer could not survive, much less evolve, in the competitive world of nature.

The terms "allelochemical" and "nutrient" are usually interpreted as referring to mutually exclusive characteristics. However, the current authors agree with Berenbaum [6] that whether a substance should be identified as a nutrient or an allelochemical is actually dependent upon the context rather than on its biosynthetic origin. Some view the release of secondary substances as a release of waste products [11, 12 p. 331]. Yet we must ask why a plant would synthesize chemicals that are not needed: Plants release secondary compounds as part of interference and defense strategies. Some secondary metabolites are known to be recycled into primary metabolism. Others have important functions such as pollination, protection from herbivory, flower coloring, allelopathy, and plant defense. Any metabolic product, including allelochemicals, may have several roles, e.g., *cis*-dihydromatricaria ester (methyl 2-decene-4,6,8-triynoate) isolated from roots of *Solidago altissima* has been shown not only to possess lethal activities against nematodes but also to inhibit seedling growth of rice [13]. An allelochemical may not be toxic or stimulative *per se*. Two important factors that determine these characteristics in the allelochemical phytotoxicity are its concentration and the substratum ecology.

Rather than discovering direct physiological effects of allelochemicals, Birkett et al. [14] highlighted the importance of signaling within rhizosphere. While studying quorum-sensing cross talk, Holden et al. [15] reported that cyclic diterpenes from *Pseudomonas aeruginosa* may allow cross talk between different signaling systems. The synthesis and bioactivity of a cell-to-cell signaling compound, 2-heptyl-3-hydroxy-4-quinone, are mediated by quorum sensing. Birkett et al. [14] opined that such signaling systems should also occur in the context of the rhizosphere of higher plants, recently shown in pea

(*Pisum sativum*) exudates [16]. In addition to the mevalonate pathway involving glyceraldehyde 3-phosphate and pyruvate, a novel biosynthetic pathway is suggested [17, 18]. This pathway generates isopentenyl diphosphate *via* 1-deoxy-D-xylulose,5-phosphate and 2-C-methyl-D-erythritol,4-phosphate. Bohlman et al. [19, 20] studied the pathway using grand fir (*Abies grandis*) as a model and reported that low-molecular-weight isoprenoids are responsible for allelopathic effects of the tree. Harley et al. [21] reported that isoprene (2-methyl-1,3-butadiene), produced generally by woody species, plays an important role in tropospheric chemistry of forest regions, i.e., it contributes to ozone formation. The various factors influenced by isoprene are (1) ozone formation in the trophosphere and OH$^-$ chemistry, (2) formation of organic nitrates, and (3) deposition of organic acids to rural sites [22–24]. Thus, recent progress in the fields of plant physiology, molecular genetics, and molecular ecology all serve to improve our understanding of allelochemistry.

As noted above, the role of allelochemicals in determining community structure is largely determined by substratum factors in the terrestrial environment. Inderjit and Weiner [2] opined that soil chemical ecology plays an important role in determining allelochemical activities in field situations. The growth suppression of a plant through the release of chemical compounds by another plant species is widely known as allelopathy. Although the phenomenon has been conceptually accepted, field evidence is largely missing [2, 25]. The aim of this chapter is not to discuss problems associated with allelopathy methodology but to discuss the influence of the substratum factor in determining allelochemical interference as well as the interactions of organic and inorganic soil components with allelochemical phytotoxicity.

Allelochemical interaction and substratum ecology

Abiotic factors

Inderjit and Dakshini [26] emphasized the significance of the interaction of soil organic and inorganic constituents in bioassays for allelopathy. Foliar leachate/root exudates contribute inhibitors (e.g., phenolic acids, methionine), promoters (e.g., nitrates), and neutral substances (e.g., glucose) into the soil environment. Blum et al. [27] investigated the role of noninhibitory concentrations of neutral substances (glucose) and promoters (nitrates) in the modification of allelopathic activities of *p*-coumaric acid on seedling biomass of morning glory (*Ipomoea hederacea*). They reported that the amount of *p*-coumaric acid required to bring 10% inhibition of morning glory biomass was influenced by the presence of methionine and nitrate. In the presence of methionine, the amount of *p*-coumaric acid required to cause 10% inhibition was reduced to 3.75 µg/g soil, compared to 7.5 µg/g soil in the absence of methionine. Higher amounts of nitrate (14 *versus* 3.5 µg/g) increased the *p*-coumaric acid concentration required to cause 10% inhibition in the biomass of morning glory (Fig. 1a, b). Pue et al. [28] reported that noninhibitory concentrations of glucose-C (≤ 72 µg carbon/g soil) enhance the inhibitory activity of *p*-coumaric acid on morning glory seedling biomass. These authors found that the increasing concentrations of glucose decreased the amount of *p*-coumaric acid required for a given level of inhibition (Fig. 2). These authors suggested the differential utilization of organic molecules by soil microorganisms. After the isolation and identification of allelochemicals, one should not attempt to correlate biological activity of isolated allelochemical(s) to that of foliar leachate or root exudates. This is primarily due to the presence of neutral substances, promoters, and inhibitors in root exudates or foliar leachates, which may modify the allelochemical activity. Allelochemicals are not likely to operate independently of organic and inorganic soil constituents, and their activity is strongly influenced by substrate factors.

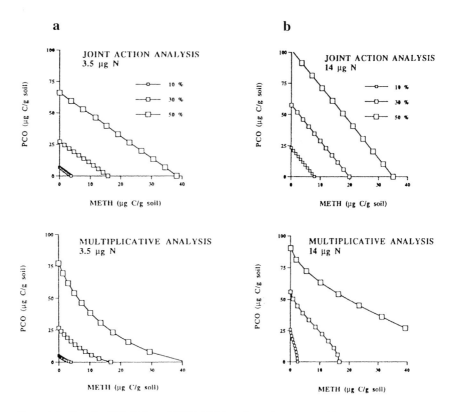

Figure 1. Isolines for 10%, 30%, and 50% inhibition of morning glory seedling biomass treated with *p*-coumaric acid C (PCO), methionine C (METH), and NO₃-N (3.5 µg/g [a]; 14 µg/g [b]). Values for isolines generated by joint action (a) and multiplicative (b) analysis. Source: [27]. Reproduced with permission from Kluwer Academic Publishers.

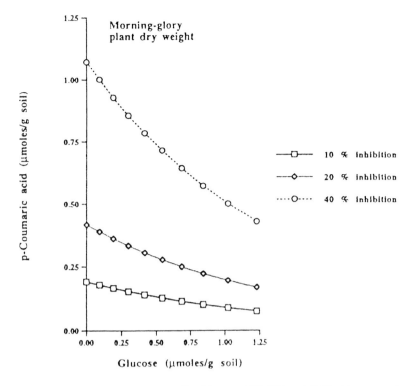

Figure 2. Inhibition of morning glory seedling biomass (10%, 20% and 40%) with *p*-coumaric acid and various concentrations of glucose. Source: [28]. Reproduced with permission from Kluwer Academic Publishers.

Biotic factors

The significance of biotic factors in the determination of allelochemical activity is often suggested [25, 29–31]. *Juglans nigra* often has been cited as an example where allelopathy is proven, and juglone (5-hydroxy-1,4-naptho-quinone) is widely cited as potent allelochemical [32]. Einhellig [33] opined that there are chances that the allelopathic effects of the walnut are actually due to the joint action of juglone and some additional, unidentified chemicals. In 1983 a gram-negative bacterium, *Pseudomonas putida*, was isolated from soil beneath black walnut trees in Germany [34], and these authors proposed that in presence of *P. putida* J1, juglone degrades into 3-hydroxyjuglone. This, they suggested, further converts into 2,3-dihydroxybenzoate and finally into 2-hydroxymucoic acid semialdehyde. Schmidt [35] then argued that juglone does not accumulate at phytotoxic levels because of its susceptibility to abiotic factors. Williamson and Weidenhamer [36], however, disagreed and proposed that walnut can maintain biotic and static availability of juglone all through the year because of its evergreen nature. Walnut soil, with and without the bacterium, however, could be expected to have different phytotoxic effects. A perennial shrub of the Florida plant community (the predominant pine species in the shrub is sand pine, *Pinus clausa*), *Ceratiola ericoides* produces ceratiolin (an inactive dihydrochalcone) which, as a result of light, heat, and acidic soil conditions, transforms into the toxic compound hydrocinnamic acid [37]. Hydrocinnamic acid further undergoes microbial degradation to form acetophenone [38]. Both hydrocinnamic acid and acetophenone suppressed the germination and growth of littlestem bluegrass (*Schizachyrium scoparium*) [37, 38]. Abiotic and biotic substratum factors, therefore, play a key role in phytoxicity of ceratiolin released from *Ceratiola ericoides*.

Schmidt et al. [39] discussed the relative distribution of soil microorganisms compared to plant distribution. It is important to find out how allelochemicals, released through root exudation, influence the spatial distribution of soil microorganisms. Since soil microorganisms use phenolics compounds as C-source [29, 31, 40], these authors hypothesized that the population of phenolics metabolizing microorganisms is higher in plants producing phenolic compounds than in plants not producing phenolic compounds. A study was conducted to investigate the effect of *Salix brachycarpa* on salicylate-mineralizing microorganisms in alpine soil. Higher levels of salicylate-mineralizing microorganisms were observed in soils beneath *S. brachycarpa* than in the surrounding meadows dominated by the sedge *Kobresia myosyroides* [39]. The above study illustrates the significance of soil microbial ecology in the expression of allelochemicals leached or exuded by the donor plant.

Conclusion

Although the chemical nature of an allelochemical is important in determining its activity, more important are substratum factors. A chemical will function as an allelochemical in a given substratum. Depending upon soil factors, a single chemical may be a potent allelochemical or may have neither toxic nor stimulatory effects. A plant may release several compounds, but the important point is to demonstrate their biological activity. Factors that determine allelochemical phytotoxicity or stimulation are (1) the nature and concentration of the compound, (2) abiotic and biotic soil factors, (3) the assay species, and (4) physical/climatic factors. Allelochemical phytotoxicity can best be conceptualized and studied in terms of the full picture of soil chemical ecology [2]. The natural product chemist often focuses on the novel compound and/or on compounds present in higher concentrations. Some compounds may not be novel and may be present at lower concentrations/amounts. Yet they may possess detectable phytotoxic effects, e.g., *p*-hydroxyacetophenone in conifer ecosystem. To understand the ecological significance of an allelochemical in the environment, the following issues must be considered:

1. Chemical analysis of the producing organism, which includes compounds present both at low and at high concentrations, is needed.

2. Compounds released into the environment must be clearly identified.
3. Data on residence time and fate of chemicals in the environment is needed.
4. Do allelochemicals, released in the substratum, influence growth of species actually growing in the vicinity of the producing organism?
5. If a compound originally contributed by the donor plant is not toxic, are its degradation products toxic?
6. What factors contribute to the phytotoxicity of an allelochemical?
7. What are the mechanisms of action of the alleged allelochemicals?
8. Are phytotoxic or stimulatory effects temporary or permanent?
9. How are allelochemical effects influenced by site, climate, and season?
10. Are joint actions of allelochemicals explained by the additive dose model or the multiplicative survival model?

Methodology is the weakest area of plant allelopathy [41, 42]. Although several workers proposed protocols [43–45], no study has ever met all criteria elaborated to demonstrate allelochemistry While considering terrestrial allelopathy, Blum et al. [25] state, *"Because no field observations or controlled field experiments have ever met the criteria elaborated by Willis (1985) in their entirety, the idealized concept of allelopathic phenomena has never actually been confirmed in the field."*

Because sophisticated biotechnological tools and mutants are available for several crop and weed species, it is possible to understand particular mechanisms underlying allelopathy. For example, certain strains of the bacterium *Escherichia coli* possess a plasmid responsible for the production of a toxin called colicin in the environment, and they also carry an immune protein to protect their cells from colicin. Other strains (colicin-sensitive), however, are without the plasmid and do not produce colicin. Iwasa et al. [46] reported that competition between colicin-producing and colicin-sensitive strains varied between a spatially structured and a completely mixed population. Later, Nakamaru and Iwasa [47] analyzed the rate of a third strain (colicin-immune) that, while it does not produce colicin, is colicin immune. It was reported that without spatial structure, this third strain inhibits colicin-producing strains and thus helps the colicin-sensitive strain to outcompete. Their study clearly demonstrates the significance of the aid of one strain for other in allelochemical interactions.

In field situations the chances of mixed populations of weed biotypes (herbicide-resistant and -sensitive) are likely, e.g., *Phalaris minor* in wheat cropping systems in northern India [48]. It is likely that allelopathic activities are due to joint action of several allelochemicals released by the donor plant and to other, unidentified compounds present in the substratum [49]. It may be possible that weed-suppressing wheat cultivars [50] respond differently to weed biotypes. It is, therefore, important to understand the ecophysiology and (agro)ecology prior to designing a study on allelochemical interactions. The above discussion illustrates that allelochemicals in terrestrial ecosystems largely do not operate independent of substratum factors. We must take a holisitic approach to integrating substratum and allelochemical activities.

References

1 Janzen DH (1978) Complications in interpreting the chemical defenses of trees against tropical arboreal plant eating vertebrates. *In*: CG Montgomery (ed): *The ecology of arboreal folivores*. Smithsonian Insititute, Washington, 73–84

2 Inderjit, Weiner J (2001) Plant allelochemical interference or soil chemical ecology? *Perspect Plant Ecol Evol Syst* 4: 3–12

3 Whittaker RH, Feeny PP (1971) Allelochemics: chemical interactions between species. *Science* 171: 757–770

4 Conn EE (ed) (1981) *Secondary plant products: the biochemistry of plants*. Vol 7. Academic Press, New York

5 Buchanan B, Grussen W, Jones R (eds): (2000) *Biochemistry and molecular biology of plants*. American Society of Plant Physiologists, Rockville, Madison

6 Berenhaum MR (1995) Turnabout is fair play: secondary roles for primary compounds. *J Chem Ecol* 21: 925–940

7 Seigler DS, Price P (1976) Secondary compounds in plants: primary functuions. *Am Nat* 110: 101–105

8 Seigler DS (1977) Primary roles for secondary compounds. *Biochem Syst Ecol* 5: 195–199

9 Green E, Berenbaum MR (1994) Phototoxicity of citral to *Trichoplausia ni* (hepidoptera: Noctuidace) and its amelioration by vitamin A. *Photochem Photobiol* 60: 459–462

10 Stahl (1888) Pflanzen und Schnecken. *Jena Z Med Naturwiss* 22: 559–684

11 Muller CH (1966) The role of chemical inhibition (allelopathy) in vegetational composition. *Bull Torr Bot Club* 93: 332–351

12 Orcutt DM, Nilsen ET (2000) *Physiology of plant under stress: soil and biotic factors*. John Wiley and Sons, New York

13 Saiki H, Yoneda K (1981) Possible dual roles of an allelopathic compound, *cis*-dehydromatricaria ester. *J Chem Ecol* 8: 185–193

14 Birkett MA, Chamberlain K, Hooper AM, Pickett JA (2001) Does allelopathy offer real promise for practical weed management and for explaining rhizosphere interactions involving higher plants? *Plant Soil* 232: 31–39

15 Holden MTG, Chhabra SR, de Nys R, Stead P, Bainton NJ, Hill PJ, Manefield M, Kumar N, Labatte M, England D et al (1999) Quorum-sensing cross talk: isolation and chemical characterization of cyclic dipeptides from *Pseudomonas aeruginosa* and other gram-negative bacteria. *Mol Microbiol* 33: 1254–1266

16 Teplitski M, Robison JB, Bauer WD (2000) Plant secrete substances that mimic bacterial *N*-acyl-homoserine lactone signal activities and affect population density-dependent behaviour in associated bacteria. *Mol Plant-Microbe Interact* 13: 637–648

17 Lichtenthaler HK, Rohmer M, Schwender J (1997) Two independent biochemical pathways for isopentenyl diphosphate and isoprenoid biosynthesis in higher plants. *Physiol Plant* 101: 643–652

18 Rohmer M (1999) The discovery for isoprenoid biosynthesis in bacteria, algae and higher plants. *Nat Prod Rep* 16: 565–574

19 Bohlman J, Crock J, Jetter R, Croteau R (1998) Terpenoid-based defenses in conifers: cDNA cloning, characterization and functional expression of wound-inducible (E)-α-bisabolene synthase from ground fir (*Abies grandis*). *Proc Natl Acad Sci USA* 95: 6756–6761

20 Bohlman J, Phillips M, Ramachandran V, Katoh S, Croteau R (1999) cDNA cloning, characterization and functional expression of four new monoterpene synthase members of Tpsd gene family from ground fir (*Abies grandis*). *Arch Biochem Biophys* 368: 232–243

21 Harley PC, Monson RK, Lerdau MT (1999) Ecological and evolutionary aspects of isoprene emission from plants. *Oecologia* 118: 109–123

22 Harley PC, Guenther A, Zimmerman P (1996) Effects of light, temperature and canopy position on net photosynthesis and isoprene emission from sweetgum (*Liquidambar styraciflua* L.) leaves. *Tree Physiol* 16: 25–32

23 Monson R, Fall R (1989) Isoprene emission from aspen leaves: influence of environment and relation to photosynthesis. *Plant Physiol* 90: 267–274

24 Sharkey TD, Singsaas EL, Vanderveer PJ, Geron C (1996) Field measurements of isoprene emission from trees in response to temperature and light. *Tree Physiol* 16: 649–654

25 Blum U, Shafer SR, Lehman ME (1999) Evidence for inhibitory allelopathic interactions involving phenolic acids in field soils: concepts *versus* experimental model. *Crit Rev Plant Sci* 18: 673–693

26 Inderjit, Dakshini KMM (1999) Bioassay for allelopathy: interactions of soil organic and inorganic constituents. *In*: Inderjit, KMM Dakshini, CL Foy (eds): *Principles and practices in plant ecology: allelochemical interactions*. CRC Press, Boca Raton, 35–44

27 Blum U, Gerig TM, Worsham AD, King LD (1993) Modification of allelopathic effects of *p*-Coumaric acid on morning-glory seedling biomass by glucose, methionine, and nitrate. *J Chem Ecol* 19: 2791–2811

28 Pue KJ, Blum U, Gerig TM, Shafer SR (1995) Mechanism by which noninhibitory concentrations of glucose increase inhibitory activity of *p*-coumaric acid on morning-glory seedling biomass accumulation. *J Chem Ecol* 21: 833–847

29 Schmidt SK, Ley RE (1999) Microbial competition and soil structure limit the expression of phytochemicals in nature. *In*: Inderjit, KMM Dakshini, CL Foy (eds): *Principles and practices in plant ecology: allelochemical interactions*. CRC Press, Boca Raton, 339–351

30 Inderjit, Cheng HH, Nishimura H (1999) Plant phenolics and terpenoids: transformation, degradation, and potential for allelopathic interactions. *In*: Inderjit, KMM Dakshini, CL Foy (eds): *Principles and practices in plant ecology: allelochemical Interactions*. CRC Press, Boca Raton, 255–266

31 Dalton BR (1999) The occurrence and behavior of plant phenolic acids in soil environment and their potential involvement in allelochemical interference interactions: methodological limitations in establishing conclusive proof of allelopathy. *In*: Inderjit, KMM Dakshini, CL Foy (eds): *Principles and practices in plant ecology: allelochemical interactions*. CRC Press, Boca Raton, 57–74

32 Willis RJ (2000) *Juglans* spp , juglone and allelopathy. *Allelo J* 7: 1–55

33 Einhellig FA (1995) Allelopathy: current status and future goals. *In*: Inderjit, KMM Dakshini, FA Einhellig (eds): *Allelopathy: organisms, processes, and applications*. ACS Sym. Ser. 582, American Chemical Society, Washington, DC, 1–24

34 Rettenmaier H, Kupas U, Lingens F (1983) Degradation of juglone by *Pseudomonas putida* J1. *FEMS Microbiol Lett* 19: 193–197

35 Schmidt SK (1990) Ecological implication of destruction of juglone (5-hydroxy-1-,4-naptho-quinone) by soil bacteria. *J Chem Ecol* 16: 3547–3549

36 Williamson GB, Weidenhamer JD (1990) Bacterial degradation of juglone: evidence against allelopathy? *J Chem Ecol* 16: 1739–1742

37 Tanrisever N, Fronczek FR, Fischer NH, Williamson GB (1987) Ceratiolin and other flavonoids from *Ceratiola ericoides*. *Phytochemistry* 26: 175–179

38 Fischer NH, Williamson GB, Weidenhamer JD, Richardson DR (1994) In search of allelopathy in the Florida scrub: the role of terpenoids. *J Chem Ecol* 20: 1355–1380

39 Schmidt SK, Lipson DA, Raab TK (2000) Effects of willows (*Salix brachycarpa*) on populations of salicylate-mineralyzing microorganisms in alpine soil. *J Chem Ecol* 26: 2049–2057

40 Inderjit (1996) Plant phenolics in allelopathy. *Bot Rev* 62: 186–202

41 Inderjit, Dakshini KMM (1995) On laboratory bioassays in allelopathy. *Bot Rev* 61: 28–44

42 Romeo JT, Weidenhamer JD (1998) Bioassays for allelopathy in terrestrial plants. *In*: KF Haynes, JG Millar (eds): *Methods in chemical ecology. Vol. 2. Bioassay methods*. Kluwer Academic Publishing, Norvell, 179–211

43 Putnam AR, Tang CS (1986) Allelopathy: state of science. *In*: AR Putnam, CS Tang (eds): *The science of allelopathy*. John Wiley and Sons, New York, 1–19

44 Horsley SB (1991) Allelopathy. *In*: ME Avery, MGR Cannell, CK Ong (eds): *Biophysical research for Asian agroforestry*. Winrock International, USA, 167–183

45 Willis RJ (1985) The historical bases of the concept of allelopathy. *J Hist Biol* 18: 71–102

46 Iwasa Y, Nakamura M, Levin SA (1998) Allelopathy of bacteria in lattice population: competition between colicin-sensitive and colicin-producing strains. *Evol Ecol* 12: 785–802

47 Nakamura M, Iwasa Y (2000) Competition by allelopathy proceeds in traveling waves: colicin-immune starins aids colicin-sensitive strains. *Theor Pop Biol* 57: 131–144

48 Singh S, Kirkwood RC, Marshall G (1999) Biology and control of *Phalaris minor* Retz. (littleseed canarygrass) in wheat. *Crop Prot* 18: 1–16

49 Blum U (1996) Allelopathic interactions involving phenolic acids. *J Nematol* 28: 259–267

50 Wu H, Pratley J, Lemerle D, Haig T (1999) Crop cultivars with allelopathic capabilities. *Weed Res* 39: 171–180

Ecological relevance of allelopathy: some considerations related to Mediterranean, subtropical, temperate, and boreal forest shrubs

Erik Tallak Nilsen

Biology Department, Virginia Polytechnic Institute and State University, Blacksburg, VA 24061, USA

Introduction

Ecological relevance, the most recalcitrant topic in chemical ecology, has hindered the acceptance of allelopathy in ecological theory. Proponents of allelopathy point to significant ecological phenomena such as bare zones around shrubs as evidence for the action of allelopathy in natural systems [1–3]. Skeptics of allelopathy in natural systems demand extensive experimental evidence on the toxic compound, its concentration in natural systems, and its pathway of delivery to target species [4]. Moreover, scientists studying allelopathy must demonstrate that resource competition is not the causative agent, while researchers who study competition are rarely held to the same standards [1, 5]. The purpose of this chapter is to probe the difficulties surrounding research on allelopathy and to present some imaginative ways scientists have investigated allelopathy in recent years. Have recent technology and scientific innovation provided enough support for allelopathy to take its place in ecological theory?

A short history of the development of allelopathy in ecological theory will lay the groundwork for our considerations. Hans Molisch [6] first mentioned allelopathy in a physiological context. However, the first serious consideration of the topic examined the effect of black and butternut walnut on wilting in several crop species [7]. Research on "toxicity" in walnuts and "soil sickness" in agricultural systems were the focus of allelopathic research until the early 1960s when C.H. Muller and W.H. Muller began publishing their research on both airborne and soil solution toxins [2]. Nonetheless, before the work of C.H. Muller, chemical toxicity among plants had been reviewed 13 times in prominent German and English journals (e.g., [8–11]). Critical reviews of allelopathy in relation to vegetation composition and coevolution among plants by C.H. Muller should be considered the foundation of allelopathy in ecological theory [2, 12, 13]. Muller's insight into allelopathy in nature was founded on a deep ecological understanding and extensive research with allelopathy in semiarid systems. The studies he, his students, and his colleagues performed on terpenes released from *Salvia* species and on phenolic acids released from chaparral plants are classics of the field. The first complete texts on allelopathy in the English language were by E.L. Rice [14, 15]. Since the publishing of these texts, there has been a revitalization of innovative research on allelopathy in many natural and crop ecosystems and a rejuvenation of the controversy about ecological relevance in natural systems [5, 16].

Allelopathy is placed in ecological theory among a matrix of plant-plant interactions (Tab. 1) based on the nature of the reciprocal effects on both species when grown together and when grown alone. Competition between species results in a negative effect on the fitness of both species, but allelopathy results in a negative influence on the "target" species and a positive influence on the fitness of the "source" species (source of the chemical toxin). In order to invoke allelopathy, the positive effect on source-species fitness must be demonstrated. Although Table 1 represents clear-cut differences among plant-plant interactions, there is considerable overlap among interactions, in nature. For example, nutrient-limiting conditions can cause an increase in allelopathic compounds in plant tissues [17], and fertilizing can negate the effects of allelopathy in some cases [18]. Therefore, under nutrient-limiting conditions (a prerequisite for most cases of competition for nutrients) allelopathic interactions can be enhanced. Moreover, the efficacy of allelopathic agents may be through inhibition of nutrient accumulation [19]. Although scientists have attempted to separate the effects of competition and allelopathy [20], these two interactions may be impossible to separate [21]. The chance for resource-chemical interaction (competition + allelopathy) in natural systems is greater than the chance for isolated interactions (competition or allelopathy alone) between plant species.

The ultimate cause of allelopathy, based on solid evolutionary theory, is another major issue in the field of allelopathy. Is there any evidence of natural selection for enhancement of allelopathy in plants? Most scientists promote

Table 1. Types of plant-plant interactions as commonly defined in the ecological literature

Name of interaction	Fitness when grown together		Fitness when grown alone	
	Species A	Species B	Species A	Species B
Competition	−*	−	0	0
Allelopathy	(source) +	(target) −	(source) 0	(target) 0
Parasitism	(parasite) +	(host) −	(parasite) −	(host) +
Mutualism	+	+	−	−
Amensalism	0 or +	−	0	0
Commensalism	+	0	−	0
Neutralism	0	0	0	0

* Effect (+,−,0) is defined here as the change in fitness of a species when grown together with the other species or when grown alone (modified from [16a]).

the idea that allelopathic capacity causes increased fitness in environments with limited resources. Those plants able to produce the most potent toxins will have the fewest competitors, the greatest availability of resources, and the highest fitness. This well-documented synergistic interaction between environmental stress (resource limitation) and enhanced plant toxicity has high importance in this theory [15]. Other scientists have emphasized the value of allelopathy to reduce fuel load and thereby to minimize the incidence of fire through reduced fuel load [22]. The lower incidence of fire increases the source plant fitness. Conversely, there may be an important link between increased fire incidence and accentuated allelopathy in chamise (*Adenostoma fasiculatum*) chaparral [23]. The burgeoning field of herbivore defense mechanisms complicates these arguments about the evolutionary significance of allelopathy. The basic classes of allelochemicals are virtually the same as the herbivore-defense compounds [24]. Under stressful conditions, leaf longevity increases and those leaves must have greater defense against herbivory [25, 26]. In fact, the term "allelochemicals" is used to represent the toxins involved in herbivore defense mechanisms [27]. The dominance of one species in a region (often ascribed to an allelopathic interaction) may be due to the action of herbivores rather than to competition or allelopathy. Moreover, in water- and nutrient-limited systems, evergreen species predominate, and these species have greater concentrations of herbivore-defense compounds than do deciduous species. Is it possible that allelopathic interference among species is a fortuitous result of evolution for herbivore defense?

The prevalence of autotoxicity is another important issue in allelopathy theory. If we assume that allelopathy (allotoxicity) confers adaptive significance for interspecific interference, then why are some species vulnerable to the compounds they have produced (autotoxicity)? Autotoxicity is frequent [24] but often is not considered in allelopathy research. Most researchers theorize that autotoxicity is a cost of defense against herbivores and pathogens [28, 29]. The allelochemical defense system can develop or intensify by natural selection as long as the cost of autotoxicity does not negate the advantage of defense. An alternative view is that autotoxicity is a self-thinning mechanism that reduces intraspecific competition and promotes outcrossing [30]. For example, pepper trees (*Schinus* species) produce a prodigious quantity of seed that falls below the tree. The seed releases allelochemicals into the upper soil layer that prohibit seed germination. Only seed that falls far from the tree will germinate after the toxins have been leached from the endocarp [31, 32]. The parent tree is unaffected by the seed-borne allelochemicals because the root system resides below the affected soil layers. Whether the ultimate significance of allelopathy is a fortuitous secondary benefit of herbivore-pathogen defense or a mechanism of self-thinning does not reduce the importance of allelopathy in community ecology.

Research on allelopathy began with natural systems and quickly evolved into a focus on agricultural systems and weeds. This is a logical trend because of the value of agriculture to humans and the monoculture nature of agricultural systems. Repeated culture and harvest of monocultures cause a legacy of residue that can quickly develop toxicity, and this has been noticed in many agricultural systems [5]. However, in natural systems the frequency of monoculture is uncommon, and succession is usually associated with a change in species composition over time. The result is a minimization of legacies from any one species. Exceptions to this general tenet include fire-dominated ecosystems (e.g., chaparral, boreal forest, western conifer forest) subcanopy shrubs, clonal plants, and some invasive species. In each of these cases, the species involved can develop large monoculture stands that recover rapidly following a disturbance (often fire). Other cases of allelopathy may be important within the immediate proximity of an individual plant (e.g., tree of heaven, black walnut, sassafras) but may have less significance at the community level. The presence of chemical interference in agricultural and natural systems is indisputable, but the key question is, to what extent is the chemical inhibition due to specific toxins released by the source plant? Allelopathic toxins, in contrast to herbivore defense compounds, travel a circuitous path from source plant through complex intervening media to the target plant. The recognition of the interface between source and target plant has promoted a series of suggestions to help support the case for allelopathy.

The research philosophy and some of the research accomplishments required for demonstrating the existence and ecological significance of allelopathy are listed in Orcutt and Nilsen [24] and are modified below.

Research philosophy:
1. Perform laboratory bioassays in a manner that closely represents field conditions.
2. Make every effort to identify possible confounding processes such as competition for limiting resources, herbivory, or environmental stressors.
3. Incorporate field experiments (particularly manipulations) before inferring that allelopathy occurs in nature.
4. Clearly identify how the chemical release results in positive fitness for the source plant and negative fitness for the target plant (this should be a priority).

Methodological accomplishments:
1. Identify the allelochemicals that are released from the source plant.
    a. Extract allelochemicals by using an aqueous nonpolar solvent.
    b. When making leachates, use a plant tissue that is appropriate for the specific growth cycle (developmental age) in amounts and duration that simulate natural leaching mechanisms.
2. Determine the active toxin species.
    a. Utilize substances measured in leachates from the source plant.
    b. Utilize more than one test species in bioassays such as *Lactuca sativa*

(lettuce), *Avena fatua* (oats), *Cucumis sativus* (cucumber), or *Lapidium sativum* (cress).

3. Determine the allelochemical dose required to cause 50% inhibition of germination or root elongation.
4. Determine the concentration and flux rates of allelochemical delivery to target plants.
5. Demonstrate inhibitory effects on native target species by laboratory bioassay.
   a. Use concentrations and flux rates that preserve natural abiotic elements.
   b. Take caution when interpreting the significance of laboratory bioassay to natural systems.
   c. When using a ratio of treatment to control, be aware that the data are likely to falsify statistical tests for normality and homoscedasticity.
6. Determine the influence of microbial populations on the source-plant allelochemical.
   a. Measure the residence time for allelochemicals in soil or atmosphere.
   b. Measure the presence of source-plant allelochemical breakdown products in natural systems and evaluate the toxicity of these compounds.
7. Perform field measurements that evaluate the likelihood of resource limitation to the fitness of target species above- or below ground.
   a. Consider target plant density and the dilution effect on the uptake of allelochemicals *versus* resource limitation.
   b. Manipulate resource availability to evaluate potential resource limitation and the interaction between allelopathy and resource limitation.
   c. Include evaluations of environmental stressors (water and heat) with evaluations of allelopathic mechanism.
   d. Determine the effect of source-plant leachates and litter fall on soil processes (chemistry and microbial populations).
8. Demonstrate the physiological mechanism by which the source allelochemical(s) inhibit target-plant fitness.

This is only a partial list of the high standards applied when ecological relevance of allelopathy is being proposed. There are very few studies (those discussed below) that utilize this philosophy and take all of these suggestions into consideration. In the rest of this chapter, I will focus on several studies of shrub systems and the efforts made to support the case for ecological relevance of allelopathy in natural ecosystems. The shrub systems considered in this chapter are from Mediterranean, subtropical, temperate, and boreal ecosystems.

Allelopathic potential of chaparral shrubs

The laboratories of Drs. C.H. Muller and W.H. Muller were vibrant with research on allelopathy during the 1960s and 1970s. Many different plant functional types were evaluated, including native evergreen species [23, 33], native deciduous shrubs [34], introduced herbaceous species [13], ferns [35], and introduced trees [31, 36]. The diversity of species and ecological processes studied in these laboratories fueled many discourses on ecological relevance of allelopathy in natural systems. The focus was on semiarid systems because of the location of the labs (Santa Barbara, CA) and the likelihood that allelopathic interactions may be particularly effective in relatively dry systems. Here I will concentrate on the research program centered on the deciduous coastal sage shrub *Salvia leucophylla* Greene.

Coastal sage scrub (or southern coastal scrub) occupies a discontinuous coastal strip from southern Oregon extending into Baja, California [37]. The species composition of this vegetation changes from north to south, and the dominants in southern California are species of *Salvia* and *Artemesia* [13]. *Salvia leucophylla* Green. and *Artemisia californica* Less. co-dominate the coastal sage scrub at the interface with coastal grassland in southern California. It was this interface in the Santa Ynez valley that sparked the interest of the Muller laboratories.

The ecological observation was a 5 m wide band of bare soil (the "bare zone") that followed the interface between *Salvia*-dominated coastal sage scrub and grassland (Fig. 1). No seedlings of grassland species survived in this zone. Moreover, grasses in an 8 m wide band adjacent to the "bare zone" were stunted compared to the unaffected grassland community (Fig. 1). The bare zone was most accentuated during dry periods and could disappear completely during a series of wet years.

I believe that allelopathy, rather than competition, was initially posed as a likely mechanism for this interference between species because of the highly aromatic nature of *S. leucophylla*. However, in order to support the case for allelopathy, the toxin had to be identified, the pathway to target plants had to be verified, and all other interference mechanisms (e.g., resource limitation, environmental stress, competition, and herbivory) had to be ruled out. Sampling the putative allelochemicals was easy because they were volatile (no soil microbial interference) and could be sampled in the air above the *Salvia* plants. A diversity of terpenes were identified from the air above *S. leucophylla* plants, but the highest concentrations were of cineole and camphor. Were these compounds the active toxic agent? In order to verify the toxicity of these terpenes, a new bioassay technique was needed that exposed test seed to an atmosphere containing the toxins and prevented any contact between test seed and the leaf material. Three species of *Salvia* that produced different amounts of cineole and camphor were used to study the relationship between terpene concentration from the source and toxicity [38]. The higher the terpenes in the leaf material, the more toxic was the atmosphere in the bioassay chamber. Moreover, filter paper soaked with cineole or camphor was the most detrimental to root elongation among all terpenes tested. These bioassays clearly showed that cineole and camphor were the toxic compounds released by *Salvia* leaves and that the toxicity increased with an increase in dose.

Although the bioassay design (Fig. 2) was very successful in showing toxicity and the atmospheric pathway, there was one drawback to the technique.

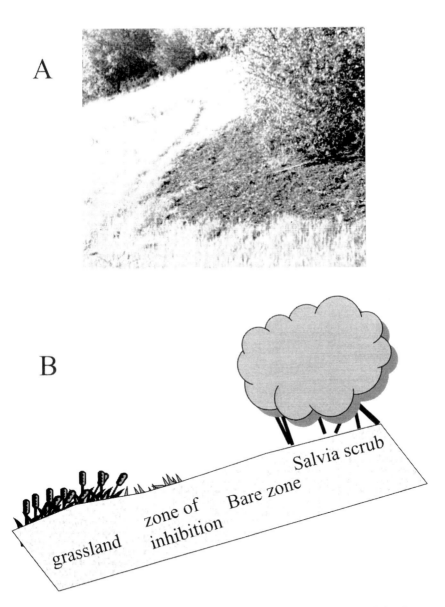

Figure 1. (A) A photograph of the ecotone between coastal sage scrub (dominated by *Salvia leuco-phylla*) and coastal grassland in southern California. (B) A diagrammatic representation of a cross-section through the ecotone between coastal sage scrub and grassland in southern California.

The concentration of terpenes in the bioassay chambers was significantly higher than that experienced in the field (a common criticism of allelopathy bioassay). The concentration of terpenes in the atmosphere above *Salvia* plants was

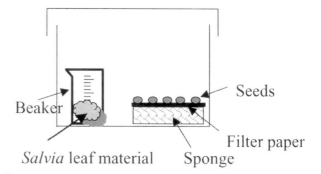

Figure 2. The bioassay design used to study atmospheric transfer of toxins from *Salvia* sp. leaves to various test species.

much lower than the concentration of terpenes that caused toxicity in bioassay. Therefore, a mechanism for concentrating terpenes at the site of action (the seed) was needed if allelopathy was to be supported. The researchers proposed that terpenes could concentrate from the atmosphere by dissolving in the cuticular wax encasing seeds. This proposition was supported because terpenes were removed from an experimental atmosphere by hard paraffin. In addition, dry soil adsorbed a significant amount of terpenes. A thin layer of soil exposed to atmosphere influenced by *Salvia* leaves became toxic to germination and radical elongation, even after the soil had been stored in an open container in a greenhouse for months [2]. Therefore, the toxins were identified, toxicity was verified, a potential route of delivery was outlined, and a concentration mechanism was proposed.

Studies on potential competition utilized the spatial gradient from the shrub margin to the "bare zone". Competition for light could not be a mechanism because light intensity was the same in all zones (bare, inhibition, grassland). Moreover, reduced radiation, such as shade from trees, promoted growth of grassland seedlings in the bare zone [13]. Competition for soil nutrients could not be a mechanism of interaction because nutrients were similar in all zones and *S. leucophylla* roots did not extend to the far edge of the bare zone [38]. In addition, natural fertilization under around cow droppings did not stimulate seed germination in the "bare zone" but did stimulate seedling growth in the grassland [13]. Seeds that fell into the bare zone would experience lower water availability and higher temperature than those that fell in the other zones. Thus, water stress in the zone of seed germination could be part of the mechanism controlling the "bare zone". In fact, the interaction between allelochemical inhibition of radical growth and water stress was cited as the main factor regulating the presence of the "bare zone" [13].

Seed and seedling herbivory was more difficult to rule out in these studies than was competition. Opponents to allelopathy suggested that birds, rodents, and rabbits residing under and in the coastal sage community could clear a zone of vegetation near the shrubs and browse the plants further out from the shrubs, causing the zone of inhibition [39]. Although no exclosure studies were performed, extensive observations of marked seedlings showed that few seedlings were completely lost due to grazing, many seedlings remained permanently stunted although they did not experience grazing, and some typical inhibition zones experienced no grazing [2]. Seed herbivory studies demonstrated a significant amount of seed loss to rodents and sparrows in the bare and inhibition zones. However, the spatially heterogeneous nature of this herbivory could not explain the uniform presence of the bare zone [13]. Herbivory by rodents (abundant under the shrubs) and sparrows (abundant in flocks at specific locals) could account for a part of the bare zone in some areas, but the entire extent of the uniform bare zone was ascribed to allelopathy.

The physiological mechanism of inhibition was also investigated. The active toxins released from *S. leucophylla* leaves caused a reduction in cell division and an increase in cuticle thickness [40–42]. Respiration rate was lower in tissue affected by the terpenes compared to a control without terpenes [43], as was the rate of mitochondrial activity [44]. It was clear that cineole and camphor from cut *S. leucophylla* leaves reduced radical growth rate by inhibiting the activity of mitochondria.

In conclusion, this study had demonstrated a strong ecological phenomenon that was correlated with the emission of terpenes from the coastal sage species. A unique bioassay technique had been devised, and dominant grassland species were used in some bioassays. The efficacy of specific terpenes was measured, a mechanism for concentrating the terpenes at the site of action was proposed, and the physiological mode of action was determined. Many possible versions of competition for resources had been ruled out, and environmental stresses were incorporated into the model. The laboratory analysis of toxicity and its mechanism of action were strong. However, field measurements were limited to observation. At the time of these studies, observation was the major mode of ecological research. The dominance of field manipulation experiments in ecological research had not developed. Consequently, the relative significance of herbivory was not settled, and the problem of field toxin concentration was never adequately solved. Moreover, the positive effect of allelopathy on source-plant fitness was suggested but not experimentally demonstrated. These three stigmas on an otherwise exemplary research program caused considerable doubt about the importance of allelopathy in this system.

Allelopathic potential of shrubs in a subtropical environment

In recent years the "bare zone" has been re-evaluated in another ecosystem. Two types of vegetation dominate the southeastern coastal plain of the United States. The sandhill community is dominated by slash pine (*Pinus elliottii* Engelm.), longleaf pine (*P. palustris* Mill), several species of oak (*Quercus* sp.), and a thick understory of grasses such as wiregrass (*Aristida stricta*

Mishx.). The sparse canopy and dense grass understory is maintained by frequent fire. Pines and grasses are well adapted to fire and continue growth rapidly after fire. Throughout the sandhill community are islands of a shrub-dominated community called the sand pine scrub. Endemic shrubs such as *Ceratiola ericoides* (A. Gray), *Chrysoma pauciflosculosa* ((Michx.) Green), and several members of the mint family dominate this community when it is young. Upon maturity, the community has a closed canopy of sand pine (*Pinus clausa* Vasey ex Sarg.) and a subcanopy of scrub oaks (*Quercus chapmanii* Sarg., *Q. germinata* Willd., and *Q. myrtifolia* Wild.). However, throughout the sand pine scrub succession there is an almost complete absence of herbaceous ground cover. In contrast to the sandhill, this community does not burn frequently and species in the community are not fire tolerant. Fires that move through the sandhill community often extinguish at the ecotone with the sand pine scrub community because there is no fuel to propagate the fire. When fire does propagate through the sand pine scrub, all species are killed and must regenerate from seed. The interface between these two communities was often described as a "fire-fighting association" [45].

The ecotone between the sandhill and sand pine scrub community is barren of vegetation, as is that between the coastal sage scrub and grassland in California [22]. Because of the presence of shrubs in the mint family (like *Salvia leucophylla*) and the presence of the bare zone, allelopathy was postulated as the mechanism of interaction between these two communities [46]. Plant ecologists and chemists, in collaboration, studied this interaction in the 1980s and 1990s.

The toxic compounds released by the scrub species were determined to be in the same class as those released by coastal sage scrub species: terpenes. However, in coastal sage scrub the delivery pathway was atmospheric and was accentuated by the semiarid environment. The environment around sand pine scrub is humid and wet enough to make an atmospheric delivery mechanism ineffective. If terpenes are not water soluble, how could they have efficacy in the sand pine scrub-sandhill ecotone? The non-water-soluble nature of terpenes was promoted by the Muller labs in the 1960s and had become the dogma in allelopathic research. Yet, leaf washes of several scrub species were found to be toxic to native grasses from the sandhills [47]. Moreover, the leaf surface waxes of the same species contained many terpenes [48], and many of these terpenes in saturated aqueous solutions were highly toxic [49]. In addition, surfaces of all three dominant sand pine scrub species contained high concentrations of the tri-terpene ursalic acid. This molecule was suggested to act as a natural surfactant that produced micelles in aqueous solution, and these micelles could be the delivery vector for monoterpenes [3]. Later studies showed that ursalic acid did not increase the solubility of oxygenated monoterpenes, but the solubility of oxygenated monoterpenes in aqueous solution was found to be much higher than previously thought [50]. In fact, the toxic monoterpenes found by the Muller laboratories (cineole, camphor) had relatively high solubility in water (Tab. 2), as did other terpenes from the sand pine

Table 2. Solubility of several terpenes found in species from the sand pine scrub vegetation in Florida and the coastal sage scrub vegetation of California

Name of compound[*]	Water solubility at saturation (ppm)
Limonene	13
α-pinene	22
Bornyl acetate	23
Camphene	27
1,8-Cineole	332
(+)-Evodone	409
(1S)-(−)-Camphor	550
(1R)-(−)-Myrtenol	1010
(−)-Carvenol	1115

[*] Compounds are listed in order of solubility. (Modified from [50].)

scrub species (e.g., carvenol, mytenol). Therefore, the toxic compounds of sand pine scrub species had been identified and a delivery mechanism proposed. However, the ecological relevance of this toxicity remained to be shown.

Soil resource availability was measured across the sand pine scrub-sand hill ecotone much as that done for the coastal sage-grassland ecotone in California. No differences in soil physical or chemical characteristics were found [45, 51, 52], and the addition of fertilizers in scrub areas did not increase germination or survival of grass species [52]. Therefore, competition for soil nutrients could not be a cause of the inhibition. Competition for water could not be a mechanism because of the high precipitation of the subtropical environment in the coastal plain of northern Florida. Irradiance is also relatively high and consistent across the ecotone, as evidenced by the high soil surface temperature in the sand pine scrub community [52] and the sparse canopy of the sandhill community. Fire seems to be the most important environmental factor separating these two species. Fire prevents the sand pine scrub vegetation from invading the sandhill community. Between fire episodes, sand pine shrubs establish and grow well in the sandhill vegetation, but the next fire kills them [53, 54]. Therefore, competition may be an important process regulating this ecotone in fire-free areas, but environmental stressors (fire and surface soil temp) and allelopathy are dominant in fire-maintained areas.

Ruling out competition is only one aspect of verifying ecological relevance of allelopathy. The *in situ* pathway for toxin transfer from source plants to target plants must be shown. Moreover, the *in situ* concentration of the toxin must be shown to have efficacy. The influence of soil biological activity on the compounds released from the source plant is critical to allelopathic function. Ceratiolin, which is found in leachates of *Ceratiola ericoides* leaves [55], breaks down in the soil to acetphenone and then hydrocinnamic acid. Hydrocinnamic acid has strong phytotoxicity, which was enhanced by nutrient limitation [56]. However, the question of soil concentration of these toxic compounds was not specifically addressed. This research group often mentioned the significance of chronic rather than acute toxin concentrations. A slow and constant delivery of toxins from the source plant could result in growth inhibition of target species. This concept is similar to the cuticular wax concentration system proposed by the Muller labs. The best way to determine whether chronic application of toxins can be effective in allelopathy is to perform field manipulation studies. The authors confirm that simulations that closely resemble the periodic release of toxins, and the complex soil chemistry, will be needed before ecological relevance of allelopathy can be verified or refuted [49].

The study on sand pine species allelochemistry demonstrated clearly that terpenes released from leaves into leaf washes were toxic to sandhill species. The studies also demonstrated that benign compounds released from leaves can be converted into active toxic compounds along the pathway to the target species. Some field evidence was used to suggest that competition was relatively insignificant. Moreover, the positive effect on source species was clearly laid out; allelopathy is a fire-prevention system for the sand pine scrub community. Yet the ecological relevance of allelopathy cannot be verified or refuted until field-based experiments are performed that take into account natural toxin delivery and concentration.

Allelopathic potential of temperate and boreal shrubs

Subcanopy evergreen ericoid (Ericaceae and Empetraceae) shrubs inhibit canopy tree seedling recruitment in many temperate and boreal forests worldwide. *Gaultheria shalon* Pursh., growing in the subcanopy of coniferous forests in the northwestern United States, can interfere with canopy tree seedling survival [57, 58]. The invasive species *Rhododendron ponticum* L. inhibits regeneration of canopy trees in the United Kingdom [59–61]. The best examples of this phenomenon in eastern North America are *Rhododendron maximum* L. in the southern Appalachian Mountains and *Kalmia angustifolia* (Small) Fernald. in northeastern Canada [62–66]. Also, *Empetrum hermaphroditum* inhibits canopy tree regeneration in boreal forest of Sweden [20].

The toxic compounds released by these species have been identified as phenolic acids (e.g., [67]) and a stilbene [20]. In most cases, no individual phenolic acid has been singled out as the exclusive toxic agent. Instead, toxicity has been ascribed to the sum total of phenolic compounds. Soil phenolic compounds are greatly increased when the soil harbors the decomposing litter of these shrubs [67], and the quantity of phenolics recovered from soils can constitute a toxic level [68, 69]. Also, there is a correlation between increased soil-borne phenolics and decreased target-plant performance both above- and below ground [67]. These studies have demonstrated the nature of the toxic substances, the pathway to the target species, the *in situ* concentration of these substances in the soil, and target species inhibition.

The mechanism of action also has been suggested as the interaction between phenolic concentration and nutrient availability. Several studies have shown that nutrient availability is reduced in regions harboring these shrubs (e.g., [70, 71]). The addition of shrub litter into soil can reduce the availability of nutrients [67], but when litter from different species is simply laid upon the surface of soils, the effect is less dramatic [72]. In a subcanopy environment, light availability may be the most important factor controlling the success of canopy tree seedlings, and several authors have claimed that nutrient variation is insignificant for seedling performance in low-light conditions [73–75]. Although phenolics from ericoid shrubs may reduce nutrient availability, this may be insignificant for subcanopy seedlings. The regulation of seedling performance by a factor other than light below a subcanopy shrub was found for *Rhododendron maximum* L. (Fig. 3). Canopy tree seedlings under *R. maximum* thickets grew less than seedlings in a forest without *R. maximum* at all light availabilities. A suppressive factor other than light availability was associated with the shrub thickets. That suppressive factor could be the phenolic acids released from *R. maximum* foliage and decomposing litter.

Field experiments have been a major theme in research on the potential allelopathic influences of ericoid shrubs. Two are particularly pertinent to the ecological relevance of allelopathy. The first case concerns the inhibition of boreal forest tree species by the shrub *Empetrum hermaphroditum*. A field experiment was designed to separate the effects of below-ground competition for resources between the shrubs and tree seedlings from the effects of allelopathy from shrubs on tree seedlings. The experiment was based on a buried "PVC tube" design (Fig. 4). Pine (*Pinus sylvestris* L.) seedlings were

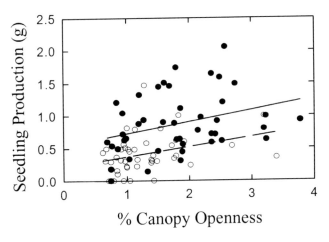

Figure 3. The relationship between seedling growth and light intensity in forest with or without a subcanopy of *Rhododendron maximum*. Open symbols represent *Quercus rubra* (red oak) seedlings growing under a thicket, and closed symbols represent *Q. rubra* seedlings growing in forest without a thicket.

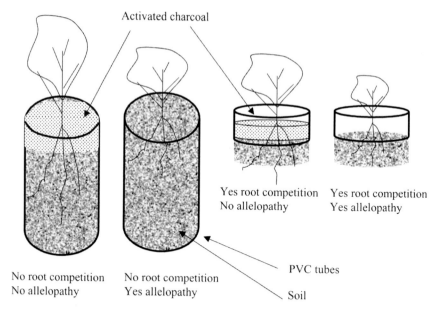

Activated charcoal

Yes root competition
No allelopathy

Yes root competition
Yes allelopathy

No root competition
No allelopathy

No root competition
Yes allelopathy

PVC tubes

Soil

Figure 4. Buried PVC tube experimental design used to separate the effects of belowground competition and allelopathy on tree seedling growth in shrub-dominated ecosystems.

planted below *E. hermaphroditum* shrubs within a PVC tube or without the tube. The tube served to inhibit competition below ground (reduced competition effect). The surface of the soil around half of the seedlings was topped with a layer of activated charcoal. The charcoal was designed to remove organic compounds (e.g., the phenolic acids and stilbene) coming from the shrubs, thereby reducing toxicity (reduced allelopathy effect). Seedlings grown in the tubes with charcoal (reduced competition and reduced allelopathy) grew the largest, while seedlings in either a tube without charcoal or charcoal without a tube grew to intermediate size. Seedlings that did not receive the tube or charcoal treatment grew the least [20]. These results indicate that both competition for resources and allelopathy contribute to *P. sylvestris* seedling inhibition by *E. hermaphroditum*.

The second study concerned the inhibitory effect of *R. maximum* on canopy tree seedlings in the southern Appalachian mountains. Bioassays of aqueous leaf and litter leachates from *R. maximum* indicated toxicity in test species germination and in root elongation [65]. Canopy throughfall (rain passing through the canopy) was not toxic in bioassay. A field experiment was designed to determine the long-term effects of litter and humus on seedling performance of a native tree species (*Quercus rubra* L.; red oak). The experimental design was a reciprocal transplant of litter and humus (in all combinations) between forest sites with and without a thicket of *R. maximum* (Fig. 5) followed by planting *Q. rubra* seeds in all plots. If transplanting humus and litter from the

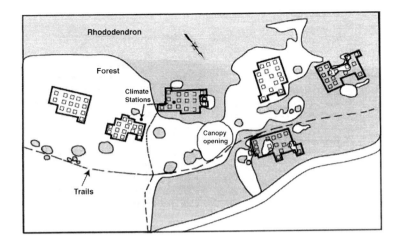

Plot treatments within blocks:

plot type	Plot location	substrate type
1*	in the *R.max* thicket	*R.max* litter + *R.max* humus
2	in the *R.max* thicket	*R.max* litter + forest humus
3	in the *R.max* thicket	Forest litter + *R.max*. humus
4	in the *R.max* thicket	Forest litter + Forest humus
5	in forest without a thicket	*R.max* litter + *R.max* humus
6	in forest without a thicket	*R.max* litter + forest humus
7	in forest without a thicket	Forest litter + *R.max*. humus
8*	in forest without a thicket	Forest litter + Forest humus

* = control plots in each vegetation type

Figure 5. Experimental design used to evaluate allelopathic potential of forest substrate below thickets of *R. maximum* compared with substrate from forest without a thicket. Each plot within each block received one of the listed treatments.

R. maximum thicket to forest sites without a thicket of *R. maximum* reduced seedling performance relative to control plots in forest without *R. maximum*, then the forest substrate from a *R. maximum* thicket contained inhibitory substances. On the other hand, if there was no effect of this forest floor transplant and transplanted substrate from forest without a thicket of *R. maximum* into a thicket had no effect compared to controls in the thicket, then there was no toxicity in the forest substrate materials. The latter case occurred, indicating no ecologically relevant toxicity in the forest substrate from the *R. maximum* thicket [65]. However, there was an inhibitory effect other than light limitation under the *R. maximum* thicket (Fig. 3). It remains to be determined whether long-term exposure to throughfall toxicity caused seedling inhibition in the thicket. Moreover, seedling inhibition could have been caused by low nutrient availability that was due to the action of phenolic acids or to efficient ericoid root nutrient accumulation [71].

Another important ecological link developed by studies on these shrubs is that between ectomycorrhizal fungi and potential allelopathic effects of the ericoid shrubs. Seedlings of *Picea mariana* Mill. (black spruce) growing close to *Kalmia angustifolia* had lower tissue nutrient concentration and lower mycorrhization than did black spruce seedling growing away from *Kalmia angustifolia* shrubs [76]. Although the diversity of ectomycorrhizal species (based on sporocarp diversity) is similar in forest types with or without a thicket of *R. maximum* [77], the association between ectomycorrhizae and tree seedling roots is inhibited below a *R. maximum* thicket [66]. That inhibition could be due to toxic effects of the shrubs, competition between ericoid mycorrhizae (on the shrub roots) and ectomycorrhizae (on the tree seedling roots), or reduced photosynthate flow to the seedling roots resulting from low light under the thicket. Bioassay of several ectomycorrhizal fungal species growth in media supplemented with *R. maximum* leaf pieces indicated toxicity [65]. However, similar or greater toxicity was found when the leaf material of other canopy tree species was incorporated into the medium. Therefore, the synthesis between seedling roots and ectomycorrhizal fungi was inhibited below a *R. maximum* thicket, and this could have contributed to reduced resource acquisition by the seedlings, but the mechanism of that inhibition was unknown. Leachates of senescent leaves from *E. hermaphroditum* also have been shown to inhibit mycorrhizal synthesis and nutrient acquisition of *P. sylvestris* roots [78].

Several of these studies of shrub toxicity have examined chemical ecology on an ecosystem scale [79]. Phenolic acids released from ericoid shrubs reduce both the availability of nitrogen in the soil and nitrogen mineralization rates. Moreover, nutrients released by decomposing litter are rapidly accumulated by ericoid mycorrhizae that inhabit these upper layers of humus. Ectomycorrhizae of canopy trees and tree seedlings tend to reside mostly in the mineral soil or the lower humus layers. Consequently, nutrients such as nitrogen that fall to the forest floor in throughfall (leaching of nutrients from leaves), stem flow (leaching of nutrients from trunks), or litter fall are quickly reaccumulated into the shrub biomass. This tight nutrient cycle reduces the availability of nutrients for canopy tree seedlings.

Conclusions

Research on allelopathy with shrubs has been extensive. Shrubs that inhabit subtropical, Mediterranean, temperate, and boreal forest have been studied. Terpenes transported through the air and water, phenolic acids transported by leaf washes, and decomposition products have been identified as the toxic substances. Imaginative bioassay designs that employ test species and/or native species have been used to identify toxic compounds and establish dose-response curves. Mechanisms for toxic substance concentration in the soil and at the site of action have been proposed and tested. Moreover, toxic compounds have been measured at active concentrations in field situations. The specific physiological mechanism of toxin action on target species has been elucidated. However, allelopathy has not been unambiguously verified or refuted in any of these cases. Why not?

Part of the problem comes from the definition of allelopathy in ecological theory. Allelopathy can be evoked only if the source plant has an increase in fitness because of toxic substances it releases that reduce the fitness of target species. Most of the researchers who work with these shrub systems take the broader view that allelopathy is simply the release of toxic chemicals that inhibit target species. This more lax view should be identified as amensalism under a strict plant-plant interaction classification (Tab. 1). Also, autotoxicity has not been studied often because autotoxicity would be a negative effect on the source species, which contradicts the definition of allelopathy. Increased fitness of the source plant is discussed only rarely in studies of shrub allelopathy, and no specific experiments have been designed to validate the increased fitness. A new definition of allelopathy may be called for now that research on this topic has progressed. One possible definition would be: Allelopathy occurs when a plant releases a chemical substance into the environment that inhibits the fitness of nearby species, releasing the source plant from interspecific competition and increasing the fitness of the source plant.

This definition recognizes the importance of interaction between chemical toxicity and resource availability. Above-ground competition has been easy to rule out, but below-ground competition between source and target species has been much more difficult to separate from allelopathy. If toxic substances cause a reduction in resource availability large enough to create limitation, then it will be impossible to separate competition from allelopathy. In fact, the interaction between chemical toxicity and resource limitation (nutrients or water) often made allelopathy effective in these shrub ecosystems. However, the reduction in resource availability may reduce the fitness of the source plant as well as of the target plant. On the contrary, if the toxic substance reduces the fitness of target species without influencing resource availability, then the source plant may have an increase in fitness.

The pathway from source species to target species is complex because of many biotic and abiotic factors. That complexity increases the burden of proof required to evoke allelopathy. Action of microbes in the soil, root symbionts, the rhizosphere, and mycorrhizae can all be involved in toxin degradation, increased efficacy, or toxicity amelioration. Verifying the pathway, stability of toxins, and toxicity of byproducts is a difficult task. Only a few of the research teams working on shrub ecosystems undertook it.

What is the best way to demonstrate ecological relevance of allelopathy? All of the bioassay and dose toxicity testing in laboratory situations can propose only the potential for allelopathy. Measurements of toxins in the soil or air around target plants can infer only the possibility of toxicity. Demonstrating toxicity on the target plant in the field verifies only half the requirement for allelopathy. Support for the ecological relevance of allelopathy is best accomplished by imaginative field manipulation experiments that demonstrate both positive fitness for the source plants and negative fitness for the target plants. The use of reciprocal transplanting of forest substrate, exclosures to remove herbivores, addition of activated charcoal, removing source species branches, working with a gradient of source species density, and varying target species density are some of the field experiments that have been used to evaluate ecological relevance of allelopathy in shrub systems. In the future, mutants of source plants may be found (or created by knocking out, or down-regulating key genes) that do not produce specific toxic substances. Controlled plantings of these mutants may be an excellent way to test for the ecological relevance of allelopathy. For example, monogenetic stands could be established at a research facility juxtaposed with California grassland. One stand type would be composed of a mutant genotype of *Salvia leucophylla* that does not synthesize cineole, while the other stand type would be composed of the wild type. If no bare zone appeared around the mutant stands, a bare zone did develop around the wild-type stands, and wild type plants had higher fitness when grown with grassland species, then allelopathy would be supported.

Analysis of chemical ecology in shrub systems has provided some of the strongest evidence for the ecological relevance of allelopathy. Yet there is still uncertainty. Continued vigilance in research on this topic with an emphasis on field experimentation is the best hope to strengthen the case for the significance of allelopathy in plant-plant interactions.

Acknowledgements
Some of the research cited in this review was supported by the NSF (grant numbers # IBN-9407129 and IBN 9630791) and the USDA/NRI forest biology program (95-37101-1902). Thanks to many conversations with colleagues and the help of the Virginia Tech Ecolunch group for editorial comments. Thanks to Jonathan Horton for reviewing this manuscript. A special thanks is given to the Muller laboratories, which provided the author with a solid appreciation for the significance of allelopathy in plant ecology.

References

1 Fuerst EP, Putnam AR (1983) Separating the competitive and allelopathic components of interference: theoretical principles. *J Chem Ecol* 9: 937–944
2 Muller CH (1966) The role of chemical inhibition (allelopathy) in vegetational composition. *Bull Torr Bot Club* 93: 332–351
3 Fischer NH, Williamson GB, Tanrisever N, De La Pena A, Weidenhamer JD, Jordan ED, Richardson DR (1989) Allelopathic actions in the Florida scrub community. *Biol Plant* 31: 471–478
4 Harper JL (1981) *Population biology of plants*. Academic Press, New York
5 Inderjit, Keating KI (1999) Allelopathy: principles, procedures, processes, and promises for biological control. *Adv Agron* 67: 141–231
6 Molisch H (1937) *Der Einfluss einer Pflanze auf die andere – Allelopathie*. Gustav Fischer, Jena
7 Massey AB (1925) Antagonism of the walnuts (*Juglans nigra* L. and *Juglans cinerea* L.) in cer-

tain plant associations. *Phytopathology* 15: 773–784

8 Evenari M (1949) Germination inhibitors. *Bot Rev* 15: 153–194

9 Bonner J (1950) The role of toxic substances in the interaction of higher plants. *Bot Rev* 16: 51–65

10 Knapp R (1954) *Experimentelle Soziologie der höheren Pflanzen*. Eugen Ulmer, Stuttgart

11 Woods FW (1960) Biological antagonisms due to phytotoxic root exudates. *Bot Rev* 26: 546–569

12 Muller CH (1969) Allelopathy as a factor in the ecological process. *Vegetatio* 18: 348–357

13 Muller CH (1970) The role of allelopathy in the evolution of vegetation. *In*: K Chambers (ed): *Biochemical coevolution, proceedings of the twenty-ninth annual biology colloquium, 1968*. Oregon State University Press, Corvalis, 13–31

14 Rice EL (1974) *Allelopathy*. Academic Press, New York

15 Rice EL (1984) *Allelopathy (2nd Ed.)* Academic Press, Orlando

16 Inderjit, Dakshini KMM, Einhellig FA (1995) *Allelopathy: organisms, processes, and applications*. American Chemical Society, Washington, DC

16a Burkholder PR (1952) Cooperation and conflict among primitive organisms. *Am Sci* 40: 601–630

17 Lehman RH, Rice EL (1972) Effects of deficiencies of nitrogen, potassium and sulfur on chlorogenic acid and scopolin in sunflower. *Am Midl Nat* 87: 71–80

18 Einhellig FA (1989) Interactive effects of allelochemicals and environmental stress. *In*: CH Chou, GR Waller (eds): *Phytochemical Ecology: allelochemicals, mycotoxins and insect phermones and allomones*. Institute of Botany, Academia Sinica, Taipei, 101–116

19 Inderjit, Mallik A (1997) Effects of *Ledum groenlandicum* amendments on soil characteristics and black spruce seedling growth. *Plant Ecol* 133: 29–36

20 Nilsson MC (1994) Separation of allelopathy and resource competition by the boreal dwarf shrub *Empetrum hermaphroditum* Hagerup. *Oecologia* 98: 1–7

21 Inderjit, Del Moral R (1997) Is separating resource competition from allelopathy realistic? *Bot Rev* 63: 221–230

22 Williamson GB (1990) Allelopathy, Koch's postulates and the neck riddle. *In*: JB Grace, D Tillman (eds): *Perspectives in plant competition*. Academic Press, New York, 143–162

23 Kaminsky R (1981) The microbial origin of the allelopathic potential of *Adenostoma fasciculatum* Han&A. *Ecol Monogr* 51: 365–382

24 Orcutt DM, Nilsen ET (2000) *The Physiology of Plants under stress: soil and biotic factors*. John Wiley and Sons, New York

25 Mooney HA, Gulmon SL (1982) Constraints on leaf structure and function in relation to herbivory. *Bioscience* 32: 198–206

26 Nilsen ET, Orcutt DM (1996) *The physiology of plants under Stress: abiotic factors*. John Wiley and Sons, New York

27 Whittaker RH (1975) *Communities and ecosystems*. Macmillan, New York

28 Stowe LG (1979) Allelopathy and its influence on the distribution of plants in an Illinois old-field. *J Ecol* 67: 1065–1085

29 Friedman J, Waller GR (1985) Allelopathy and autotoxicity. *Trends Biochem Sci* 10: 47–50

30 Seigler DS (1996) Chemistry and mechanisms of allelopathic interactions. *Agron J* 88: 876–885

31 Nilsen ET, Muller WH (1980) The relative naturalization ability of two *Schinus* species in southern California, II: Seedling establishment. *Bull Torr Bot Club* 107: 232–237

32 Nilsen ET, Muller WH (1980) The relative naturalization ability of two *Schinus* species in southern California, I: seed germination. *Bull Torr Bot Club* 107: 51–56

33 McPherson JK, Muller CH (1969) Allelopathic effects of *Adenostoma fasciculatum*, "chamise", in the California chaparral. *Ecol Monogr* 39: 177–198

34 Muller CH, Muller WH, Haines BL (1964) Volatile growth inhibitors produced by aromatic shrubs. *Science* 143: 471–473

35 Gliessman SR, Muller CH (1978) The allelopathic mechanisms of dominance in bracken (*Pteridium aquilinum*) in southern California. *J Chem Ecol* 4: 337–362

36 Del Moral R, Muller CH (1972) Fog drip: A mechanism of toxin transport from *Eucalyptus globulus*. *Bull Torr Bot Club* 96: 467–475

37 Mooney HA (1977) Southern Coastal Scrub. *In*: MG Barbour, J Major (eds): *Terrestrial vegetation of California*. John Wiley and Sons, New York, 471–489

38 Muller WH, Muller CH (1964) Volatile growth inhibitors produced by *Salvia* species. *Bull Torr Bot Club* 91: 327–330

39 Bartholomew B (1970) Bare zone between California shrub and grassland communities: the role of animals. *Science* 170: 1210–1212

40 Muller WH, Hague R (1967) Volatile growth inhibitors produced by *Salvia leucophylla*: Effect on seedling anatomy. *Bull Torr Bot Club* 94: 182–191

41 Lorber P, Muller WH (1976) Volatile growth inhibitors produced by *Salvia leucophylla*: Effects on seedling root tip ultrastructure. *Am J Bot* 63: 196–200

42 Lorber P, Muller WH (1980) Volatile growth inhibitors produced by *Salvia leucophylla*: Effects on cytological activity in *Allium cepa*. *Comp Physiol Ecol* 5: 60–67

43 Muller WH, Lorber P, Haley B (1968) Volatile growth inhibitors produced by *Salvia leucophylla*: Effect on seedling growth and respiration. *Bull Torr Bot Club* 95: 415–422

44 Lorber P, Muller WH (1980) Volatile growth inhibitors produced by *Salvia leucophylla*: effects on metabolic activity in mitochondrial suspensions. *Comp Physiol Ecol* 5: 68–75

45 Webber HJ (1935) The Florida scrub, a fire fighting association. *Am J Bot* 22: 344–361

46 Williamson GB, Richardson DR (1989) Bioassays for allelopathy: measuring treatment responses with independent controls. *J Chem Ecol* 14: 181–187

47 Richardson DR, Williamson GB (1988) Allelopathic effects of shrubs of the sand pine scrub on pines and grasses of the sandhills. *For Sci* 34: 592–605

48 Fischer NH, Tanrisever N, Williamson GB (1988) Allelopathy in the Florida scrub community as a model for natural herbicide actions. *In*: JB Harborne, FA Tomas-Barberan (eds): *Ecological Chemistry and Biochemistry of Plant Terpenoids*. Oxford University Press, Oxford, 233–249

49 Fischer NH, Williamson GB, Tanrisever N, Weidenhamer JD, Richardson DR (1994) In search of allelopathy in the Florida scrub: the role of terpenoids. *J Chem Ecol* 20: 1355–1380

50 Weidenhamer JD, Macias FA, Fischer NH, Williamson GB (1993) Just how insoluble are monoterpenes? *J Chem Ecol* 19: 1827–1835

51 Laessle AM (1958) The origin and successional relationship of sandhill vegetation and sand pine scrub. *Ecol Monogr* 28: 361–387

52 Richardson DR (1985) Allelopathic effects of species in the sand pine scrub of Florida. Ph.D. dissertation, University of South Florida, Tampa, Florida

53 Veno PA (1976) Successional relationships of five Florida plant communities. *Ecology* 57 : 498–508

54 Hebb EA (1982) Sand pine performs well in Georgia-Carolina sandhills. *South J Appl For* 6: 144–147

55 Williamson GB, Obee EM, WeidenhamerJD (1992) Inhibition of *Schizachyrium scoparium* (Poaceae) by the allelochemical hydrocinnamic acid. *J Chem Ecol* 18: 2095–2105

56 Williamson GB, Richardson DR, Fischer NH (1992) Allelopathic mechanisms in fire-prone communities. *In*: SJH. Rizvi, V Rizvi (eds): *Allelopathy: basic and applied aspects*. Chapman and Hall, London, 58–75

57 Klinka K, Carter RE, Feller MC, Wang Q (1989) Relation between site index, salal, plant communities, and sites in coastal Douglas-fir ecosystems. *Northwest Sci* 63: 19–29

58 Messier C (1993) Factors limiting early growth of western redcedar, western hemlock, and Sitka spruce seedlings on ericaceous-dominated clearcut sites in coastal British Columbia. *For Ecol Manage* 60: 181–206

59 Fuller RM, Boorman LA (1977) The spread and development of *Rhododendron ponticum* L. on the dunes at Winterton, Norfolk, in comparison with invasion by *Hippophae rhamnoider* at salt fleetby, Lilcolnshire. *Biol Conserv* 12: 82–94

60 Cross JR (1981) The establishment of *Rhododendron ponticum* L. in the Killarney oakwoods, S.W. Ireland. *J Ecol* 69: 807–824

61 Mitchell RJ, Marrs RH, LeDuc RH (1997) A study of succession on lowland heaths in Dorset, south England: changes in vegetation and soil chemical properties. *J Appl Ecol* 34: 1426–1444

62 Clinton BD, Boring LR, Swank WT (1994) Regeneration patterns in canopy gaps of mixed-oak forests of the southern Appalachians: influences of topographic position and evergreen understory. *Am Midl Nat* 32: 308–320

63 Mallik AU (1995) Conversion of temperate forests into heaths- role of ecosystem disturbance and ericaceous plants. *Environ Manage* 9: 675–684

64 Baker TT, Van Lear DH (1998) Relations between density of *Rhododendron* thickets and a diversity of riparian forests. *For Ecol Manage* 109: 21–32

65 Nilsen ET, Walker JF, Miller OK, Semones SW, Lei TT, Clinton BD (1999) Inhibition of seedling survival under *Rhododendron maximum* L. (Ericaceae): could allelopathy be a cause? *Am J Bot* 86: 1597–1605

66 Walker JF, Miller OK, Lei TT, Semones SW, Nilsen ET, Clinton BD (1999) Suppression of ecto-

mycorrhizae on canopy tree seedlings in *Rhododendron maximum* L. (Ericaceae) thickets in the southern Appalachian mountains. *Mycologia* 9: 49–56

67 Inderjit, Mallik A (1996) The nature of interference potential of *Kalmia angustifolia*. *Can J For Res* 26: 1899–1904

68 Gallet C, Nilsson MC, Zackrisson O (1999) Phenolic metabolites of ecological significance in *Empetrum hermaphroditum* leaves and associated humus. *Plant Soil* 210: 1–9

69 Wallstedt A, Nilsson MC, Zackrisson O, Odham G (2000) A link in the study of chemical interference exerted by *Empetrum hermaphroditum*: Quantification of batatasin-III in the soil solution. *J Chem Ecol* 26: 1311–1323

70 Damman AWH (1971) Effects of vegetation changes on the fertility of a Newfoundland forest site. *Ecol Monogr* 41: 243–270

71 Nilsen ET, Lei TT, Semones SW, Walker JF, Miller OK, Clinton BD (2001) Resource availability for canopy tree seedlings of the Appalachian mountains in the presence or absence of *Rhododendron maximum*. *Am Midl Nat* 145: 325–343

72 Nilsson MC, Wardle DA, Dahlberg A (1999) Effect of plant litter species composition and diversity on the boreal forest plant-soil system. *Oikos* 86: 16–26

73 Walters MB, Reich PB (1997) Growth of *Acer saccharum* seedlings in deeply shaded understories of northern Wisconsin: effects of nitrogen and water availability. *Can J For Res* 27: 237–247

74 Pacala SW, Canham CD, Saponara J, Silander Jr, A, Kobe RK, Ribbens E (1996) Forest models defined by field measurements: estimation, error analysis and dynamics. *Ecol Monogr* 66: 1–43

75 Coomes DA, Grubb PJ (2000) Impacts of root competition in forests and woodlands: a theoretical framework and review of experiments. *Ecol Monogr* 70: 171–207

76 Walker JF (1998) *The inhibitory effect of Rhododendron maximum L. (Ericaceae) thickets on mycorrhizal colonization of canopy tree seedlings*. MS thesis, Virginia Tech, Blacksburg

77 Yamasaki SH, Fyles JW, Egger KN, Titus BD (1998) The effect of *Kalmia angustifolia* on the growth, nutrition, and ectomycorrhizal symbiont community of black spruce. *For Ecol Manage* 105: 197–207

78 Nilsson MC, Hogberg P, Zackrisson O, Wang FY (1993) Allelopathic effects by *Empetrum hermaphroditum* on development and nitrogen uptake by roots and mycorrhizae of *Pinus sylvestris*. *Can J Bot* 71: 620–628

79 Wardle DA, Nilsson MC, Gallet C, Zackrisson O (1998) An ecosystem-level perspective of allelopathy. *Biol Rev Cambridge Phil Soc* 73: 305–319

Linking ecosystem disturbance with changes in keystone species, humus properties and soil chemical ecology: implications for conifer regeneration with ericaceous understory

Azim U. Mallik

Biology Department, Lakehead University, Thunder Bay Ontario, Canada P7B 5E1

Introduction

Plant-plant interaction mediated through allelochemicals in the environment is complex. In a recent paper, Inderjit and Weiner [1] articulated the complexity of the phenomenon quite nicely. At the ecosystem level, this complexity reaches another height when the ecosystem processes are regulated by natural and anthropogenic disturbance. Under natural disturbance regime, abiotic and biotic interactions such as competition, symbiosis, and allelopathy operate simultaneously and/or sequentially at various levels of temporal and spatial scales in organizing and maintaining plant community structure and diversity. Human interference of natural disturbance cycle can easily break the balance of community functioning, resulting in large-scale vegetation change and habitat degradation. Using examples of fire-adapted boreal forest dynamics, one can identify the various factors of ecological interdependencies. The dynamic equilibrium of boreal forest structure and composition is maintained largely by natural fires and, to a lesser extent, by insect defoliation and wind throw. Periodic fires rejuvenate plant communities by reducing competition and by consuming polyphenol-rich humus that create a suitable seedbed for canopy trees and provide a regeneration niche for understory species. The combined effect of frequency and intensity of natural fire regulates the speed and direction of secondary succession by controlling (1) the creation of differential regeneration niche for canopy trees and understory plants and (2) competition, allelopathy, symbiosis, and substratum factors such as pH change, cation exchange capacity (CEC), and nutrient cycling.

Absence of high-severity fire in Northern boreal forest with ericaceous understory can result in a large accumulation of humus that is rich in polyphynols and other secondary compounds, leading to long-term habitat change [2, 3]. In this ecosystem the thickness of forest floor humus determines whether the post-disturbance community will be regenerated to a conifer-dominated forest with ericaceous understory or to a treeless heath community dominated by ericaceous plants. A thick layer of humus, consisting mainly of conifer and ericaceous litter of the pre-disturbed forest or forest after clearcutting or that following a non-severe fire, is not conducive to conifer regeneration because of its phytotoxicity [4–8]. On the other hand, a thick humus layer provides an excellent medium for rhizomatous growth and sprouting of the understory ericaceous plants [9]. Severe fires consume the humus layer and create favorable mineral soil seedbeds for conifers [10, 11] and at the same time reduce the ericaceous understory by killing their vegetative parts [12]. Thus, the physical and chemical nature of the forest floor humus left after disturbance plays a predominant role in this ecosystem in structuring the plant community by preempting competition as a mechanism of community organization. The aim of this chapter is to highlight the causes and consequences of vegetation change by linking the effect of natural and anthropogenic disturbance with the dynamics of dominant species, their humus quality, and soil chemical ecology. It has been argued that disturbance-mediated keystone species change is the principal cause of ecosystem-level allelochemical effect and subsequent habitat degradation.

Conifer-ericaceous humus and allelopathy

Plant litter influences the physical and chemical property of forest floor (humus), which in turn may lead to long-term changes in soil chemical property and productivity [2]. Unfavorable humus chemistry and allelopathy have been attributed to natural regeneration failure of Norway spruce (*Picea abies* L. Karst) in sub-alpine spruce forests with bilberry (*Vaccinium myrtilus* L., hereafter referred to as *Vaccinium*) understory [5–7]. Gallet and Lebreton [13] studied the patterns of phenolic polymers (tannins) and monomers (phenolic acids and flavonoids) in the living leaf and root tissue and their litter and humus. They found high abundance of tannin in bilberry leaves, whereas the spruce needles had high abundance of *p*-hydroxyacetophenone. Compared to the green foliage, the brown foliage (litter) of bilberry exhibited greater loss of monomeric compounds but had high tanning activity. The amount of protocatechuic and vanillic acids (the degradation intermediates) were increased with increasing litter decomposition. Spruce-dominated organic soil under bilberry had high tanning activity and higher levels of phenolic acids. From a laboratory bioassay with natural leachates of bilberry and spruce, Gallet [6] concluded that *p*-hydroxyacetophenone, a spruce-tree-specific metabolite, and phenolic acids and tannins associated with bilberry reduce and often stop natural regeneration of spruce by inhibiting primary root growth of spruce seedlings. Jaderlund et al. [14] found that water leachate of green leaves of billberry was more inhibitory to germination and root growth of Norway spruce than that of brown leaves. While spruce germination in control (no *Vaccinium*, no litter removed) and above-ground *Vaccinium*-removed plots was

2% and 3%, respectively, it was 27% in plots that received humus-removal treatment [15]. Two phenolic compounds, caffeic and p-coumaric acids were found in rather high quantities, 1.7×10^{-3} M and 0.02×10^{-3} M respectively, in the humus. Pure compounds of both the phenolic acids were strongly inhibitory to germination and primary root growth of Norway spruce [15]. High concentration of phenolic acids in the humus layer of this forest has been suggested as a cause of Norway spruce regeneration failure.

Similarly, natural regeneration failure of Scotch pine (*Pinus sylvestris* L.) has been reported in northern Sweden after clearcutting when another understory Ericales, northern crowberry (*Empetrum hermaphorditum* Hagerup), dominates the post-harvest habitats [16, 17] in the absence of natural fire [18]. A phenolic derivative batatasin III (3,3-dihydroxy-5-methoxy-dihydrostilbene) has been found in high concentration in the forest floor humus and also in leaf hair glands of crowberry [19, 20]. This compound has been found in water leachate of humus that inhibited seed germination and seedling growth of the conifers [8, 21, 22]. An increase in batatasin with the increasing abundance of crowberry after clearcutting and seasonal variation in concentration of batatasin has been found to be positively correlated with poor seedling regeneration of the conifers [23]. Fire can reduce the levels of allelochemical, and charcoal of the post-fire habitat can adsorb the remaining phenolic compounds, rendering the habitat more conducive to conifer seed regeneration [18].

Inadequate natural regeneration and poor growth of planted black spruce (*Picea mariana*) in the presence of another ericaceous understory shrub, sheep laurel (*Kalmia angustifolia* L. var *angustifolia*, hereafter referred to as *Kalmia*), has been reported from nutrient-poor sites of eastern Canada [4, 24–27]. The combined effects of allelopathic inhibition of *Kalmia* humus on primary root growth of black spruce, resource competition from rapidly sprouting *Kalmia* on post-disturbed habitat, and soil nutrient imbalance created by the *Kalmia* humus phenolics have been suggested as reasons for conifer regeneration failure in this ecosystem [4, 28–30]. High rates of organic matter accumulation have been reported under *Kalmia*, and pH of *Kalmia* humus can be quite acidic, ranging from 2.4 to 3.0 [31]. In a controlled experiment, addition of *Kalmia* leaf leachate to the non-*Kalmia* organic and mineral soil caused a significant decrease in N and P and an increase in pH, total phenol, and metallic ions such as Fe, Al, Ca, K, Mg, Mn, Zn, Na, and Ba [28]. An increase in phytotoxicity of phenolic compounds with an increase in nutrient-poor soils has been reported [32, 33]. Phenolic allelochemicals are more phytotoxic at low soil N and P concentrations than at high soil nutrient concentrations [34]. The concentrations of phenolic compounds are correlated with soil pH, soil moisture, total carbon, and total soil N [35]. The oceanic and high-altitude temperate forests with ericaceous understory are characterized by high rates of organic accumulation, high soil moisture, and low pH and often are deficient in soil-available N and P.

Quality of seedbed

Seedbed quality plays a vital role in natural regeneration of conifers. Suitability of seedbed is determined by the physical and chemical nature of the seedbed substratum. Conifers require mineral soil seedbed for successful natural regeneration. In boreal ecosystem, severe fire creates such conditions by consuming the thick humus layer and exposing the mineral soil [10, 11]. In the absence of such a disturbance, the seedbed remains inhospitable for conifer regeneration because of the physical and chemical nature of the seedbed substrate. Post-logging, lightly burned, or insect-defoliated forest seedbed consists of thick humus originating from the partially decomposed litter of the canopy trees and understory plants, which is often rich in germination and growth-inhibitory allelochemicals [5, 7, 8]. Periodic fires remove these compounds by consuming the forest floor humus. The current method of forest management by fire suppression and clearcut logging creates a shift in dominance of keystone species by restricting the creation of favorable mineral soil seedbed for conifer regeneration. On the other hand, it allows prolific growth of understory ericaceous shrubs that regenerate vegetatively mainly by stem-base sprouting and rhizomatous growth under the protection of humus [9]. In nutrient-poor sites, the conifer-ericaceous communities thus transform into ericaceous heath following clearcut logging and non-severe fires [36, 37].

Keystone species and humus property

Keystone species in a habitat are those species that provide the unique structure and function of the ecosystem by performing essential ecosystem services [38, 39]. The removal of keystone species would result in a significant change in community structure and function [40]. The keystone species act as "ecosystem engineers" by controlling directly or indirectly the availability of resources to other species and by causing state changes on biotic and abiotic materials [40, 41]. In this section, I would like to relate the concept of keystone species and their functional roles with the physical and chemical changes of the seedbed for conifer regeneration. In the context of boreal forest allelopathy, it would be relevant to examine the role of different forms of disturbance in changing the keystone species, which in turn change the structure and function of the ecosystem (Tab. 1).

Forest floor allelopathy in northern temperate forest is regulated by the chemical nature of the humus. These forests often have one or two keystone species and a dominant ericaceous understory species that play the most predominant role in controlling above- and below-ground ecological functions by capturing most of the photosynthates and by their "afterlife effects" through the addition of leaf litter and fine roots to forest floor humus [42, 43]. A disproportionately large amount of the humus develops from the litter of keystone species, and the physical and chemical nature of the decomposing litter makes the seedbed for newly regenerating conifers from seeds. Changes in keystone species following forest disturbance bring about changes in forest floor humus chemistry and in its physical property. The cool, moist oceanic and high-altitude temperate forests generally have two layers of predominant species: conifer in the forest-canopy layer, forming the keystone species and the understory, or ground-level species consisting of ericaceous plants. For example, in the sub-alpine forest the canopy keystone species is Norway spruce or Scotch pine and the dominant understory species is *Vaccinium* [15, 44–46]. Similarly the boreal forests of eastern Canada have black spruce as a keystone species in the forest canopy and *Kalmia* as the dominant understory [36, 37]. In northern Sweden the keystone species at the forest-canopy level is Norway spruce or Scotch pine, with *Empetrum* as the dominant ground-level species [8]. Forest structure and function are regulated by vegetation dynamics as well as by humus chemistry of these keystone species and the understory species, which controls the above- and below-ground processes. Humus developing largely from conifer and the ericaceous understory litter contains an array of allelochemicals that interfere with natural regeneration of conifers by inhibiting their seed germination and seedling growth [4, 7, 8, 28, 29, 47, 48].

Logging disturbance, fire suppression, and shift in keystone species

Natural fire has been a major force in maintaining the characteristic structure and composition of boreal forest [49–51]. Periodic fires remove competition, reduce allelochemicals, and release nutrients by thermal decomposition of humus, creating a favorable seedbed for conifers and ephemerals by resetting the progressive secondary succession leading to forest development. The current forest-management practice of fire suppression and clearcut harvesting does not provide the ecological services necessary for resetting the characteristic secondary succession. The thick humus layer that develops over time from the accumulated litter of keystone species and dominant understory species is rich in germination- and growth-inhibitory allelochemicals, which interfere with natural regeneration of the canopy keystone species, the conifers. Furthermore, the physical characteristics of the accumulated humus in the absence of periodic natural fires act as a barrier of root establishment into the mineral soil. The delicate primary roots of the germinating seedlings in the loosely packed, partially decomposed humus get desiccated during the periodic hot spells of summer [52]. This condition, on the other hand, is favorable for the fire-survived resprouting species. All the dominant understory species mentioned above regenerate mainly by stem-base sprouting and rhizomatous growth [9, 45]. Increased light at the understory and warmer temperature at the humus layer following forest canopy removal by logging, insect defoliation, mild surface fire, and wind throw in over mature *Kalmia*-black spruce forest [52] stimulate vegetative growth of understory plants. In the absence of adequate seed regeneration of the canopy keystone species, the vegetatively regen-

Table 1. Regeneration conditions for keystone and understory species following natural and anthropogenic disturbance in nutrient-poor conifer forests with ericaceous understory

Disturbance type	Seedbed condition	Mode of regeneration	Spatial distribution
Natural fire (uncontrolled): Often large and severe fire burn with smoldering combustion	Mineral soil exposed seedbed, thin organic matter, flush of available nutrients	Favorable natural regeneration of keystone conifer species from seeds	Fairly uniform seedling regeneration of conifers
	Damaged seedbank and budbank of ericaceous plants	Limited vegetative regeneration of ericaceous plants	Patchy regeneration of ericaceous plants
Natural fire (suppressed): Smaller mild surface fire put out before it becomes a severe fire	Thick layer of charred humus, poor medium for seedling establishment, except small patches of high-severity burns	Insufficient conifer seedling regeneration	Scattered patches of conifer regeneration from seeds
	Favorable regeneration of ericaceous plants from fire-protected vegetative buds	Rapid vegetative regeneration of ericaceous plants	High and uniform cover of ericaceous plants
			Patchy conifer cover
Clearcut harvesting: Selective removal of conifer keystone species	Unfavorable seedbed in the presence of thick humus and understory ericaceous plants	Lack of seed regeneration of conifers; some advanced regeneration by layering	High and uniform cover of ericaceous plants
	Favorable regeneration of ericaceous plants from vegetative buds	Prolific vegetative regeneration of ericaceous plants	Patchy conifer cover
			High and uniform cover of ericaceous plants

(continued on next page)

Table 1. (Continued)

Disturbance type	Seedbed condition	Mode of regeneration	Spatial distribution
Selective harvesting	Unfavorable seedbed in presence of thick humus and understory ericaceous plants	Lack of seed regeneration of conifers; some advanced regeneration by layering	High and uniform cover of ericaceous plants
	Favorable regeneration of ericaceous plants from vegetative buds	Prolific vegetative regeneration of ericaceous plants	
Large-scale defoliation of insects such as spruce budworm and hemlock looper	Unfavorable seedbed in presence of thick humus and understory ericaceous plants	Lack of seed regeneration of conifers; some advanced regeneration by layering; dead standing snags	High cover of ericaceous plants, often patchy.
	Favorable regeneration of ericaceous plants from vegetative buds	Prolific vegetative regeneration of ericaceous plants	

erating understory ericaceous species soon become the predominant plant in the post-disturbance habitat and become the new keystone species. From then on, humus development occurs from litter accumulation of the ericaceous plants rich in allelochemicals. Phenolic compounds of ericaceous plants [53] are found to create soil nutrient imbalance by decreasing available N and P and increasing metallic ions [28]. Occupancy of a habitat by ericaceous plants for an extended time can bring about irreversible habitat degradation through high rates of organic accumulation and soil chemical changes [2, 3, 54, 55].

Seedbed allelopathy preempting competition

Competition is believed to be the predominant force in the structuring of an ecosystem. However, inhospitable seedbed condition, resulting from unfavorable physical and chemical characteristics of the humus, can preempt competition by limiting natural regeneration of conifers. It is clear from the preceding discussion that logging and absence of natural fire in conifer-ericaceous ecosystems can stimulate the growth of understory species, which can replace the canopy keystone species by limiting their natural regeneration [4, 5, 8, 29, 56, 57]. Consequently, the complex, multi-layered forest structure is replaced by simple ericaceous shrub-dominated heath, which further deteriorates the habitat by inducing changes in soil chemical ecology [28, 37, 47, 58, 59]. Recent studies indicate that even fire can create similar conditions of retrogressive succession from forest to heath, if it is not severe enough to burn off the ericaceous humus and form favorable seedbed for conifers [11]. Bloom [11] assessed favorable seedbed types (based on fire severity classes) by mapping six 20 m × 10 m plots in three recently burned (three years after fire) forests in Terra Nova National Park, Newfoundland, where an active fire-suppression program has been enforced for the last four decades [52]. Mappings were done by following the choropleth method [60] with an *a priori* assignment of four seedbed classes: (1) mineral soil exposed, (2) partially burned organic matter (25% left unburned), (3) charred humus, and (4) *Kalmia* vegetative regrowth. The results showed that mineral-soil-exposed seedbed created by the fire constitutes only 5% to 8% of the burned area and that the bulk of the post-fire habitat is made up of charred humus, which is unfavorable for conifer seed regeneration. Results of seeding experiments in these seedbed types showed that mineral-soil-exposed condition provides the most favorable seedbed for germination and establishment of black spruce; other seedbed substrate types are not conducive to black spruce seed regeneration. Areas other than the one experiencing high-severity fire had rapid vegetative growth of *Kalmia* from fire-survived rhizomes under the protective cover of charred humus. It was concluded from the study that the successful fire-suppression program in the Terra Nova National Park put out the fire before it became a large and high-intensity natural fire. Consequently, it did not produce enough high-severity burn to consume *Kalmia* humus and create the mineral soil seedbed necessary for black spruce regeneration. Many areas of the park previously dominated by black spruce with *Kalmia* understory that burned 1 to 40 years ago are now being dominated by *Kalmia* heath and have practically no black spruce regeneration. Such *Kalmia*-dominated heath is also common outside the park boundary, where the provincial ministry of forest maintains an aggressive fire-suppression program.

Keystone species as ecosystem engineers

In the absence of natural regeneration of black spruce (the canopy keystone species), the understory ericaceous species, predominantly *Kalmia*, becomes the new keystone species. At this stage, the multi-layered forest structure is replaced by a relatively uniform low-growing heath community. The rapid vegetative growth of *Kalmia* accumulates large amounts of humus that are characteristically different from that of black spruce. The "afterlife effects" [42] of *Kalmia* litter make the soil even more acidic; its phenolic allelochemicals bind N in protein-phenol complexes and the habitat thus becomes even more deficient in available N [61–63]. In the presence of a large array of phenolic acids, metallic cations such as Fe, Al, Ca, Zn, and Mn precipitate to the lower soil horizon, forming extremely hard iron pan and changing the soil-plant water relation. With the rapid buildup of acidic humus and a high rate of paludification, occupancy of the ericaceous heath community brings about long-term change in the habitat, making it less suitable for conifer regeneration [2, 3, 54, 55]. Ericaceous plants such as *Kalmia*, *Vaccinium*, and *Empetrum* thus act as ecosystem engineers. With their autogenic properties they bring about allogenic habitat changes [40].

Allelochemicals, microorganisms, ectomycorrhizae, and conifer regeneration

Litter decomposition in organic soil is mediated by litter-dwelling macro- and microorganisms. These organisms can decompose humus allelochemicals as well as produce new allelochemicals from their decomposition products [64–66]. Microorganisms are able to metabolize large amounts of phenolic compounds rather quickly [67–71]. The amount of available N and P in the polyphenol-rich humus is dependent on the combined effects of the activity of a vast array of soil organisms and their ability to influence the carbon and nitrogen cycles [72]. Post-disturbance ericaceous-dominated forests experiencing high rates of humus accumulation and conifer regeneration failure usually exhibit deficiency of available N in soil. An increase in population of nitrifying bacteria such as *Nitrosomonas* and *Nitrobacter* [73]; competition for ammonium between heterotrophic microbes, plants, and nitrifying bacteria [74, 75]; allelopathic plant exudates; and humus decomposition products all inhibit nitrification [76–80]. Both nitrifiers and other groups of soil microorganisms are associated with the allelopathic effects of forest soils [81]. In humus with low pH, fungi play a significant role in litter decomposition. Some soil fungi are negatively affected by soil allelochemicals, whereas others show stimulatory effects [70, 82, 83].

In organic soils, mycorrhizal fungi play vital roles in the functioning of higher plants of temperate region [84, 85]. Handley [86] and Robinson [87] demonstrated that certain conifer ectomycorrhizae were negatively affected by ericaceous (*Calluna vulgaris*) root exudates. Others have found that the effect can be both inhibitory and stimulatory, depending on the ectomycorrhizae, conifer species, and the type and concentration of allelochemicals involved [88–91]. Response of ectomycorrhizal fungi to allelochemicals is complex and often is related to chemical structure, chemical mixture, concentration, and fungal species [80, 91, 92]. Andre [44] suggested that high phenolic contents of a *Vaccinium*-dominated humus layer on the forest floor restrict the development of mycorrhizae of Norway spruce seedlings in sub-alpine forests. Four forest floor humus and humic solution allelochemicals (catechol, *p*-hydroxyacetophenone, *p*-hydroxybenzoic acid, and protocatechuic acid) of sub-alpine Norway spruce-*Vaccinium* forest were found to cause significant reduction in respiration of two conifer ectomycorrhizae, *Laccaria laccata* and *Cenococcum graniforme* [5, 92]. These authors concluded that conifer regeneration failure observed in this forest can be explained at least partly by the inhibitory effects of the humus allelochemicals that threaten the symbiotic relationship between the conifer and their ectomycorrhizae.

Yamasaki et al. [93] examined mycorrhizal infection, plant height and diameter growth, and foliar N of black spruce seedlings growing in the field close to (<1 m) and away (>1 m) from *Kalmia*. Black spruce seedlings close to *Kalmia* have been found to contain significantly fewer mycorrhizal short roots, less foliar N, and less height and diameter growth compared to those growing further away. In a field survey, black spruce seedlings were found to be relatively small and had about 50% less mycorrhizal infection in a site dominated by *Kalmia* compared to a contiguous non-*Kalmia* site.

Ericoid mycorrhizae

Ericaceous plants are in symbiotic association with a variety of mycorrhizae that are specifically adapted to nutrient-poor habitats [94, 95]. Ericaceous plants produce large quantities of polyphenolic materials (e.g., tannins, humic acids, melanins, and quinines) that can bind soil organic N as calcitrant protein-phenol complexes [96, 97]. Bending and Read [61] have shown that ericoid mycorrhizae are able to utilize protein N that is complexed with tannic acid by means of enzymatic degradation, whereas ectomycorrhizal fungi associated with conifers could not obtain N from the same source. They also demonstrated that ericoid mycorrhizae can utilize tannin as a carbon source, a feature other mycorrhizae do not have. Studies of these and other authors suggest that ericoid mycorrhizal associations may have resulted from the selective force of low available nitrogen environments [62, 98, 99]. Largent et al. [100] and Xiao and Berch [101] found that roots of salal *(Gaultheria shalon* Pursh), another ericaceous plant, are associated with three types of mycorrhizae—ericoid, arbutoid, and ectomycorrhizae—making them very efficient in obtaining N and P in acidic soils as well as in obtaining nutrients from complex organic forms. Thus, ericaceous plants seem to be much better equipped than conifers, which are typically associated with only ectomycorrhizae, for nutrient acquisition in polyphenol-rich, shrub-dominated conditions occurring after disturbance.

Ecological consequence of fire suppression and clearcutting

Clearcut harvesting and active fire suppression in nutrient-poor boreal forests with ericaceous understory can transform forest communities into ericaceous heath [37]. If scattered black spruce can be established in the limited mineral soil seedbed and subsequently can regenerate vegetatively by layering, over a long period of time (70–100 years) the forest may develop into a wood savanna-type community, as observed in some parts of Terra Nova National Park in Newfoundland. Replacement of multi-layered forest with shrub-dominated heath community or wood savanna-type community has important biodiversity implications. With the loss of forest structure in such communities, habitat for many understory plant and wildlife species such as birds, large and small game and non-game animals, insects, and butterflies is lost.

It appears that periodic removal of forest floor humus by wild fires is a precondition for rejuvenating forest community by creating the necessary conditions for keystone species regeneration and resetting the secondary succession. Removal of forest floor humus requires fire, but high-intensity smoldering combustion consumes humus along with the underground perennating structures of the ericaceous plants. This ensures the much-needed mineral soil seedbed for successful conifer regeneration as well as the relatively competition-free and allelochemical-free initial stage of progressive secondary succession. Any other form of canopy disturbance, such as clearcutting, insect defoliation, wind throw or even mild fire (under a successful fire-suppression regime), in the conifer-ericaceous community tends to replace the canopy keystone conifer species with ericaceous species. The rate and degree of biotic and abiotic changes are dependent on the type of canopy tree, the dominant understory plant, soil type, nutrient, pH, and climatic condition. For example, mild fires, clearcutting, or heavy spruce budworm defoliation in nutrient-poor black spruce or balsam fir *(Abies balsamea* (L.) Mill.) *Kalmia* forest can convert the forest community into *Kalmia* heath, and it can last for a very long time. If, on the other hand, the understory is predominantly Labrador tea *(Ledum groenlandicum* Oeder.), the growth inhibition of black spruce may last for 6–10 years, and after that the black spruce may regain the canopy dominance [47].

Ecosystem-level perspective of allelopathy

Ecosystem-level perspective of allelopathy has been studied very little. Citing two examples, (1) nodding thistle (*Carduus nutans*) containing New Zealand pasture and (2) crowberry containing Swedish boreal forest, Wardle et al. [43] argued that secondary metabolites of these invading species can cause ecosystem-level changes by negatively influencing the regeneration of dominant plants and key ecological processes. They suggested that the concept of allelopathy is more applicable to ecosystem- than to population-level processes.

Accumulation of large amounts of litter in the northern environment can be explained by the dominant plants, adaptation to produce polyphenol-rich litter in nutrient-stressed environments [102, 103] and by reduced litter decomposition that is due to poor litter quality and cool climatic conditions [37, 104]. Fire adaptation and anti-herbivory also can be considered as reasons for the production of polyphenol-rich litter by the dominant plants in this ecosystem [105–107]. It is logical to think that the subtle changes of an ecosystem are continually being brought about by the secondary compounds of living and dead remains of the dominant plants that influence the biotic and abiotic processes of the ecosystem. However, our research efforts in allelopathy so far have been almost exclusively directed toward the more dramatic effects of plant secondary compounds on the neighboring plants at the individual and population levels. We must focus our attention to the study of a broader landscape-level perspective of disturbance and allelopathy. We need to study the combined effects of many subtle physical and biochemical changes involving allelochemicals at a range of temporal and spatial scales.

Conclusions

Allelopathic phenomenon can be better appreciated at the ecosystem level in combination with other biotic and abiotic processes. By linking the concept of keystone species and their role as ecosystem engineers, one can detect the habitat change resulting from allelochemical nutrient interactions. This approach is described by using examples of conifer regeneration failure in the presence of ericaceous plants in boreal and sub-alpine temperate forests. The change of keystone species following ecosystem disturbance results from a combined effect of biotic processes such as competition and species regeneration strategies that influence productivity and litter accumulation, which in turn control the rate and direction of habitat changes and succession. To understand the mechanism(s) of community structuring following disturbance, one must understand and relate the sequence of major biotic and abiotic events and their roles in ecosystem functioning. In most cases, not one but several factors working at the ecosystem level can be identified as the cause and consequence of vegetation change.

Acknowledgments

The research was supported by a Natural Science and Engineering Research Council of Canada grant. Comments and editorial suggestions of Drs. Inderjit, J. Wang, F. Pellissier, and Steve Murphy were useful in revising the paper.

References

1 Inderjit, Weiner J (2001) Plant allelochemical interference or soil chemical ecology? *Perspect Plant Ecol Evol Syst* 4: 3–12

2 Damman AWH (1971) Effects of vegetation changes on the fertility of a Newfoundland forest site. *Ecol Monogr* 41: 253–270

3 Damman AWH (1975) Permanent changes in the chronosequence of a boreal forest habitat induced by natural disturbances. *In*: W Schmidt (ed): *Sukzessionsforschung.* J. Cramer, Vaduz, 499–515

4 Mallik AU (1987) Allelopathic potential of *Kalmia angustifolia* to black spruce. *For Ecol Manage* 20: 43–51

5 Pellissier F (1993) Allelopathic inhibition of spruce germination. *Acta Oecol* 14: 211–218

6 Gallet C (1994) Allelopathic potential in bilberry-spruce forests: influence of phenolic compounds on spruce seedlings. *J Chem Ecol* 20: 1009–1024

7 Gallet C, Pellissier F (1997) Phenolic compounds in natural solutions of coniferous forest. *J Chem Ecol* 22: 2401–2412

8 Zackrisson O, Nilsson M-C (1992) Allelopathic effects by *Empetrum hermaphroditum* on seed germination of two boreal tree species. *Can J For Res* 22: 1310–1319

9 Mallik AU (1993) Ecology of a forest weed of Newfoundland: vegetative regeneration strategies of *Kalmia angustifolia. Can J Bot* 71: 161–166

10 Mallik AU, Roberts BA (1994) Natural regeneration of red pine on burned and unburned sites in Newfoundland. *J Veg Sci* 5: 179–186

11 Bloom R (2001) *Disturbance conditions, species composition and functional diversity in* Kalmia-*black spruce forests.* M.Sc. thesis, Lakehead University, Thunder Bay

12 Flinn MA, Wein RW (1977) Depth of underground plant organs and theoretical survival during fire. *Can J Bot* 55: 2550–2554

13 Gallet C, Lebreton P (1995) Evolution of phenolic patterns and associated litter and humus of a mountain forest ecosystem. *Soil Biol Biochem* 27: 157–165

14 Jaderlund A, Zackrisson O, Nilsson M-C (1996) Effects of bilberry (*Vaccinium myrtillus* L.) litter on seed germination and early seedling growth of four boreal tree species. *J Chem Ecol* 22: 973–986

15 Mallik AU, Pellissier F (2000) Effects of *Vaccinium myrtillus* on spruce regeneration: testing the notion of coevolutionary significance of allelopathy. *J Chem Ecol* 26: 2197–2209

16 Sarvas R (1950) Metsatieteellisen tutkimuslaitoksen. Julkasisuja. *Commun Inst For Fenn* 38: 85–95 (English summary)

17 Hagner M (1984) Hur paverkas plantoran av sin narmaste miljo pa hygget? *In*: P-O Backstrom (ed): Skogsforyngring i fjallnara skogar. Sveriges Lantbruksuniversitet, Umea

18 Zackrisson O, Nilsson M-C, Wardle DA (1996) Key ecological function of charcoal from wildfire in the boreal forest. *Oikos* 7: 10–19

19 Oden PC, Brandtberg PO, Andersson R, Gref R, Zackrisson O, Nilsson M-C (1992) Isolation and characterization of a germination inhibitor from leaves of *Empetrum harmaphroditum* Hagerup. *Scand J For Res* 7: 497–502

20 Wallstedt A (1998) *Temporal variation and phytotoxicity of Batatasin-III produced by Empetrum hermaphroditum.* Ph.D. thesis, Lund University, Lund, Sweden

21 Nilsson M-C (1992) *Mechanisms of biological interference by Empetrum harmaphroditum on tree seedling establishment in boreal forest ecosystem.* Ph.D. dissertation, Swedish University of Agricultural Science, Umea

22 Gallet C, Nilsson M-C, Zackrisson O (1999) Phenolic metabolites of allelopathic significance in *Empetrum hermaphroditum* leaves and associated humus. *Plant Soil* 210: 1–9

23 Nilsson M-C, Gallet C, Wallstedt A (1998) Temporal variability of phenolics and batatasin-III in *Empetrum harmaphroditum* leaves over an eight year period: interpretation of ecosystem function. *Oikos* 81: 6–16

24 Candy RH (1951) *Reproduction on cutover and burned-over land in Canada.* Can. Dept. Res. and Dev., Forest Res. Div., Silv. Res. Note 92. 224p

25 Page G (1970) *The development of Kalmia angustifolia on black spruce cutover in central Newfoundland.* Forest Research Laboratory, St John's, Newfoundland, Internal Report N-27, 7 p

26 Wall RE (1977) *Ericaceous ground cover on cutover sites in southwestern Nova Scotia.* Can. Dep. Fish. Environ, Can. For. Serv., Inf. Rep. N-X-71, 55 p

27 Thompson ID, Mallik AU (1989) Moose browsing and allelopathic effects of *Kalmia angustifolia* on balsam fir regeneration in central Newfoundland. *Can J For Res* 19: 524–526

28 Inderjit, Mallik AU (1996) The nature of interference potential of *Kalmia angustifolia. Can J For Res* 26: 1899–1904

29 Mallik AU (1992) Possible role of allelopathy in growth inhibition of softwood seedlings in Newfoundland. *In:* SJH Rizvi, V Rizvi (eds): *Allelopathy: Basic and Applied Aspects.* Chapman and Hall, London, 321–341

30 Mallik AU (1998) Allelopathy and competition in coniferous forests. *In:* K Sassa (ed): *Environmental forest science.* Kluwer Academic Publishers, London, 309–315

31 Inderjit, Mallik AU (1999) Nutrient status of black spruce forest soils dominated by *Kalmia angustifolia* L. *Acta Oecol* 20: 87–92

32 Putnam AR (1985) Weed allelopathy. *In:* SO Duke (ed): *Weed physiology,* Vol. I. CRC Press, Boca Raton, 131–155

33 Einhellig FA (1995) Allelopathy: current status and future goals. *In:* Inderjit, KMM Dakshini, FA Einhellig (eds): *Allelopathy: organisms, processes, and applications.* ACS Symposium Series 582, American Chemical Society, Washington, DC, 1–25

34 Stowe LG, Osborn A (1980) The influence of nitrogen and phosphorus levels on the phytotoxicity of phenolic compounds. *Can J Bot* 58: 1149–1153

35 Blum U, Gerig TM, Worsham AD, King LD (1993) Modification of allelopathic effects of *p*-coumaric acid on morningglory seedling biomass by glucose, methionine, and nitrate. *J Chem Ecol* 19: 2791–2811

36 Mallik AU (1994) Autecological response of *Kalmia angustifolia* to forest types and disturbance regimes. *For Ecol Manage* 65: 231–249

37 Mallik AU (1995) Conversion of temperate forests into heaths: role of ecosystem disturbance and ericaceous plants. *Environ Manage* 19: 675–684

38 Ehrlich PR (1986) *The Machinery of Nature.* Simon and Schuster, New York

39 Ehrlich PR, Wilson EO (1991) Biodiversity studies: science and policy. *Science* 253: 758–762

40 Lawton JH, Jones CG (1995) Linking species and ecosystems: organisms as ecosystem engineers. *In:* CG Jones, JH Lawton (eds): *Linking species and ecosystems.* Chapman and Hall, New York, 141–150

41 Jones CG, Lawton JH, Shachak M (1994) Organisms as ecosystem engineers. *Oikos* 69: 373–389

42 Wardle DA, Bonner KI, Nicholson KS (1997) Biodiversity and plant litter: experimental evidence which does not support the view that enhanced species richness improves ecosystem function. *Oikos* 79: 247–258

43 Wardle DA, Bonner KI, Nicholson KS (1998) An ecosystem-level perspective of allelopathy. *Biol Rev* 73: 305–319

44 Andre J (1994) Régénération de la pessiere a myrtle: allelopathie, humus et mycorhizes. *Acta Bot Gallica* 141: 551–558

45 Andre J, Gensac P, Pellissier F, Trosset L (1987) Régénération des peuplements d'epicea en altitude: researches preliminaries sur le l'allelopathie et de la mycorhization dans les premiers stades du developpement. *Rev Ecol Biol Sol* 24: 301–310

46 Ponge JF, Andre J, Bernier N, Gallet C (1994) La régénération naturelle: connaissances actuelles. Le cas de l'epicea en forêt de Macot (Savoie). *Rev For Fr* 46: 25–45

47 Inderjit, Mallik AU (1996) Growth and physiological responses of black spruce (*Picea mariana*) to sites dominated by *Ledum groenlandicum. J Chem Ecol* 22: 575–585

48 Zackrisson O, Nilsson M-C, Dahlberg A, Jaderlund A (1997) Interference mechanisms in conifer-Ericaceae-feathermoss communities. *Oikos* 78: 209–220

49 Wein RW, MacLean DA (1983) An overview of fire in northern ecosystems. *In:* RW Wein, DA MacLean (eds): *The role of fire in northern circumpolar ecosystems,* John Wiley and Sons Ltd., Toronto, 1–18

50 Rowe S (1983) Concepts of fire effects on plant individuals and species. *In:* RW Wein, DA MacLean (eds): *The role of fire in northern circumpolar ecosystems,* John Wiley and Sons Ltd.

Toronto, 135–1554

51 Mallik AU (1982) *Post-fire microhabitat and plant regeneration in Calluna heathland*. Ph.D. thesis, Aberdeen University, Aberdeen

52 Power R (2000) Terra Nova National Park Vegetation Management Plan. Department of Canadian Heritage, Parks Canada, Ottawa

53 Zhu H, Mallik AU (1994) Interactions between *Kalmia* and black spruce: isolation and identification of allelopathic compounds. *J Chem Ecol* 20: 407–421

54 Meades WJ (1983) The origin and successional status of anthropogenic dwarf shrub heath in Newfoundland. *Advanced Space Res* 2: 97–101

55 Meades WJ (1986) *Successional status of ericaceous dwarf-shrub heath in eastern Newfoundland*. Ph.D. thesis, University of Connecticut, Connecticut

56 Pellissier F (1994) Effect of phenolic compounds in humus on the natural regeneration of spruce. *Phytochemistry* 36: 865–867

57 Pellissier F (1998) The role of soil community in plant population dynamics: is allelopathy a key component? *Trends Ecol Evol* 13: 407

58 Inderjit, Mallik AU (1997) Effect of *Ledum groenlandicum* amendment on soil characteristics and black spruce seedling growth. *Plant Ecol* 133: 29–36

59 Inderjit, Mallik AU (1997) Effect of phenolic compounds on selected soil properties. *For Ecol Manage* 92: 11–18

60 Burrough PA (1995) Spatial aspects of ecological data. *In*: RHG Jongman, CFJ ter Braak, OFR can Tongeren (eds): *Data analysis in community and landscape ecology*. Cambridge University Press, Cambridge, 213–251

61 Bending GD, Read JR (1996) Effects of soluble polyphenol tannic acid on the activities of ectomycorrhizal fungi. *Soil Biol Biochem* 28: 1595–1602

62 Bending GD, Read JR (1996) Nitrogen mobilization from protein-polyphenol complex by ericoid and ectomycorrhizal fungi. *Soil Biol Biochem* 28: 1603–1612

63 Mallik AU (2001) Black spruce growth and understory species diversity in contiguous plots with and without sheep laurel (*Kalmia angustifolia*). *Agron J* 93: 92–98

64 Kaminsky R (1981) The microbial origin of the allelopathic potential of *Adenostoma fasciculatum* (HandA). *Ecol Monogr* 51: 365–382

65 Rice EL (1984) *Allelopathy*. Academic Press, Orlando

66 Inderjit (1996) Plant phenolics in allelopathy. *Bot Rev* 62: 186–202

67 Blum U (1998) Effect of microbial utilization of phenolic acid and their phenolic acid breakdown products on allelopathic interactions. *J Chem Ecol* 24: 685–708

68 Blum U, Shafer SR (1988) Microbial populations and phenolic acids in soil. *Soil Biol Biochem* 20: 793–800

69 Lockwood JL, Filonow AB (1981) Response of fungi to nutrient-limiting conditions and to inhibitory substances in natural habitats. *In*: M Alexander (ed): *Advances in microbial ecology*. Plenum Press, New York, 1–61

70 Souto XC, Chiapusio G, Pellissioer F (1998) Soil microorganisms and plant phenolics: their implications in natural forest regeneration. *In*: K Sassa (ed): *Environmental forest science*. Kluwer Academic Publishers, Dorchdrecht, 301–308

71 Inderjit, Cheng HH, Nishimura H (1999) Plant phenolics and terpenoids: transformation, degradation and potential for allelopathic interactions. *In*: Inderjit, KMM Dakshini, CL Foy (eds): *Principles and practices in plant ecology: allelochemical interactions*. CRC Press LLC. Becona Raton, 225–285

72 Bradley RL, Fyles JW, Titus B (1997) Interactions between *Kalmia* humus quality and chronic low C inputs in controlling microbial and soil nutrient dynamics. *Soil Biol Biochem* 29: 1275–1283

73 Smith W, Bormann FH, Likens GE (1968) Response of chemoautotrophic nitrifiers to forest cutting. *Soil Sci* 106: 471–473

74 Robertson GP, Vitousek PM (1981) Nitrification potentials in primary and secondary succession. *Ecology* 62: 376–386

75 Stienstra AW, Gunnewiek PK, Laanbroek HJ (1994) Repression of nitrification by balsam poplar and balsam fir. *FEMS Microbiol Ecol* 14: 45–52

76 Thaibault JR, Fortin JA, Smirnoff WA (1982) *In vitro* allelopathic inhibition of nitrification by balsam poplar and balsam fir. *Am J Bot* 69: 676–679

77 Baldwin IT, Olson RK, Reiners WA (1983) Protein binding phenolics and their inhibition of nitrification in subalpine balsam fir soils. *Soil Biol Biochem* 15: 419–423

78 Olson RA, Reiners WA (1983) Nitrification in subalpine balsam fir soils: tests for inhibitory factors. *Soil Biol Biochem* 15: 413–418

79 Jobidon R, Thaibault JR, Fortin JA (1989) Phytotoxic effects of barley, oat, and wheat straw mulches in eastern Quebec forest plantations, II. Effects on nitrification and black spruce (*Picea mariana*) seedling growth. *For Ecol Manage* 29: 295–310

80 Pellissier F, Souto C (1999) Allelopathy in northern temperate and boreal semi-natural woodland. *Crit Rev Plant Sci* 18: 637–652

81 Souto XC, Gonzalez L, Reigosa MJ (1994) Comparative analysis of allelopathic effects produced by four forestry species during decomposition process in their soils in Galecia (NW Spain). *J Chem Ecol* 11: 3005–3015

82 Zhang Q (1997) Effects of soil extracts from repeated plantation woodland of Chinese fir on microbial activities and soil nitrogen mineralization dynamics. *Plant Soil* 191: 205–212

83 Lindeberg G, Lindeberg M, Lindeberg L, Popoff T, Theander O (1980) Simulation of litter-decomposing basidiomycetes by flavonoids. *Trans Br Mycol Soc* 75: 455–459

84 Malloch D, Malloch B (1981) The mycorrhizal status of boreal plants: species from northeastern Ontario. *Can J Bot* 59: 2167–2172

85 Malloch D, Malloch B (1982) The mycorrhizal status of boreal plants: additional species from northwestern Ontario. *Can J Bot* 60: 1035–1040

86 Handley WRC (1963) *Mycorrhizal* associations and *Calluna* heathland afforestation. *For Common Bull* No. 36

87 Robinnson RK (1972) The production by *Calluna vulgaris* of a factor inhibitory to growth of some mycorrhizal fungi. *J Ecol* 60: 219–224

88 Rose SL, Perry DA, Pilz D, Schoeneberger MM (1983) Allelopathic effects of litter on growth and colonization of mycorrhizal fungi. *J Chem Ecol* 9: 1153–1162

89 Cote JF, Thibault JR (1988) Allelopathic potential of raspberry foliar leachates on growth of ectomycorrhizal fungi associated with black spruce. *Am J Bot* 75: 966–970

90 Nilsson M-C, Hogberg P, Zackrisson O, Fengyou W (1993) Allelopathic effects of *Empetrum hermaphroditum* on development and nitrogen uptake by roots and mycorrhizas of *Pinus silvestris*. *Can J Bot* 71: 620–628

91 Mallik AU, Zhu H (1995) Overcoming allelopathic growth inhibition by mycorrhizal inoculation. *In*: Inderjit, KM Dakshini, FA Einhellig (eds): *Allelopathy: organisms, processes and prospects*. ACS Books, Washington, DC, 39–57

92 Bouflis A, Pellissier F (1994) Allelopathic effects of phenolic mixtures on respiration of two spruce mycorrhizal fungi. *J Chem Ecol* 20: 2283–2289

93 Yamasaki SH, Fyles JW, Egger NE, Titus BD (1998) The effect of *Kalmia angustifolia* on growth, nutrition, and ectomycorrhizal symbiont community of black spruce. *For Ecol Manage* 105: 197–207

94 Read DJ (1983) The biology of mycorrhiza in the Ericales. *Can J Bot* 61: 985–1004

95 Read DJ (1991) Mycorrhizas in ecosystems. *Experientia* 47: 31–36

96 Tackechi M, Tanaka Y (1987) Binding of 1,2,3,4,6-pentagalloylglucase to proteins, lipids, nucleic acids and sugars. *Phytochemistry* 26: 95–97

97 Mole S, Waterman PG (1987) A critical analysis of techniques for measuring tannins in ecological studies I. techniques for chemically defining tannins. *Oecologia* 72: 137–147

98 Leak JR, Read DJ (1989) The effects of phenolic compounds on nitrogen mobilization by ericoid mycorrhizal system. *Agric Ecosyst Environ* 29: 225–236

99 Leak JR, Read DJ (1991) Experiments with ericoid mycorrhiza. *In*: JR Norris, DJ Read, AK Varma (eds): *Techniques for the study of mycorrhiza: methods in microbiology*. Vol. 23, Academic Press, London, 435–459

100 Largent DL, Sugihara N, Wishner C (1980) Occurrence of mycorrhizae on ericaceous and pyrolaceous plants in northern California. *Can J Bot* 58: 2274–2279

101 Xiao G, Berch SM (1992) Ericoid mycorrhizal fungi of *Gaultheria shallon*. *Mycologia* 84: 470–471

102 Muller RN, Kalisz PJ, Kimmerer TW (1987) Interspecific variation in production of astringent phenolics over a vegetation-resource gradient. *Oecologia* 72: 211–215

103 del Moral R (1972) On variability of chlorogenic acid concentration. *Oecologia* 9: 289–300

104 Facelli JM, Pickett STA (1991) Plant litter: its dynamics and effects on plant community structure. *Bot Rev* 57: 1–32

105 Willamson GB, Black EM (1981) High temperature forest fires under pines as a selective advan-

tage over oaks. *Nature* 293: 643–644
106 Coley PD (1988) Effects of plant growth rate and leaf lifetime on the amount and type of anti-herbibore defense. *Oecologia* 74: 531–536
107 Coley PD, Bryant JP, Chapin IIIFS (1985) Resource availability and plant herbivore defense. *Science* 230: 895–899

Black walnut allelopathy: current state of the science

Shibu Jose

School of Forest Resources and Conservation, University of Florida, 5988 Hwy 90, Building 4900, Milton, FL 32583, USA

Introduction

Allelopathy, generally defined as any direct or indirect effect by one plant species on another through the production of chemical compounds that are released to the environment [1–3], has important widespread implications in natural communities and in artificial plant assemblages such as agriculture, horticulture, and forestry [2, 4–8]. Allelopathic influences may be subtle in natural communities, but they can be more pronounced in artificial communities among plant species whose evolutionary history is seldom connected. Black walnut (*Juglans nigra*), a tree species often cited as an example of allelopathy, is one of the most valuable hardwood timber species in the United States. It is often found in natural forests and grown in plantations. Anecdotal stories abound of the allelopathic effects of black walnut on associated vegetation, dating back to the first century A.D. [9]. Our knowledge base of black walnut allelopathy has increased steadily over the last two decades, allowing us to distinguish between myths and science. This chapter examines the different facets of black walnut allelopathy in the light of recent advances in chemical ecology.

Black walnut

Black walnut, the first tree species to receive a U.S. patent, is widely distributed throughout the eastern half of the United States, from Kansas to the Atlantic seaboard, in mixed forests on moist alluvial soils [10, 11] (Fig. 1). Although black walnut is associated with a number of tree species, it is not dominant in most forests but occurs as scattered single trees or as small isolated groups within hardwood stands. It is also grown in plantations because of the high demand for black walnut veneer in domestic and overseas markets. Individual trees can fetch as much as $10,000, with the highest reported value of a single tree being over $50,000. While the majority of black walnut culture has focused on timber production, black walnut is also valued for its nuts. Over

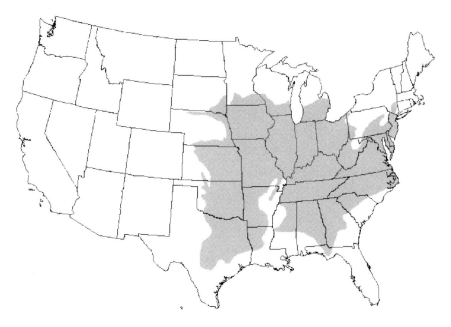

Figure 1. Natural range of black walnut (*Juglans nigra*) in the United States.

10,000 metric tons of black walnut nuts are harvested annually in the United States. The majority of this production comes from hand-harvested nuts from trees growing in native stands throughout its range. Despite its high value and wide acceptance among growers, black walnut is still regarded as a potential threat to many companion species.

Considerable interest has surfaced over the years in black walnut's controversial role as a potential allelopathic tree species. While some authors consider it the most notorious of allelopathic trees [12], others have questioned its alleopathic nature is a result of the contradictory claims of species affected [13–16]. The difficulty in allelopathic research is to ascertain the relative importance of allelopathy within the complexities of other direct and indirect causes of plant interference [17]. Plant-plant interaction can be either competitive or complementary. Most often, trees compete with associated herbaceous vegetation or crop plants for soil water, nutrients, and light. This can result in reduced growth and vigor of the associated plant species. Black walnut has been shown to exert intense competition on intercropped species for both above- and below-ground resources, thereby reducing crop yield [18, 19]. Allelopathy may play a major role in the observed yield reductions; however, it is often difficult to separate it from other competitive interactions. In fact, allelopathic influences may be modified significantly by environmental stress factors. Although the allelopathic phenomenon is intertwined with environmental factors and cannot be isolated from the environmental matrix, Willis [20, 21] has recommended a set of criteria to help identify allelopathy. Thus, in order to suggest that allelopathy is operative at a particular site or in a particular system, the following six criteria must be met.

1. A pattern of inhibition (or association) of one species by another species must be present.
2. The putative donor plant must produce biologically active substances.
3. The putative donor must have a mode of release of allelochemicals into the environment.
4. There must be a mode of allelochemical transport and/or accumulation in the environment.
5. The receiver plant must have some means of allelochemical uptake.
6. The observed symptoms and pattern of growth cannot be explained solely by physical or other biotic factors, such as competition or herbivory.

Black walnut has been shown to meet several of the aforementioned criteria but is in need of experimental evidence for others. After two millennia of identifying *Juglans* as an allelopathic genus, controversy still continues on the specifics of the synthesis, storage, and release of the allelochemical; its persistence in the soil; and its effect on other plants.

History of black walnut allelopathy

The history of walnut allelopathy dates back to the first century A.D. In about 1 A.D., Pliny the Elder wrote in his *Naturalis Historia* as follows: *"The shade of the walnut even caused headaches in man and injury to anything planted in the vicinity"*.

Similar observations have been documented by several writers during the medieval period and in modern history. A detailed historic account of walnut allelopathy is given elsewhere [21].

The American species black walnut was noted to have injurious effects on other plants during the latter part of the nineteenth century. Initial reports pointed to the poisonous nature of black walnut that killed apple trees in orchards [22, 23]. Further evidence came from experiments in which tomato plants exhibited growth failure when grown in water and soil media supplemented with black walnut bark [24]. Field observations by Massey [24] also confirmed toxic effects of black walnut on tomatoes and alfalfa within the rooting zone of the trees. Controversy started when Greene [13] questioned the toxic effects of black walnut. After several years, MacDaniels and Muenscher [14] reported the results of a 3-year greenhouse study in which they contradicted the results of Massey. According to these authors, no significant difference was found in the growth of tomato plants, alfalfa, and small apple trees when grown with small walnut trees or walnut roots in containers. Tomato plants watered with leachate from ground up black walnut roots also did not exhibit any signs of toxicity. The controversy continued when Brown [25] reported growth inhibition of tomato and alfalfa by black walnut bark and fresh roots. In 1948, the United States Department of Agriculture cleared the name of black walnut by publishing a clip sheet under the title "Test Clears Walnut's Reputation" [26]. This publication described the results of a field experiment in which tomato plants were planted under mature black walnut trees as follows: *The results of the tests seemed to clear the walnut trees. There was no evidence of a poisonous effect. When the tomatoes were supplied with water and nutrients they produced as good a crop as could be expected under conditions of poor sunlight under the walnut trees.*

However, numerous reports have been published since then documenting the inhibitory effects of black walnut on associated vegetation (e.g., [8, 27–30]). The controversy still remains unresolved [15, 16, 21].

The allelochemical: juglone

It is now well known that a phenolic compound called juglone (5-hydroxy-1,4-napthoquinone) is the causal agent in black walnut allelopathy (Fig. 2). Although it was isolated from *Juglans* spp. and identified in the latter half of the nineteenth century [31, 32], it was not until 1925 that juglone was suggested as the allelotoxin in black walnut allelopathy [24]. Later, Davis [33] supported this idea through experimental research in which he demonstrated the toxic effects of synthetic juglone on tomato and alfalfa plants.

The occurrence of juglone is a characteristic of the Juglandaceae family. While some scientists argue that juglone is restricted to Juglandaceae (e.g., [34]), others have reported its occurrence in unrelated families such as Proteaceae [35], Caesalpiniaceae [36], and Fabaceae [37]. In living tissues, juglone occurs mainly in a reduced nontoxic form called hydrojuglone [9, 38]. When exposed to the air, hydrojuglone is oxidized to its toxic form [9]. Small quantities of juglone also can be found in plant tissues along with hydroju-glone and other precursors and byproducts [21].

Dansette and Azerad [39] have proposed the most widely accepted biosyn-thetic pathway of juglone. It involves the condensation of succinyl-seminalde-

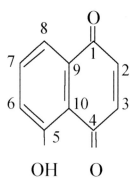

Figure 2. Structure of juglone (5-hydroxy-1,4-napthoquinone).

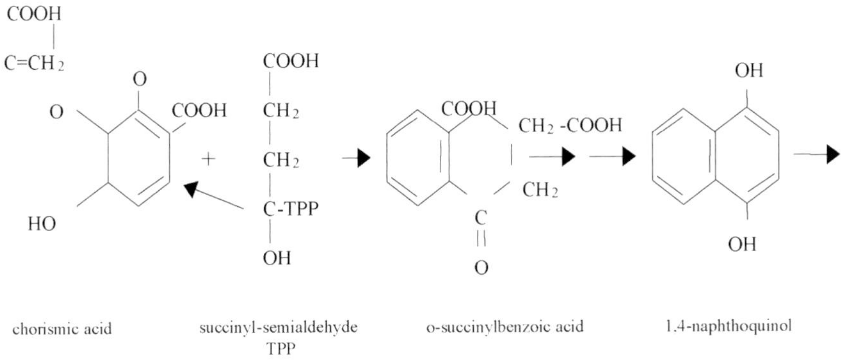

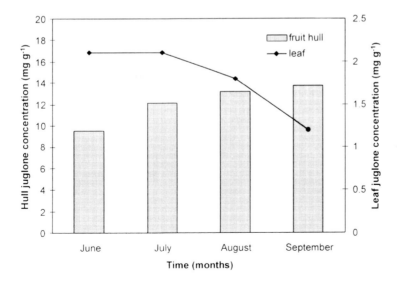

Figure 3. Biosynthetic pathway of juglone as proposed by Dansette and Azerad [39].

hyde thiamine-pyrophosphate carbanion with chorismate to form o-succinyl-benzoic acid. Upon ring closure, the immediate precursors of juglone, 1,4-napthoquinol or 1,4-naphthoquinone, are formed (Fig. 3).

Juglone is present in all parts of the black walnut, including roots, bark, twigs, leaf parts, buds, pollen, and fruit parts [38, 40, 41], except for the sap-wood and heartwood [42]. The concentration of juglone and/or its precursors shows great variation among cultivars and even individuals. Temporal varia-tions in juglone levels in different plant parts also have been reported (Fig. 4). In general, there is an increase in juglone in the fruit hulls and a decrease in

Figure 4. Temporal variation in juglone in fruit hull and leaf samples of black walnut (based on data published in Lee and Campbell [38]). It is suggested that juglone is produced in leaves and translo-cated to other plant parts including hull during fruit development.

leaves throughout the growing season. It also appears from the work of Lee and Campbell [38] that the highest level of juglone is present in the roots despite evidence of its production in the leaves [43]. Evidently, juglone produced in the leaves is translocated to other plant parts, and the roots act as a major sink for juglone and its precursors. But how much of the juglone is released into the soil? How long will it take to accumulate enough juglone in the soil to cause inhibitory effect on other plants? These questions remained unanswered until scientists recently began exploring the soil juglone dynamics [6, 7, 30].

Quantification of soil juglone

De Scisciolo et al. [6] reported the most reliable procedure available today for quantifying soil juglone using high-performance liquid cromatography (HPLC). Jose and Gillespie [7] have successfully used this procedure and observed recovery rates as high as 90%. In brief, this procedure involves extracting soil samples with chloroform (Fig. 5). One hundred grams sieved soil (through a 2 mm screen) is shaken with 100 mL chloroform for one hour. The mixture is vacuum-filtered through Whatman No. 1 filter paper. The soil is then reextracted with 50 mL chloroform for 30 min. The filtrates are combined and reduced *in vacuo* to a volume of 3 mL by using a rotary evaporator at 30°C. The concentrated extract is further dried under a jet of N_2. The dried extract is eluted with 0.5 mL chloroform. Aliquots (0.02 mL) are streaked onto silica gel G TLC plates and developed with a solvent mixture of hexane, chloroform, and glacial acetic acid (140:40:20). A juglone standard is also run on the same plate. The putative juglone band is scraped into a test tube and resuspended in 0.7 mL chloroform to which 0.3 mL 0.1 N acetic acid is added after one minute. This solution is centrifuged through 0.45 µm nylon filters, and the filtrate is used to quantify juglone by using HPLC on a 5 µm ODS column (4.6 × 150 mm). Elution is isocratic, with methanol containing 30% 0.1 N acetic acid at a rate of 0.8 mL min^{-1}. Juglone quantification is done based on a standard curve of peak areas monitored spectrophotometrically at 427 nm. The identity of the juglone peak is later confirmed by using a gas chromatography (GC)-mass spectrometry system. The GC is performed on a DB-1 fused silica capillary column (30 m × 0.25 mm) with a film thickness of 0.25 µm. The oven temperature is held at 50°C for 0.1 minute and raised 15°C per minute to 280°C. Mass spectrometry is by electron impact at 70 eV.

The soil juglone concentrations determined by HPLC have to be corrected using standard curves of juglone recoveries from soil. For juglone recovery experiments, known concentrations of commercial juglone in solution should be added to 100 g sieved soil collected from a juglone free site. After 10 minutes, the soil can be extracted with chloroform, and juglone can be quantified as previously described.

Recovery rates of juglone from soil can vary depending on soil and site conditions. Even within the same soil, recovery rates may vary among different

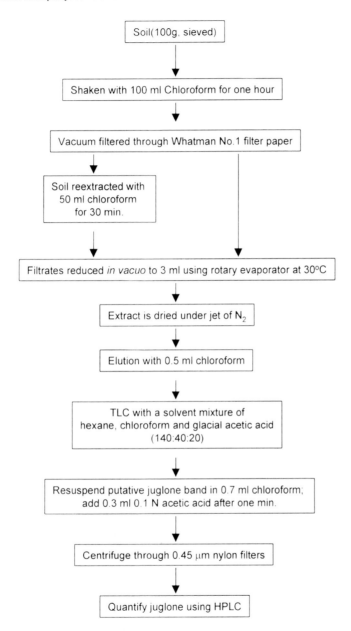

Figure 5. A flow chart of the procedure involved in extracting and quantifying juglone from soil samples.

seasons. In the recovery experiments by Jose and Gillespie [7], the highest recovery of added juglone was observed for spring soil samples (91.1%) and the lowest for fall samples (79.8%), with the summer samples being intermediate (83.4%). However, these differences were not significant (p = 0.4456). A strong linear relationship ($R^2 = 0.998$) was observed between the quantity of juglone added and that recovered from soil after 10 minutes (Fig. 6). Implicit in this observation is perhaps the complex chemical and physical interaction of juglone with soil. In nature, juglone may be removed from the soil in a number of ways, including leaching, chemical processes, microbial breakdown (discussed later), and uptake by plants. Further, juglone can be bound and made unavailable by soil organic matter and clays. These mechanisms operating in nature can take place in recovery experiments as well. When juglone is added to soil in low quantities, most of it is broken down or tied up in the soil. However, in a limited amount of soil (as in most recovery experiments), when the quantity added is increased more juglone becomes active or available (e.g., for extraction) because of the saturation of reaction sites or the slowing down of breakdown processes.

The standard juglone recovery curves based on 10 minutes of soil incubation may lead to overestimation of juglone in the soil. It is reasonable to assume that soil fixation may take more than 10 minutes. Juglone recovery rates following a temporal sequence of soil incubation have to be determined to correct for this potential error of overestimating soil juglone levels.

Spatio-temporal variation in soil juglone

An allelopathic role for juglone would require its release from tissues and accumulation in the soil in sufficient quantities to inhibit the growth of other plants. As with any allelochemical, juglone may enter the soil through process-

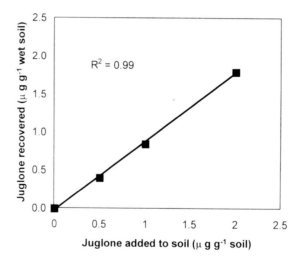

Figure 6. Juglone recovery as a function of quantity of juglone added to soil (source: Jose and Gillespie [7], used with kind permission from Kluwer Academic Publishers).

es such as root exudation, litter decay, and throughfall. The persistence and accumulation of juglone in soil will depend on local site and soil conditions. Early studies suggested juglone's presence only in the immediate vicinity of roots and its lack of persistence [24, 44]. Recent reports indicate a gradient of juglone levels with respect to depth and/or distance from black walnut trees [30, 45]. For example, in a 14-year-old black walnut plantation, Ponder and Tadros [30] reported juglone concentrations (at a depth of 0–8 cm) ranging from 1.85 to 3.95 μg g^{-1} soil at a distance of 0.9 m from the tree rows. However, sampling at a distance of 1.8 m resulted in lower concentrations that ranged from 0.70 to 1.55 μg g^{-1} soil. In another study, De Scisciolo et al. [6] reported juglone concentrations of up to 1.88 μg g^{-1} soil in a 73-year-old black walnut plantation mixed with other hardwoods (within a distance of 0.5 to 1.0 m from tree base and to a depth of 10 cm). Jose and Gillespie [7] have reported the most recent work in this regard. These authors measured soil juglone levels in a 10-year-old black walnut plantation and reported a significant decrease in juglone levels with distance from trees (Fig. 7). When the distance increased to 4.25 m from the tree row, juglone concentration decreased as much as 80% (0.31 μg g^{-1} soil) as compared to the within-tree row concentration (1.63 μg g^{-1} soil).

Jose [46] and Jose and Gillespie [7] also revealed the role of roots in contributing to the soil juglone pool. They inserted polyethylene root barriers at a distance of 1.2 m from tree rows to a depth of 1.2 m. This root-barrier treatment had a significant effect on soil juglone concentration (Fig. 8). Juglone

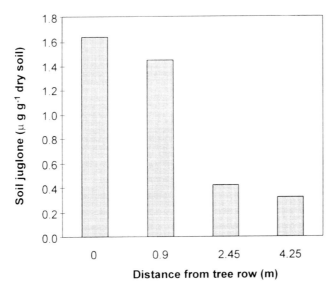

Figure 7. Soil juglone concentration as influenced by distance from tree row (based on data published in Jose and Gillespie [7]).

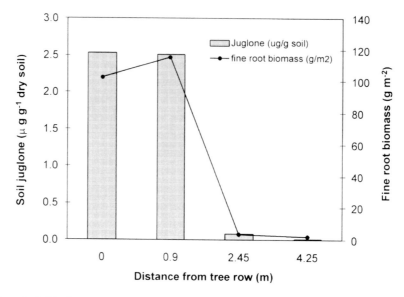

Figure 8. Soil juglone concentration as influenced by polyethylene root barrier inserted at a distance of 1.2 m from tree rows. The barrier prevented tree roots from growing beyond 1.2 m. Soil juglone concentration was also negligible beyond the root barrier (based on data published in Jose [46]).

concentration beyond the root barrier decreased to trace levels of 0.08 and 0.01 μg g^{-1} soil (at a distance of 2.45 and 4.25 m, respectively) in the root-barrier treatment as compared to 0.42 and 0.32 μg g^{-1} soil in the no-barrier control treatment. However, juglone levels were significantly higher at 0 and 0.9 m (2.53 and 2.51 μg g^{-1} soil) in the root-barrier than in the no-barrier (1.63 and 1.44 μg g^{-1} soil) treatment. This result supports the idea that roots release most of the juglone present in the soil. The higher concentration of juglone within the root barrier (i.e., at 0 and 0.9 m) may be due to enhanced root proliferation following root pruning, as shown by Jose et al. [18]. Confining the root mass within a restricted volume of soil (by installing the polyethylene barrier) also could cause juglone to build up in greater concentrations as rooting density increases. Beyond the barrier, one would expect either no roots or only a negligible rooting density, and that was reflected in the significantly lower juglone levels observed.

The temporal changes in soil juglone levels were also examined by Jose and Gillespie [7] in a black walnut plantation in Indiana, USA. Soil juglone did not show any significant differences between seasons. However, on average, fall samples had the lowest juglone concentration (1.03 μg g^{-1} soil), followed by spring (1.16 μg g^{-1} soil) and summer (1.17 μg g^{-1} soil). De Scisciolo et al. [6] observed a seasonal variation in juglone concentration, with the summer samples exhibiting the lowest concentrations as compared to the fall and spring samples. Although they attributed this to the variation in soil moisture, no significant correlation was observed. Soil moisture in the Indiana study was lowest during summer (16.3%) and highest during spring (20.1%), with the fall season being intermediate (16.7%). It has been suggested that, under well-aerated soil conditions, chemical breakdown of juglone increases and, therefore, soil juglone levels decrease [12, 14]. However, Jose and Gillespie [7] did not observe any correlation between soil juglone and soil moisture.

Although temporal patterns of juglone in various black walnut tissues are well documented [38, 41, 47], the contribution of juglone to soil from each plant part remains unknown [6]. As suggested by Williamson and Weidenhamer [48], toxicity by juglone is likely a function of both concentration in the soil and the flux or renewal rate from plant tissues. The results of the aforementioned experiment in Indiana show that live roots are the major source of "renewal" to juglone in the soil. For example, it is reasonable to expect a higher concentration of soil juglone during fall because of the presence of leaves and fruits on the soil surface. Similar juglone concentrations for all three seasons suggest that juglone released from leaves and fruits is probably negligible or nonpersistent (readily oxidized to other nontoxic substances, see [9]) on the soil surface as compared to the continuous leakage of juglone from live roots.

Bacterial degradation of soil juglone

The effects of biotic processes, including microbial breakdown, on the bioavailability of juglone have received much attention in the recent past. A bacterium (*Pseudomonas putida*) isolated from soil beneath black walnut trees has been found to metabolize juglone [49–51]. Juglone is rapidly converted to 3-hydroxyjuglone, then slowly to 2,3-dihydroxybenzoate, and further to 2-hydroxymuconic acid semialdehyde (Fig. 9). The allelopathic potential of these juglone degradation products still remains unknown.

Some authors have used the microbial breakdown of juglone as an evidence against black walnut allelopathy. For example, Schmidt [50] concluded that

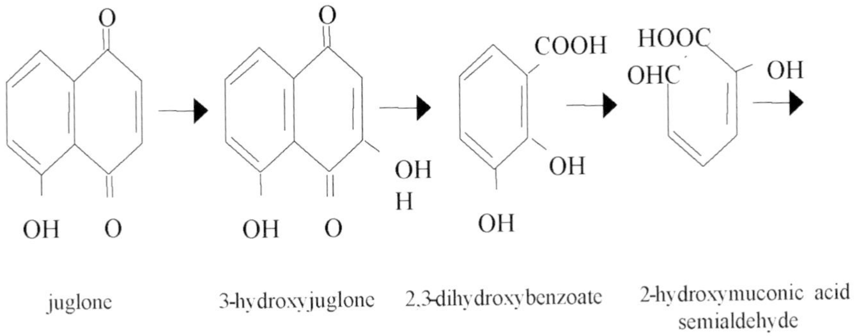

juglone	3-hydroxyjuglone	2,3-dihydroxybenzoate	2-hydroxymuconic acid semialdehyde

Figure 9. Degradation of juglone by *Pseudomonas putida* (Retenmaier et al. [49]).

"rapid degradation of juglone and other suspected allelochemicals by soil bacteria make it unlikely that these compounds are important mediators of plant-plant interactions under natural conditions". This was challenged by Williamson and Weidenhamer [48], who argued that the very existence of such specialized bacteria confirmed the frequent, if not continuous, presence of juglone and such chemicals in the soil. Contrary to Schmidt's conclusion that juglone concentrations in soil are maintained below phytotoxic levels by soil bacteria, several studies have found (as explained earlier) high concentrations of juglone in the soil beneath black walnut.

It is difficult to compare chloroform-extracted soil juglone concentrations with solution (hydroponic) juglone concentrations. Many of the reported soil juglone concentrations are sufficient to produce toxic solution concentrations. In general, a solution juglone concentration of 10^{-5} M (0.0017416 g juglone per liter) has been found to inhibit seed germination and dry weight accumulation of various herbaceous and woody species in hydroponic systems [7, 12, 29]. Further, shoot elongation of certain plants was inhibited by concentrations as low as 10^{-6} M (0.00017416 g juglone per liter) [29]. The only reported work showing a relationship between chloroform-extractable soil juglone and a noticeable allelopathic effect in the field is that by Ponder and Tadros [30]. They observed European black alder (*Alnus glutinosa* L. Gaertn.) mortality in a mixed plantation with black walnut where the soil juglone levels had reached 3.95 µg g^{-1} soil within a 14-year period. This level (0.00395 g juglone per 1000 g soil) is more than twice the concentration (0.0017416 g juglone per liter) used in the hydroponic systems mentioned earlier. Although it was speculated that a 12- to 25-year period would be required to accumulate toxic amounts of juglone in the soil [29], data by Jose and Gillespe [7] (0.00144 g to 0.00253 g juglone per 1000 g soil for 0 and 0.9 m from the tree stem) suggest a large amount of juglone buildup in the soil within a 10-year period. It is reasonable to believe that any juglone-sensitive species would be adversely affected by such concentrations since black walnut allelopathy under field conditions can occur as a result of long-term exposure to even moderate levels of juglone [29].

Effects of black walnut on other plants

Four major types of plant associations involving black walnut can be identified where allelopathy may be operative (modified from [12]). They are
1. natural forest stands containing black walnut as a component,
2. various species planted within the influence zone of trees in yards or fields,
3. mixed plantations of black walnut and other tree species, and
4. black walnut intercropping systems with food or forage crops.
Studies conducted under field conditions are rare [6, 21]. However, extensive screening of various herbaceous and woody plants for juglone sensitivity has been carried out using bioassays. The majority of existing literature on black walnut allelopathy involves the second type of association. Although an exten-

sive review of the available literature (see [12, 21, 28, 52]) is beyond the scope of this chapter, some of the most relevant works are discussed below.

Field studies

As noted earlier, the inhibitory effects of *Juglans* on other plants have been observed by many for at least two millennia. Surprisingly, hard evidence from field studies where allelopathy was investigated under natural conditions is rare in the scientific literature. Observations in the early part of the twentieth century by Massey [24] confirmed wilting and dying of alfalfa, tomato, and potato in the vicinity of black walnut trees. Similarly, death of apple trees [22, 23, 53] and of a white pine stand [54] was reported in response to black walnut invasion or proximity. Another incident of white pine mortality took place at the Cunningham Experimental forest owned by Purdue University in Indiana, USA (W. Beineke, personal communication). A white pine plantation planted in 1948 was invaded by black walnut. By 1962, the white pines were losing vigor and dying, and by 1971 all the white pines had been killed.

Althen [55] observed that red oak (*Quercus rubra*) interplanted with black walnut appeared healthy and grew normally, while red pine (*Pinus resinosa*) began to die after 15 years. Black walnut toxicity to red pine has also been reported by Brooks [52] under field conditions. Other species such as black alder [12, 30] and white birch [56] also have been found to be sensitive when interplanted with black walnut trees.

Black walnut allelopathy has been implicated for altering the course of old field succession. In a study by Bratton [57], American elm (*Ulmus americana*) and ash (*Fraxinus pennsylvanica*) were found to be sensitive to black walnut. These species either declined or were killed in the vicinity of black walnut trees.

De Scisciolo et al. [6] compared the herbaceous and woody understory of black walnut to that of sugar maple (*Acer saccharinum*) and red oak (*Quercus rubra*) in mixed hardwood plantations in Syracuse, New York. Despite significant soil juglone beneath black walnut trees, no significant difference in understory vegetation was observed between black walnut and other tree species in three of the four stands studied. These authors concluded that the allelopathic nature of juglone under field conditions was questionable. However, a recent field investigation has shown a decline in the abundance of a woody climber, *Rhus radicans*, near black walnut trees [58]. Laboratory research with black walnut bark extract has exhibited inhibitory effects on seed germination of this species.

Bioassays

Bioassays are an integral part of all allelopathic studies [59]. Although they fail to demonstrate allelopathy in natural or field conditions, they are necessary for evaluation of the allelopathic potential of a species. Most of the work to examine the effects of extracted plant materials on test plants has been under laboratory or greenhouse conditions (Tab. 1). In of the earliest bioassays conducted by Massey [24], he added black walnut root bark to tomato plants in water culture. Within 48 hours plants exhibited wilting and browning of the root system. Brown [25] corroborated these findings by conducting a similar experiment with tomato and alfalfa plants.

Some of the conflicting results of nontoxicity by black walnut (e.g., [13, 14]) may be explained by the findings of Weidenhamer et al. [60] and Romeo and Weidenhamer [61]. These authors have found that phytotoxicity is density-dependent. For example, Weidenhamer et al. [60] grew tomato plants at different densities in the greenhouse in soil collected from beneath black walnut and from an adjacent field. Tomato growth was reduced much more in the walnut soil at low density than at high density. The results provide evidence in

Table 1. Percent change in shoot and root dry weights of seedlings grown at different juglone concentrations in hydroponic cultures. A negative change means growth was reduced, whereas a positive change means growth was enhanced compared to control seedlings

Species	Shoot dry weight (%)			Root dry weight (%)		
	10^{-6}	10^{-5}	10^{-4}	10^{-6}	10^{-5}	10^{-4}
Herbs [a]						
Crimson clover (*Trifolium incarnatum*)	−15	−50	−81	11	−50	−78
Crown vetch (*Coronilla varia*)	2	−83	−94	−1	−82	−97
Hairy vetch (*Vicia villosa*)	2	−29	−67	−10	−10	−57
Korean lespedeza (*Lespedeza stipulacea*)	−9	−47	−27	14	−29	−14
Sericea lespedeza (*L. cuneata*)	−30	−72	−92	−6	−63	−88
Shrubs [a]						
Ginnala maple (*Acer ginnala*)	67	64	−35	−35	−28	−83
Siberian peashrub (*Caragana arborescens*)	24	−46	−83	−14	−72	−91
Russian olive (*Elaegnus angustifolia*)	−16	32	−92	8	99	−75
Autumn olive (*E. umbellate*)	−45	−65	−94	−18	−41	−88
Amur honeysuckle (*Lonicera maackii*)	−41	−61	−91	−55	−61	−94
Trees [b]						
White pine (*Pinus strobus*)	−7	−3	−31	−33	−29	−50
Scotch pine (*P. sylvestris*)	−38	0	−63	−50	0	−25
Japanese larch (*Larix leptolepis*)	−36	−14	−71	−20	0	−60
Norway spruce (*Picea abies*)	20	−20	−20	17	−17	−16
Whire oak (*Quercus alba*)	−23	−41	−53	−21	−27	−20
White ash (*Fraxinus Americana*)	−20	−58	−83	−7	−31	−71
European black alder (*Alnus glutinosa*)	−33	−86	−94	−26	−87	−94
Yellow poplar (*Liriodendron tulipifera*)	19	8	−72	4	−20	−77
Row crops[c]						
Corn (*Zea mays*)	4	−29	−56	6	−39	−61
Soybean (*Glycine max*)	−11	−37	−33	−9	−48	−56

[a] Seedlings were grown for 4 to 6 weeks [29]
[b] Seedlings were grown for 8 to 10 weeks for white pine, Scotch pine, Japanese larch, and Norway spruce [62]; the rest of the tree seedlings were grown for 4 to 6 weeks [29]
[c] Seedlings were grown for 3 days [46]

favor of the allelopathic potential of walnut soil in that reduced growth at low but not at high plant densities is very difficult to explain on the basis of resource competition. One implication of this experiment is that even when juglone is present in phytotoxic amounts, toxicity may not be manifested if the target plant density is high enough. The size of bioassay pots also may be a factor, i.e., in small pots, no effect may be seen, while in large pots (with more toxin available to the assay plant), toxicity will be manifested.

Bioassays using tree species also have been carried out. Pines are often considered sensitive to juglone. In addition to field observations where a number of other factors can cause pine decline or mortality, lab experiments have confirmed juglone as a causal agent. For example, Fisher [16] observed reduced radicle elongation in red pine in response to juglone. He concluded that the effect was more prominent under wet soil conditions than under dry soil conditions. Funk et al. [62] reported the results of a hydroponic culture with four coniferous species including two pines. Seedlings of Japanese larch (*Larix leptolepis*), Norway spruce (*Picea abies*), white pine (*Pinus strobus*), and Scotch pine (*P. sylvestris*) were grown for 8 to 10 weeks in nutrient solutions with juglone concentrations ranging from 10^{-2} to 10^{-10} M. Juglone was lethal to all species at concentrations of 10^{-2} and 10^{-3} M. Scotch pine and Japanese larch also were killed at a concentration of 10^{-4} M. Seedling growth of all species was inhibited at varying degrees by solutions as dilute as 10^{-7} M.

Rietveld [12, 29] investigated the sensitivity of 16 species of herbs, shrubs, and trees to juglone in solution culture with juglone concentrations varying from 10^{-3} M to 10^{-6} M (Tab. 1). Although seed germination and radicle elongation were not affected in all species, shoot elongation and dry weight accumulation were affected. Many species were sensitive to concentrations as low as 10^{-6} M. Seedlings of all species were severely wilted and eventually killed by 10^{-3} M juglone concentration.

Interest in quantifying the effects of juglone on field crops has resulted in recent hydroponic experiments using corn (*Zea mays*) and soybean (*Glycine max*) [8]. These are two of the major field crops often planted in black-walnut-based alley-cropping agroforestry systems of temperate North America [63–66]. Three different concentrations of juglone (10^{-4}, 10^{-5}, and 10^{-6} M) along with a control were applied to corn and soybean in solution culture. Within three days, juglone exhibited significant inhibitory effects on shoot and root relative growth rates (Fig. 10). In general, soybean was found to be more sensitive to juglone than was corn. Root relative growth rate was the most inhibited variable for both species, and reductions of 86.5% and 99% were observed in corn and soybean, respectively, at 10^{-4} M juglone concentration. Actual juglone concentrations in the field may not be as high as 10^{-4} M. As stated earlier, concentrations as high as 2×10^{-5} M have been reported under field conditions [7, 30]. Although reduced growth of corn in the field under black walnut has been attributed to shading and competition for water [52], the results of Jose and Gillespie [8] indicate a possibility of juglone phytotoxicity. Similarly, soybean yield reduction under black walnut [67] also may be partly

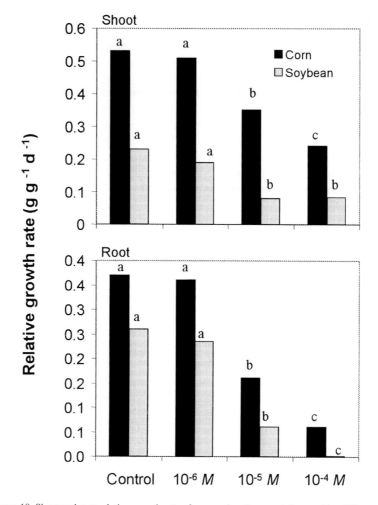

Figure 10. Shoot and root relative growth rate of corn and soybean as influenced by different juglone concentrations. Bars with the same letters are not significantly different within each species (based on data given in Jose and Gillespie [8]).

explained by the toxic effects of juglone. Further, soybean growth also may be reduced by any toxic effects of juglone on the nitrogen-fixing symbiotic microorganisms [68, 69].

Physiological basis of black walnut allelopathy

Despite documented evidence of black walnut allelopathy in several species, little is known about the actual physiological basis for observed growth reductions in response to juglone. The specific physiological modes of action for allelochemicals are difficult to determine because the interruption of one plant process usually affects other processes as well [70]. There have been two independent studies where oxygen uptake by corn and soybean mitochondria was measured in response to juglone. Koeppe [71] studied respiration rates of isolated corn mitochondria and of excised corn roots and concluded that juglone had inhibitory effects on respiration. Although Perry [72] reported no inhibition of respiration by juglone in tomato (*Lycopersicon esculentum*) and bean (*Phaseolus vulgaris*), Lang [73] observed inhibition of cyanide-sensitive respiration in the same two species. Similarly, Hejl et al. [74] reported respiratory inhibition of soybean mitochondria in response to juglone. They further observed photosynthetic reductions in soybean leaf disks and suggested that changes in normal oxygen uptake by mitochondria would work in concert with impairment of photosynthesis to cause reductions in plant growth. Although the physiological basis for potential growth reduction was thus proposed, whether whole-plant physiology and growth would, in fact, be affected remained unclear until recently.

In an effort to elucidate plant physiological response to juglone, Jose [46] and Jose and Gillespie [7] examined whole-plant net photosynthesis, respiration, transpiration, and stomatal conductance of hydroponically grown corn and soybean seedlings at various (10^{-4}, 10^{-5}, and 10^{-6} M) juglone concentrations. Net photosynthetic rates (P_{net}) were significantly affected by juglone in both species (Fig. 11). The rate of decrease was much greater for soybean than for corn. The lowest concentration did not have any significant effects on P_{net} for either species. Though not significant, a higher P_{net} was observed for corn at the 10^{-5} M compared to the control or 10^{-6} M treatments. This was not seen in soybean, where P_{net} was reduced by 72.9% by 10^{-5} M treatment. Relative to the control, the greatest reduction in P_{net} was observed in both corn (67.8%) and soybean (87.1%) at the highest juglone concentration (10^{-4} M).

As mentioned earlier, using soybean leaf disks, Hejl et al. [74] observed up to a 93% reduction in photosynthesis with 30 μM (3×10^{-5} M) juglone treatment. They found similar inhibition of photosynthesis in *Lemna minor*, which was correlated with reduction in growth. Jose and Gillespie [7] also observed similar correlations between P_{net} and growth rate for both species. Growth reduction in soybean was much better explained by differences in P_{net} ($R^2 = 0.73$, p = 0.0001) than was that in corn ($R^2 = 0.53$, p = 0.0001).

Transpiration rates (E) in both species were also significantly affected by juglone (Fig. 11). The trend observed for E in response to different juglone levels was similar to that exhibited by P_{net}. However, the effects were less in magnitude as compared to P_{net}. For example, E inhibition was only 37.7% for corn and 66.2% for soybean in the 10^{-4} M juglone treatment as compared to 67.8% (corn) and 87.1% (soybean) reductions in P_{net}. The observed reductions in E may be associated with reductions in root mass and the resultant decrease in absorbing area. As mentioned earlier, relative root growth rate exhibited greater reductions in both species than did relative shoot growth rates. E may further be reduced by the influence of juglone on stomatal functioning.

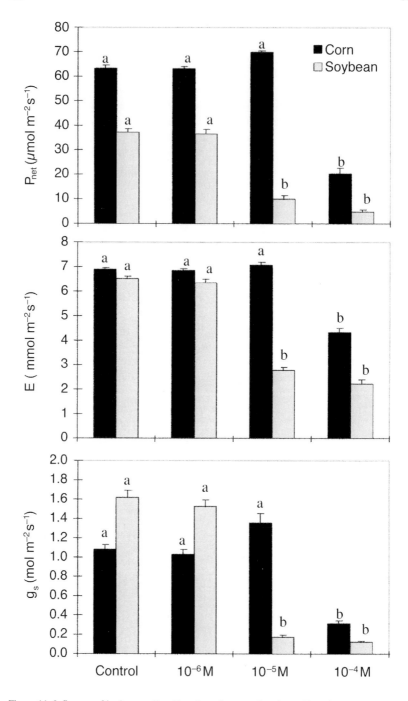

Figure 11. Influence of juglone on P_{net}, E, and g_s of corn and soybean. Error bars represent one standard error of the mean. Bars with the same letters are not significantly different (source: Jose and Gillespie [8], used with kind permission from Kluwer Academic Publishers).

Allelochemicals have been reported to interfere with stomatal functioning and thus alter plant water relationships [75]. According to Jose and Gillespie [8], stomatal conductance (g_s) followed patterns similar to those seen for E. These authors suggested that impairment of normal stomatal functioning might be occurring in both corn and soybean as a response to juglone. Alternately, one can argue that stomatal closure is a result of reduced E caused by reductions in root mass. However, considering the concurrent reductions in shoot mass where water is not limiting growth, direct action of juglone on stomatal functioning is likely. And, although it is suggested that juglone acts directly on the mechanisms of photosynthesis [74, 76], stomatal disruption by juglone also may have indirect inhibitory effects on photosynthesis. Thus, it is reasonable to assume both direct and indirect effects of juglone on the photosynthetic process, which in turn cause reductions in plant growth.

Juglone was also found to cause significant reductions in leaf and root respiration in both species [8] (Fig. 12). Leaf respiration rates (in both species) at the 10^{-6} M juglone concentration did not vary significantly from those of the control plants. Similarly, leaf respiration rates at 10^{-5} and 10^{-4} M concentrations were similar but much lower than those of plants in the control and 10^{-6} M concentrations. A maximum of 50.2% reduction was observed in corn leaf respiration at 10^{-4} M juglone concentration, whereas reduction was 46.9% for soybean. Perry [72] found similar reduction in leaf respiration (50%) in bean and tomato at 10^{-4} M juglone concentration.

For corn, root respiration (including that of any microbial life supported by roots) was less inhibited by juglone than was leaf respiration. Although all three juglone treatment concentrations had significantly lower respiration rates than the control, a maximum reduction of only 27.3% was found at the highest juglone concentration. Root respiration in soybean exhibited a much greater respiratory inhibition (52.1%) in the 10^{-4} M treatment. However, the medium and low concentrations of juglone had no effect on root respiration of soybean.

Attempts have been made to identify the causes for *in vitro* juglone-induced reduction in respiration [71, 74]. Koeppe [71] attributed the reduction in isolated mitochondrial respiration to the inhibition of some of the coupled intermediates of oxidative phosphorylation and to the resultant slowing down of electron flow to oxygen. It is reasonable to believe that *in vivo* juglone moves into cells and inhibits mitochondrial respiration. However, at the tissue level, as suggested by Koeppe [71], juglone-induced reduction in respiration may involve many mitochondrial or extramitochondrial respiratory activities. Whatever the mechanism of action, juglone altered the respiratory activities of both leaves and roots in studies by Jose [46] and Jose and Gillespie [8]. Although not as dramatic as P_{net}, the reductions in respiratory rates of both leaves and roots may complement the reductions in P_{net} to cause the observed reductions in shoot and root growth rates in both species. Similar physiological responses are likely to occur in other juglone sensitive species.

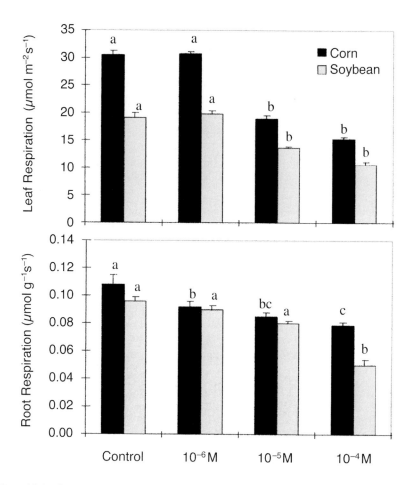

Figure 12. Leaf and root respiration of corn and soybean as influenced by different juglone concentrations. Error bars represent one standard error of the mean. Bars with the same letters are not significantly different (source: Jose and Gillespie [8], used with kind permission from Kluwer Academic Publishers).

Conclusions

Anecdotal stories of black walnut allelopathy have gained the attention of the general public and the scientific community for several decades. Through rigorous scientific research it has been well established that black walnut contains the allelochemical juglone, which is proven to be toxic to several plant species. Considerable progress has been made during the last two decades in extracting and quantifying juglone from black walnut plant parts and from the soil. Juglone has been shown to accumulate in the soil at high enough concentrations to be toxic to associated plant species. As a result, the process of bacterial degradation does not seem to remove soil juglone completely. However, it has been proven that an active source, particularly live roots, is needed to maintain high juglone levels in the soil. Hydroponic cultures provide evidence of juglone uptake and disruption of normal physiological functions in sensitive plant species. This has been directly correlated with reductions in whole-plant growth. However, we still lack direct evidence for juglone uptake by plants in the field. As mentioned earlier, it is difficult to compare chloroform-extractable juglone levels to those actually experienced by plants in the field. Soil juglone bound to organic matter and clays, which may not be available to plants, may be extracted in the chloroform procedure. More studies are needed to further distinguish the soil solution-phase juglone, which plants are likely to experience, from soil solid-phase juglone, which is likely to be unavailable to plants. A closer examination of allelopathy in black walnut's natural habitats is also warranted.

Acknowledgements
I wish to acknowledge and express my sincere thanks to Drs. Inderjit, J. Weidenhamer, P. Linehan, C. Ramsey, and an anonymous reviewer for helpful comments on an earlier version of the manuscript. Florida Agricultural Experiment Station Journal Series No. R 08344

References

1 Molisch H (1937) *Der Einfluss einer Pflanze auf die andere – Allelopathie.* Gustav Fischer, Jena
2 Rice EL (1984) *Allelopathy* (2nd ed.). Academic Press, Orlando
3 Putnam AR (1987) Introduction. *In*: GR Waller (ed): *Allelochemicals: role in forestry and agriculture.* ACS Symposium Series No. 330, American Chemical Society, Washington, DC, xiii–xviii
4 Thompson A (ed) (1985) *The chemistry of allelopathy: biochemical interactions among plants.* ACS Symposium Series No. 268, American Chemical Society, Washington, DC
5 Waller GR (1987) *Allelochemicals: role in forestry in and agriculture.* ACS Symposium Series No. 330, American Chemical Society, Washington, DC
6 De Scisciolo B, Leopold DJ, Walton DC (1990) Seasonal patterns of juglone in soil beneath *Juglans nigra* (black walnut) and influence of *J. nigra* on understory vegetation. *J Chem Ecol* 16: 1111–1130
7 Jose S, Gillespie AR (1998) Allelopathy in black walnut (*Juglans nigra* L.) alley cropping, I. Spatio-temporal variation in soil juglone in a black walnut-corn (*Zea mays* L.) alley cropping system in the Midwestern USA. *Plant Soil* 203: 191–197
8 Jose S, Gillespie AR (1998) Allelopathy in black walnut (*Juglans nigra* L.) alley cropping, II. Effects of jugone on hydroponically grown corn (*Zea mays* L.) and soybean (*Glycine max* L. Merr.) growth and physiology. *Plant Soil* 203: 199–205
9 Gries GA (1943) Juglone, the active agent in walnut toxicity. *North Nut Grow Assoc Annu Rep* 32: 52–55
10 Harlow WM, Harrar ES (1969) *Textbook of dendrology.* 5th edition. McGraw-Hill, Inc., New York
11 Williams RD (1990) *Juglans nigra* L. Black walnut. *In*: RM Burns, BH Honkala (eds): *Silvics of North America: Vol. 2. hardwoods.* Agric. Handbook 654. US Department of Agriculture Forest Service, 391–399
12 Rietveld WJ (1981) The significance of allelopathy in black walnut cultural systems. *North Nut Grow Assoc Annu Rep* 72: 117–134
13 Greene KW (1930) The toxic (?) effect of the black walnut. *North Nut Grow Assoc Proc* 1929: 152–157
14 MacDaniels LH, Muenscher WC (1941) Black walnut toxicity. *North Nut Grow Assoc Annu Rep* 31: 172–179

15 Camp RF (1986) Walnuts and white pine can be grown together successfully. *Tree Plant Notes* US Department of Agriculture Forest Service, 37: 29–31

16 Fisher RF (1978) Juglone inhibits pine growth under certain moisture regimes. *Soil Sci Soc Am J* 42: 801–803

17 Putnam AR (1983) Allelopathic chemicals Nature's herbicides in action. *Chem Eng News* 61: 34–45

18 Jose S, Gillespie AR, Seifert JR, Biehle DJ (2000) Defining competition vectors on a temperate alley cropping system in the midwestern USA 2: competition for water. *Agrofor Syst* 48: 41–59

19 Jose S, Gillespie AR, Seifert JE, Pope PE (2000) Defining competition vectors in a temperate alleycropping system in the midwestern USA: III. competition for nitrogen and litter decomposition dynamics. *Agrofor Syst* 48: 61–77

20 Willis RJ (1985) The historical bases of the concept of allelopathy. *J Hist Biol* 18: 71–102

21 Willis RJ (2000) *Juglans* spp., juglone and allelopathy. *Allelopathy J* 7: 1–55

22 Galusha OB (1870) O.B. Galusha report ad-interim – apples. *Trans Illnois State Hort Soc, New Series* 4: 78–86

23 McWhorter F, Douglas B, Schroeder D (1874) Discussion. *Tran Illinois State Hort Soc, New Series* 8: 66–67

24 Massey AB (1925) Antagonism of the walnuts (*Juglans nigra* L. and *J. cinerea* L.) in certain plant associations. *Phytopathology* 15: 773–784

25 Brown BL (1942) Injurious influence of bark on black walnut roots on seedlings of tomato and alfalfa. *North Nut Grow Assoc Annu Rep* 33: 97–102

26 Anonymous (1948) Test clears walnut's reputation. US Department of Agriculture, Clip Sheet 2375-48

27 Sherman RR (1971) *An ecological evaluation of the allelopathic influence of Juglans nigra on Lycopersicon esculentum*. M.S. Thesis, Michigan State University, East Lansing

28 MacDaniels LH, Pinnow D L (1976) Walnut toxicity, an unsolved problem. *North Nut Grow Assoc Annu Rep* 67: 114–122

29 Rietveld WJ (1983) Allelopathic effects of juglone on germination and growth of several herbaceous and woody species. *J Chem Ecol* 9: 295–308

30 Ponder FJr, Tadros SH (1985) Juglone concentration in soil beneath black walnut interplanted with nitrogen-fixing species. *J Chem Ecol* 11: 937–942

31 Vogel A, Reischauer (1856) Buchner Neues. *Rep Fur Pharm* 5: 106

32 Bernthsen A, Semper A (1887) Über die Konstitution des Juglons und seine Synthese aus Naphtalin. *Ber Dtsch Chem Ges* 20: 934

33 Davis EF (1928) The toxic principle of *Juglans nigra* as identified with synthetic juglone and its toxic effects on tomato and alfalfa plants. *Am J Bot* 15: 620

34 Daughtery W, Smith S, Wigal C, Verhoek S, Williams S (1995) Distribution of 5-hydroxy-1,4-naphthoquinone and other naphthoquinone derivatives in the Juglandaceae (walnut family) and related families. *Curr Top Plant Physiol* 15: 335–337

35 Moir M, Thomson RH (1973) Naphthoquinones in *Lomatia* species. *Phytochemistry* 12: 1351–1353

36 Nageshwar G, Radhakrishnainah M, Narayana LL (1984) Chemotaxonomy of Caesalpin. *Curr Sci* 53: 813–814

37 Marichkova L, Kumanova B (1981) Isolation of flavonoids and some accompanying substances from the above-ground part of some *Astragalus centralpinus* family Leguminosae (Russian). *Problemy Farm* 9: 63–74

38 Lee KC, Campbell RW (1969) Nature and occurrence of juglone in *Juglans nigra* L. *Hort Sci* 4: 297–298

39 Dansette PM, Azerad RG (1970) A new intermediate in naphthoquinone and menaquinone biosynthesis. *Biochem Biophys Res Commun* 40: 1090–1095

40 Borazjani A, Graves CH, Hedin PA (1983) A survey of juglone levels among walnuts and hickories. *Pecan Quart* 17: 9–13

41 Cline S, Neely D (1984) Relationship between juvenile-leaf resistance to nthracnose and the presence of juglone and hydrojuglone glucoside in black walnut. *Phytopathology* 74: 185–188

42 Gupta SR, Ravindranath B, Seshadri TR (1972) Polyphenols of *Juglans nigra*. *Phytochemistry* 11: 2634–2636

43 Borazjani A, Graves CH, Hedin PA (1985) Ocurence of juglone in various tissus of pecan and related species. *Phytopathology* 75: 1419–1421

44 Pirone PP (1938) The detrimental effect of walnut to rhododendrons and other ornamentals. *Nurs Dis Notes* 11: 13–16

45 Ponder F Jr, (1987) Allelopathic interference of black walnut trees with nitrogen-fixing plants in mixed plantings. *In*: GR Waller (ed): *Allelochemicals: role in forestry and agriculture.* ACS Symposium Series No. 330, American Society Washington, DC, 195–204

46 Jose S (1997) Interspecific interactions in alley cropping: the physiology and biogeochemistry. Ph.D. Thesis, Purdue University, West Lafayette

47 Coder KD (1983) Seasonal changes of juglone potential in leaves of black walnut. *J Chem Ecol* 9: 1203–1212

48 Williamson GB, Weidenhamer JD (1990) Bacterial degradation of juglone: evidence against allelopathy? *J Chem Ecol* 16: 1739–1742

49 Retenmaier HJ, Kupas E, Lingens F (1983) Degradation of juglone by *Pseudomonas putida* J 1. *FEMS Microbiol Lett* 19: 193–195

50 Schmidt SK (1988) Degradation of juglone by soil bacteria. *J Chem Ecol* 14: 1561–1571

51 Schmidt SK, Ley RE (1999) Microbial competition and soil structure limit the expression of allelochemicals in nature. *In*: Inderjit, KMM Dakshini, CL Foy (eds): *Principles and practices in plant ecology: allelochemical interactions.* CRC Press, Boca Raton

52 Brooks MG (1951) Effects of black walnut trees and their products on other vegetation. Agricultural Extension State Bulletin no. 347, West Virginia University, Morgantown

53 Schneiderhan FJ (1927) The black walnut (*Juglans nigra* L.) as a cause of death of apple trees. *Phytopathology* 17: 529–540

54 Perry GS (1932) Some tree antagonisms. *Proc Penn Acad Sci* 6: 136–141

55 Althen FW (1968) Incompatibility of black walnut and red pine. *Can. Dept. of Fishery and Forestry Bimonthly Research Notes* 24: 19

56 Gabriel WJ (1975) Allelopathic effects of black walnut on white birches. *J For* 73: 234–237

57 Bratton GF (1974) The role and allelopathic effects of *Juglans nigra* in old field succession. M.S. Thesis, Emporia State College, Emporia

58 Talley SM, Lawton RO, Setzer WN (1996) Host preferences of *Rhus radicans* (Anacardiaceae) in a southern deciduous hardwood forest. *Ecology* 77: 1271–1276

59 Alves PLCA, Toledo REB, Gusman AB (1999) Allelopathic potential of *Eucalyptus* spp. *In*: SS Narwal (ed): *Allelopathy update: Vol 2. basic and applied aspects.* Scientific Publishers Inc., New Hampshire, 131–148

60 Weidenhamer JD, Hartnett DC, Romeo JT (1989) Density-dependent phytotoxicity: distinguishing resource competition and allelopathic interference in plants. *J Appl Ecol* 26: 613–624

61 Romeo J, Weidenhamer J (1998) Bioassays for allelopathy in terrestrial plants. *In*: KF Haynes, JG Millar (eds): *Methods of chemical ecology, Volume 2: Bioassay methods.* Kluwer Academic Publishers, Norvell, 179–211

62 Funk DT, Case PJ, Rietveld WJ, Phares RE (1979) Effects of juglone on the growth of coniferous seedlings. *For Sci* 25: 452–454

63 Garrett HE, Kurtz WB (1983) Silvicultural and economic relationships of integrated forestry-farming with black walnut. *Agrofor Syst* 1: 245–256

64 Bandolin TH, Fisher RF (1991) Agroforestry systems in North America. *Agrofor Syst* 16: 95–118

65 Jose S, Gillespie AR (1996) Assessing the synergistic and competitive interactions in alley cropping: Lessons from temperate systems involving fine hardwoods and corn. *In*: RK Kohli, KS Arya, K Atul (eds): *Resource inventory techniques to support agroforestry and environment: Proceedings of the IUFRO-DNAES International meet, Chandigarh, India.* HKT Publications, Chandigarh, India, 147–150

66 Garrett HE, Harper LS (1999) The science and practice of black walnut agroforestry in Missouri, USA: A temperate zone assessment. *In*: LE Buck, JP Lassoie, ECM Fernades (eds): *Agroforestry in sustainable agricultural systems.* CRC Press, Boca Raton, 97–110

67 Seifert JR (1991) Agroforestry: The southern Indiana experience of growing oak, walnut, corn, and soybeans. *In*: P Williams (ed): *Agroforestry in North America.* Ministry of Agriculture and Food, Ontario, Canada, 70

68 Dawson JO and Seymour PE (1983) Effects of juglone concentration on growth *in vitro* of *Frankia* ArI3 and *Rhizobium japonicum* strain 71. *J Chem Ecol* 9: 1175–1183

69 Neave IA, Dawson JO (1989) Juglone reduces growth, nitrogenase activity, and root respiration of actinorhizal black alder seedlings. *J Chem Ecol* 15: 1823–1836

70 Boes TK 1986 Allelopathy: chemical interactions between plants. *Amer Nurs* 163: 67–72

71 Koeppe DE 1972 Some reactions of isolated corn mitochondria influenced by juglone. *Physiol Plant* 27: 89–94

72 Perry SF (1967) Inhibition of respiration by juglone in *Phaseolus* and *Lycopersicon*. *Bull Torr Bot Club* 94: 26–30

73 Lang HJ (1983) Physiological effects of juglone on sensitive and insensitive plants. *Hort Sci* 17: 536

74 Hejl AM, Einhellig FA, Rasmussen JA (1993) Effects of juglone on growth, photosynthesis, and respiration. *J Chem Ecol* 19: 559–568

75 Einhellig FA (1987) Interaction among allelochemicals and other stress factors of the plant environment. *In*: GR Waller (ed): *Allelochemicals: role in forestry and agriculture*. ACS Symposium Series No. 330, American Chemical Society, Washington, DC, 343–357

76 Einhellig FA (1995) Mechanism of action of allelochemicals in allelopathy. *In*: Inderjit, KMM Dakshini, FA Einhellig (eds): *Allelopathy: organisms, processes, and applications*. American Chemical Society, Washington, DC, 96–116

Chemical Ecology of Plants: Allelopathy in Aquatic and Terrestrial Ecosystems
ed. by Inderjit and Azim U. Mallik
© 2002 Birkhäuser Verlag/Switzerland

Allelopathy and agroecology

Stephen R. Gliessman

Center for Agroecology, Department of Environmental Studies, University of California, Santa Cruz, California 95064, USA

Introduction

During the last half of the twentieth century, agricultural production came to depend on a wide array of chemical inputs that were designed to control a complex of weed, insect, and disease organisms. The use of these chemicals is most often evaluated in terms of their efficacy in controlling pests, lowering yield losses, and increasing the output and profitability of the cropping system. But during the last decade, evidence has accumulated that indicates that the long-term and intensive use of these materials has become a major threat to our ability to sustain agricultural production into the future [1, 2].

The constant and intensive use of synthetic agrichemicals can be linked to a variety of on- and off-farm environmental and health problems [3, 4]. Evidence has accumulated that improper or excessive use of these chemicals is associated with the following problems:

- surface- and groundwater contamination,
- deleterious impacts on nontarget organisms,
- development of secondary pests,
- development of pesticide resistance and loss of effectiveness of chemicals on target organisms,
- consumer concerns for residues on food,
- increasing evidence for links between agrichemicals and human-health problems,
- risk for the applicator and other farm workers,
- increasing cost of use of agrichemicals, and
- increasing regulation of chemical use.

The growing awareness of these problems has stimulated a search for ways to reduce or eliminate the use of synthetic agricultural chemicals, especially for weed control. This has brought recent attention to allelopathy, especially where the impacts of allelochemicals in agroecosystems can be positive and contribute to the development of alternative weed-management strategies [5–9]. These strategies have been categorized as (1) avoidance of negative impacts of allelochemicals, (2) exploitation of positive impacts, (3) management and development of allelopathic crops to suppress weeds, (4) development of allelochemicals as herbicides or growth regulators, and (5) combinations of these approaches [10]. A further criterion for the development of these strategies is that they should not bring about some of the same problems as the chemicals that they are designed to replace. The use of allelopathy for weed management must not be considered merely as a way of replacing a synthetic chemical with one that is naturally produced. The entire plant, and all of the possible interactions that it can have in the context of the entire agroecosystem, also must be considered. Agroecology becomes the means of determining these interactions [1] and, subsequently, evaluating the use of allelopathy in the broader context of agricultural sustainability [11].

Agroecology

Agricultural sustainability depends greatly on the development of strategies that reduce the need for costly external inputs and reduce the environmental impacts often associated with the excessive use of these inputs. Sustainability requires an approach to agriculture and agricultural research that builds on modern ecological knowledge and methods. This approach is embodied in the science of agroecology, which is defined as the application of ecological concepts and principles to the design and management of sustainable agroecosystems [1].

Agroecology provides the knowledge and methodology necessary for developing an agriculture that is, on the one hand, environmentally sound and, on the other hand, highly productive and economically viable. Ecological methods and principles form the foundation of agroecology. They are essential for determining (1) whether a particular agricultural practice, input, or management decision is sustainable; (2) the ecological basis for the functioning of the chosen management strategy over the long term, and (3) how to integrate each management strategy into an overall integrated farming system. Once these are known, practices can be developed that reduce purchased external inputs, lessen the impacts of such inputs when they are used, and establish a basis for designing systems that help farmers sustain their farms and their farming communities.

Even though an agroecological approach begins by focusing on particular components of a cropping system and the ecology of alternative management strategies, it establishes in the process the basis for much more. Applied more broadly, it can help us to examine the historical development of agricultural activities in a region and determine the ecological basis for selecting more sustainable practices adapted to that region. It also can trace the causes of problems that have arisen as a result of unsustainable practices. Even more broadly, an agroecological approach helps us to explore the theoretical basis for developing models that can facilitate the design, testing, and evaluation of sustainable agroecosystems. Ultimately, ecological knowledge of agroecosystem sustainability can help reshape humanity's approach to growing and raising food in order for sustainable food production to be achieved worldwide.

Sustainable agriculture

Sustainability means different things to different people, but there is general agreement that it has an ecological basis [1, 2]. In the most general sense, sustainability is a version of the concept of sustained yield–the condition of being able to harvest biomass from a system in perpetuity because the ability of the system to renew itself or be renewed is not compromised.

Because "perpetuity" can never be demonstrated in the present, the proof of sustainability always remains in the future and out of reach. Thus, it is impossible to know for sure whether a particular practice is in fact sustainable or whether a particular set of practices constitutes sustainability. However, it is possible to demonstrate that a practice is moving away from, or towards, sustainability.

Based on our present knowledge, we can suggest that a sustainable agriculture would, as primary goals,

- have minimal negative effects on the environment and release no toxic or damaging substances into the atmosphere, surface water, or groundwater;
- preserve and rebuild soil fertility, prevent soil erosion, and maintain the soil's ecological health;
- incorporate natural ecosystem processes rather than work against them;
- use water in a way that allows aquifers to be recharged and the water needs of the environment and people to be met;
- rely mainly on resources within the agroecosystem, including nearby communities, by replacing external inputs with nutrient cycling, better conservation, a reliance on renewable resources, and an expanded base of ecological knowledge;
- work to value and conserve biological diversity, both in the wild and in domesticated landscapes;
- farm in ways that are regenerative and restorative, with ecosystem principles driving agroecosystem recovery from the disturbances that management and harvest cause each season;
- guarantee equality of access to appropriate agricultural practices, knowledge, and technologies and enable local control of agricultural resources;
- create an agriculture that is truly productive, rather than extractive; and
- provide the greatest benefit for as many people as possible, from farmers to consumers, as well as those in between.

Allelopathy and sustainability

The tools of agroecology, together with our knowledge of allelopathy and phytochemical ecology, can play very important roles in helping to meet the criteria for a sustainable agriculture listed above. Recent reviews on allelopathy have outlined many of the potential applications that either an allelopathic plant or the chemical that the plant produces can have in agriculture [9, 12, 13].

The challenge, then, becomes attempting to evaluate the contribution that each of these applications may have to the overall sustainability of the system involved. This cannot be done by simply comparing the effectiveness of a plant-produced allelochemical in replacing a synthetic chemical. An overly narrow analysis would probably dismiss the use of the plant for weed control because it might be less cost-effective when compared to conventional methods. We need to look beyond the chemicals to the complex of interactions that the plant has with other components of the agroecosystem. Allelopathic abilities need to be examined as one of the complex set of impacts that the presence of the allelochemical-producing plant has on the agroecosystem [11, 14]. Examples of research that employ this *agroecosystem-level approach* are presented below.

Allelopathy in intercropped agroecosystems

Intercropped or mixed-crop agroecosystems can contribute greatly to increasing agricultural sustainability [11, 15, 16]. In such systems, more than one crop occupies the same piece of land either simultaneously or in some type of rotational sequence during the season. A primary goal of the system is to maximize the potential of beneficial interactions between the component species and to minimize the negative. Yields can be increased, more efficient use of resources takes place, and the land can be occupied productively more continuously. Of course, tradeoffs such as higher labor requirements or more complex management needs can occur, but the importance of multiple cropping has long been recognized [16], and there continues to be a need for intensive agroecological studies of such mixed-crop systems.

A traditional Mesoamerican multiple cropping system in which allelopathy plays a part is the maize, bean, and squash polyculture [15, 17]. Intercropping of maize (*Zea mays*), beans (*Phaseolus vulgaris*), and squash (*Cucurbita* spp.) has been practiced in Central America since prehispanic times and continues to be important in food production today [16]. A series of studies done in Tabasco, Mexico, showed that maize, the principal crop in the mixture from the farmer's point of view, had yields increased by as much as 50% beyond monoculture yields when the three crops were planted together [15]. There was some yield reduction for the two associated crop species, but the total yield for the three crops together was higher than what would have been obtained in an equivalent land area of the three monocultures planted separately (a process known as *overyielding*). Two of the three crops have shown significant allelopathic potential [5, 18, 19].

One of the mechanisms of the yield increase in the polyculture is thought to be suppression of potentially interfering weeds by the release of either inhibitory root exudates or leaf wash from the squash [5, 20]. Table 1 shows how leaf leachate from senesced leaves of *Cucurbita pepo* var. "small sugar" can be inhibitory to germination and initial plant development of the common

Table 1. Germinated seeds, initial radicle elongation, and initial hypocotyl length of *Amaranthus retroflexus* expressed as a percent of distilled water controls, in standard petri dish bioassays of *Cucurbita pepo* var. "small sugar" senesced leaf leachate

	5% Leachate	2.5% Leachate
Germination	17.1*	97.7
Radicle elongation	73.3	135.5
Hypocotyl length	21.3*	87.0

Osmotic concentration of the 5% extract was measured at 59 mOsm, a concentration that did not significantly alter germination or initial growth of *A. retroflexus* as compared to distilled water controls when tested in a similar concentration of mannitol solution (adapted from [20]).
An asterisk (*) signifies statistical difference from control at the 0.05 level.

red root pigweed (*Amaranthus retroflexus*). Farmers consider pigweed to be a very bothersome weed in crops like corn and beans, requiring considerable cultural or agrichemical control. Crop pollen can also be allelopathic [21], with corn pollen having been shown to be inhibitory to weeds in traditional corn farming systems in Mexico [19, 22]. However, experimental results that attempt to pinpoint these allelopathic interactions as the primary reasons for improved crop performance rarely are very successful. Returning to the data in Table 1, we can see just how variable inhibition can be and that statistically significant trends are very difficult to demonstrate. We must therefore place allelopathy in the context of the other ecological mechanisms of plant interaction in order to understand the crop-yield increase and to establish a stronger basis for recommending continued or more widespread use of such cropping systems.

In polycultures with maize, beans nodulate more and potentially are more active in biological fixation of nitrogen, which could be made directly available to the maize [23, 24]. Total agroecosystem gains of nitrogen have been observed when the crops are associated, despite its removal with the harvest [25]. This contributes both to the future reductions in dependence on purchased external inputs of fertilizer and to a sounder ecological basis for managing resources within the system. At the same time, studies of the management practices employed in this polyculture have demonstrated the ecological basis upon which the practices can function. For example, despite the lower squash yields in the mixed planting, farmers insist that the crop system benefits from the squash presence through the control of weeds [5]. The thick, broad, horizontal leaves cast a dense shade that blocks sunlight that can combine with even slight inhibitory effects from leaf leachates to more strongly inhibit weed growth [20]. Herbivorous insects are at a disadvantage in the intercrop system because food sources are less concentrated and more difficult to find in the mixture [26], and the presence of beneficial insects is often promoted because of factors such as the availability of more attractive microclimatic conditions or the presence of more diverse pollen and nectar sources

[27]. The presence of allelopathic plants in the cropping system forms a component in a complex set of interactions that benefit the entire system. We would lose many of these benefits if we maintain our focus only on the inhibition of weeds or the synthesis of the active chemical agent in order to use it merely as an herbicide analog. By expanding our focus, we stand to gain understanding of both the specifics of the allelochemicals and the broader ecological impacts of the plant in the agroecosystem.

Integrating crop, soil, and weed management

A larger challenge for weed management is to find cost-effective, ecologically sound ways of integrating weed-management objectives with the needs of sustainable soil and crop management [1, 2]. To do so successfully, we must fully understand the agroecological implications of the fact that whatever weed-control strategy is employed, there will be accompanying impacts on the soil and crop components of the agroecosystem. One of the best examples of such an integrated focus in agroecosystem management, and the development of relevant farming practices, is the use of crop rotations that employ cover crops or green manure crops as part of the system [1, 2, 28]. The benefits of cover-cropping have been known to agriculture for a long time [29, 30]. The use of cover crops has declined greatly over the past 50 years, but as farmers have become much more concerned with the need to reduce inputs and protect the quality of their farm environments, cover-crop use has increased in popularity again. Some of the benefits of cover cropping include reduction of soil erosion, reduced tillage or herbicide use, reduced fertilizer needs, control of soil-borne diseases, and improved soil quality. While all of these cover-crop reactions on the crop environment together can contribute to the sustainability of a farm, it is in weed control that allelopathy plays its greatest role [7]. For this reason, allelopathic cover crops that also provide multiple impacts on the agroecosystem would be preferred. Agroecology offers us the tools to determine the ecological processes behind such impacts as well as to incorporate their use into more sustainable farming systems.

A region where a cover crop with allelopathic species has proven to be quite useful is on the central coast of California. Located in a Mediterranean climate with cool, wet winters, cover crops of different kinds have proven to be extremely useful as a rotation with annual vegetable or fruit crops such as lettuce or strawberries. In comparative field trials, two different allelopathic grasses (cereal rye [*Secale cereale*] and barley [*Hordeum vulgare*]) have been combined with a legume (a fava bean variety known as bellbean [*Vicia faba*]) to provide winter cover that is disked under in the spring. Crops are then planted for harvest during the summer. Table 2 shows the impact of the cover crops on weed and crop yields after several years of rotation. Rye and barley are prolific producers of carbon-rich biomass, and they significantly reduce weed biomass and diversity, especially when compared to the planting of bellbeans

Table 2. Impact of legume and grass cover-crop combinations on various agroecosystem factors in the central coast of California. Data are four year averages (adapted from [31])

Cover crop treatment	Total biomass dry grams/m²	Weed biomass dry grams/m²	Cabbage yield fresh kg/100 m²
Bellbeans alone	279.8ᵃ	51.3ᵃ	759.5ᵃ
Rye alone	614.5ᵇ	8.5ᵇ	406.0ᵇᶜ
Barley alone	514.5ᵇ	12.8ᵇ	524.4ᵇᶜ
Rye/bellbeans	574.0ᵇ	1.8ᶜ	634.6ᵃᵇ
Barley/bellbeans	572.0ᵇ	6.8ᵇᶜ	645.4ᵃᵇ
None (control)	194.0ᵃ	194.0ᵃ	562.8ᵇᶜ

Values in each column followed by a different letter are significantly different at the 5% level.

alone. Bellbeans bring nitrogen into the system through a symbiotic relationship with nitrogen-fixing bacteria. The fast growing grasses provide early erosion control, while the more slowly developing legume becomes established. Research has indicated that root exudates produced by the grasses [32] and inhibitory breakdown products from the incorporated grass residues [33] are the probable factors in weed reduction.

When cabbage yields are compared between treatments, the plots with bellbean cover crop alone gave greatest harvests, especially compared to the control plots or the plots planted to grass cover crops alone. Although less yield was obtained with the grass/legume mixtures, yield reductions were not nearly as severe. The highest yields were in the monoculture bellbean treatment. Interestingly, the lowest yields occurred in the monoculture grass treatments; these were even lower than in the non-cover-cropped control. In order to try to understand these yield differences, an agroecological analysis of a complex of factors must be undertaken. A focus on allelopathy alone would miss many potentially important aspects of the system. For example, total nodule biomass, nodule number, and nitrogen fixation rates need to be studied, especially when the legume is associated with an allelopathic grass. The importance of biomass quality as well as quantity must be analyzed. Bellbeans are rich in nitrogen in relation to carbon, promoting rapid decomposition following plowdown and making nitrogen more readily available to the crop. The grasses, on the other hand, are high in carbon in relation to nitrogen. Their decomposition would probably be much slower, potentially tying up nitrogen in the process but adding more soil-building compounds over the long term. The grass/legume combination may provide an advantage because of a better carbon/nitrogen ratio, but the impacts of this difference on actual soil quality may not show up in either the crop or the soil for several years. The potential buildup of toxic products from the breakdown of the grass residues and their impact on the following crop must be examined as well [34].

In another example, wild mustard (*Brassica kaber*) is used as a cover crop in apple orchards in central California [35]. Upon observation of the dominant mustard cover between the rows of apple trees in the spring, it would appear that the well-known allelopathic potential of mustards [36, 37] was clearly being expressed. Comparisons of several cover crops (wild mustard, white clover, and different annual grasses) showed the mustard to be the only cover crop that controlled weeds as effectively as conventional herbicides or the non-chemical use of plastic trap. Within 45 days following the mustard germination, it had established a dominance that accounted for 99% of the total weed biomass present. None of the other cover crops were able to replace more than 42% of the weed biomass in experimental plots. Laboratory bioassays demonstrate the presence of inhibitory compounds for mustard [11, 35, 38, 39], glucosinolates and their breakdown products being some of the most important [40].

Perhaps even more important than weeds being inhibited allelopathically was that apple production was increased over several years in the plots that had the mustard cover crop. Trees in the mustard plots produced almost three times as many apples per tree than did those in conventional herbicide plots. Trees grown with mustard also achieved greater girth, with stems and branches achieving as much as 50% greater diameter after two years than the trees in conventional plots. It appears that at least a part of the yield advantage gained in mustard plots is due to better nutrient cycling. Analysis showed that the mustard cover took up significant amounts of nitrogen during the winter cover period, lowering concentrations in the soil. Nitrogen in the soil of the bare, herbicided plots was susceptible to leaching and loss from the system, whereas nitrogen stored in the mustard biomass was available for return to the soil when it was cut and incorporated in the spring, becoming available for spring and summer growth by the apples. It appears that this nutrient advantage was especially important for the younger, establishing trees characteristic of the relatively young orchard being studied—the time that the developing trees might be most susceptible to interference from other weed species. Similar interactions would be possible with other mineral nutrients, as would longer-term impacts on soil structure, organic matter content, water-holding capacity, and other qualities. Allelopathy becomes one of many interactions that we must understand and a farmer must effectively manage.

Integrating weeds with arthropod and disease management

At the same time that an allelopathic plant is releasing phytotoxins that inhibit weeds, it may be releasing compounds that can be important in disease or arthropod pest management as well as compounds that are attractants for beneficial biological control agents [13, 41]. Allelopathic plants may provide resources, such as pollen or nectar, that also can attract beneficial organisms, modify microclimatic conditions, and diversify the agroecosystem in ways that can favor the presence of beneficial organisms or decrease the presence of

pests [41]. Again, an agroecological focus on the complex of potential interactions in the agroecosystem becomes very important.

An example of an allelopathic weed that can be managed in an agroecosystem is another wild mustard, *Brassica campestris*. In a study where specific
densities of wild mustard were intercropped into a standard crop spacing of
organically grown broccoli (*Brassica oleracea* var. Premium crop), crop yields
were stimulated as much as 50% above those observed in broccoli plots kept
free of the weed [38]. In order to understand the ecological mechanisms of the
yield stimulation, various types of agroecosystem interactions were studied. It
was found that at the same time that wild mustard was expressing allelopathic
potential (Tab. 3), it could be releasing chemicals that repel herbivores from
the crop or function as a trap crop and preferentially attract certain insect pests
that would otherwise feed on the desired crop [39, 42]. There was a great
reduction in total herbivore load on broccoli with the associated mustard as
compared to the weedy intercrop, especially in the early part of the season
(Tab. 4). The reduction of herbivore damage on broccoli is especially important in the first weeks following transplanting of the young broccoli plants
when they are especially susceptible to pest-induced growth reduction.

Table 3. Initial radicle elongation of barley (*Hordeum vulgare*), rye (*Lolium mulitflorum*), and vetch
(*Vicia atropurpurea*) at two mustard growth stages, expressed as a percent of distilled water controls
in standard petri dish bioassays of *Brassica campestris* dried whole plant leachates (5 g/100 mL)

Indicator species	Full vegetative stage	At flower initiation
Barley	34.8*	25.0*
Rye	49.8*	51.0*
Vetch	98.7	35.0*

An asterisk (*) signifies statistical difference from control at the 0.05 level.
N = 30 (adapted from [39]).

Table 4. Flea beetle (*Phyllotreta* sp.) load per dm^2 of broccoli leaf surface in plots with or without
associated weedy mustard (*Brassica campestris*)

	Flea beetle load per dm^2	
Sample date	Broccoli monoculture	Broccoli/8 mustard plants per m^2
1	33.9[a]	9.8[b]
2	21.9[a]	10.3[b]
3	12.8[a]	5.9[a]
4	11.1[a]	1.6[b]
5	0.1[a]	0.2[a]

Paired samples with the same letter not significantly different at the 0.05 level.
N = 30 plants for each mean. Sampling done at approximately 10-day intervals (adapted from [39]).

Table 5. Total population of plant parasitic nematodes recovered from 100 cc of rhizosphere soil or 5 grams of corn roots

Treatment	Plant parasitic nematodes in soil		Plant parasitic nematodes in roots	
Sampling date	30 days	45 days	30 days	45 days
Corn monoculture	2300[a]	2800[a]	560[a]	700[a]
Corn/mustard	1860[b]	1720[b]	380[b]	400[b]

Values in each column followed by a different letter are significantly different at the 5% level.
N = 6 replicates. Samples taken at 30 and 45 days after corn seedling emergence (adapted from [43]).

When another species of wild mustard (*B. kaber*) was selectively inter-cropped with sweet corn, not only was the allelopathic potential of the weed brought to bear upon other weed species but also allelochemicals from the weed reduced the incidence of soil-borne diseases [43]. In plots with mustard, parasitic nematodes recovered from both the soil and from corn roots showed much lower numbers as compared to corn planted alone (Tab. 5). The percent of recovery of several fungal root pathogens (i.e., *Fusarium* spp., *Pythium aphanidermatum*, and *Rhizoctonia solani*) from corn roots was as much as 35% lower in the crop planted in the intercrop with mustard as compared to the crop planted alone. Combining these impacts with others such as modifi-cation of soil conditions, impact on associated insect populations, and allelo-pathic control of other weed species, the need for an ecologically based man-agement plan becomes obvious. In this study, it was found that if the density of wild mustard passed a critical threshold, significant reductions in corn yield would occur, despite other possible benefits to the system. Proper management is definitely required.

Like any other allelopathic plant, be it weed, crop, or cover crop, wild mus-tard has a complex effect on the crop environment. This could reduce the need for external inputs for both weed control and pest management. The complex of glucosinolates and their derivatives common to most of the genus *Brassica* are possible active agents, both for allelopathic inhibition of other weeds and for altering insect and disease populations [38, 41]. Ultimately, we look for the final response of the crop community in yield increases, but our goal is to see this come about as a result of the multiple interactions that are possible with the presence of the weed in the system.

Conclusions

In all of the cases described above, allelopathy plays a potential role in the management of weeds in agroecosystems. There is enough evidence in the lit-erature supporting its effectiveness to expand the use of allelopathy for the benefit of crop production. The replacement of costly and damaging synthetic agrichemicals is certainly a goal of a sustainable agriculture. But it is doubtful that strict replacement will provide the incentive or the means by which allelopathy will find its greatest utility in agroecosystem management. Replacing one chemical with another, regardless of its source, probably only exacerbates the problems that threaten the sustainability of agriculture as described earlier in this chapter. We must investigate agricultural practices involving allelopathy in order to help farmers increase their options for improving the sustainability of their farms. They need not only to be environmentally sound but also to contribute to the long-term maintenance of the natural resource base of the farm. They need to be incorporated into farm management in ways that focus on avoiding the buildup of problems rather than be looked to as cures for problems that are created because of farm mismanagement. They need to contribute to the overall productive capacity of the agroecosystem.

Allelopathy is one component of the complexity of agroecosystems. In the case of cover crops, allelopathy can combine with factors such as soil-erosion control, soil organic matter and nutrient improvement, reduced inputs of fertilizers and pesticides, and the long-term maintenance of productivity. In intercropping with an allelopathic weed, phytotoxins are part of yield stimulation, biological insect pest management, and input reduction. In the case of mixed crop systems, allelopathy plays a role in erosion control, weed suppression, soil nutrient and organic matter management, and input reduction. In all cases, the reduction of agrochemical inputs has both economic and ecological benefits for the farmer, the farm worker, consumers, and the environment. The challenge, then, is to use this broader perspective as we expand our research on allelopathy as applied to the current problems that face agriculture. By doing so, we then have the opportunity to place our research in the context of long-term sustainability.

Acknowledgements
This paper is the result of the long-term collaboration with a wonderful group of colleagues and students who now have become colleagues. This group includes Miguel Altieri, Matt Liebman, Roberto Garcia-Espinosa, Moises Amador, Phillip Fujiyoshi, Jim Paulus, Juan Jose Jimenez-Osornio, Rob Kluson, Martha Rosemeyer, and Francisco Rosado-May. Each and every one of them has effectively demonstrated how important agroecology is for developing an understanding of the complex interactions involved in sustainable agroecosystems. In addition, a large group of assistants, collaborators, and students in our agroecology courses, too large to list here, contributed to the complex (and often tedious) field and laboratory work needed to show the value of an agroecological approach. To all of them, we researchers are deeply indebted. Financial support from the UCSC Agroecology Program, the UCSC Committee on Research, the Halliday Foundation, and the Alfred Heller Endowed Chair in Agroecology is gratefully acknowledged.

References

1 Gliessman SR (2000) *Agroecology: ecological processes in sustainable agriculture.* CRC Press, Boca Raton
2 Altieri MA (1995) *Agroecology: the science of sustainable agriculture.* Westview Press, Boulder

3 Pimentel D, McLaughlin L, Zepp A, Latikan B, Kraus T, Kleinman P, Vancini F, Roach W, Graap
   E, Keeton W et al (1991) Environmental and economic effects of reducing pesticide use.
   Bioscience 41: 402–409

4 National Research Council (1989) Problems in U.S. agriculture. *In*: National Research Council,
   Alternative Agriculture, National Academy Press, Washington, 89–134

5 Gliessman SR (1983) Allelopathic interactions in crop-weed mixtures: applications for weed management. *J Chem Ecol* 9: 991–999

6 Putnam AR, Tang CS (eds) (1986) *The Science of Allelopathy*. John Wiley and Sons, New York

7 Einhellig FA, Leather GR (1988) Potentials for exploiting allelopathy to enhance crop production.
   J Chem Ecol 14: 1829–1844

8 Narwal SS (1994) *Allelopathy in crop production*. International Allelopathy Foundation, Haryana,
   India

9 Weston LA (1996) Utilization of allelopathy for weed management in agroecosystems. *Agron J*
   88: 860–866

10 Einhellig FA (1995) Allelopathy: current status and future goals. *In*: Inderjit, KMM Dakshini, FA
   Einhellig (eds): *Allelopathy: Organisms, Processes, and Applications*, American Chemical
   Society Symposium Series 582, Washington, DC, 1–24

11 Gliessman SR (1987) Species interactions and community ecology in low external-input agriculture. *Am J Alternative Agric* 2: 160–165

12 Waller GR (ed) (1987) *Allelochemicals: role in agriculture and forestry*, ACS Symposium Series
   330, American Chemical Society, Washington, DC

13 Rice EL (1995) *Biological control of weeds and plant diseases: advances in applied allelopathy*,
   University of Oklahoma Press, Norman

14 Gliessman SR (1986) Plant interactions in multiple cropping systems. *In*: CA Francis (ed):
   Multiple cropping systems. MacMillan, New York,

15 Amador MF, Gliessman SR (1990) An ecological approach to reducing external inputs through
   intercropping. *In*: SR Gliessman (ed): *Agroecology: researching the ecological basis for sustainable agriculture*, Springer-Verlag, New York, 146–159

16 Francis CA (ed) (1986) *Multiple cropping systems*, MacMillan, New York

17 Gliessman SR (1992) Agroecology in the tropics: achieving a balance between land use and
   preservation. *Environ Manage* 16: 681–689

18 Fujiyoshi PT, Gliessman SR, Langenheim JH (2002) Inhibitory potential of compounds released
   by squash (*Cucurbita* spp) under natural conditions. *Allelopathy J* 9: 1–8

19 Jimenez JJ, Schultz K, Anaya AL, Hernandez J, Espejo O (1983) Allelopathic potential of corn
   pollen. *J Chem Ecol* 9: 1011–1025

20 Fujiyoshi PT (1998) Mechanisms of weed suppression by squash (*Cucurbita* spp) intercropped
   with corn (*Zea mays* L.). PhD dissertation, Biology, University of California, Santa Cruz

21 Murphy SD (1999) Pollen allelopathy. *In*: Inderjit, KMM Dakshini, CL Foy (eds): *Principles and
   practices in plant ecology: allelochemical interactions*. CRC Press, Boca Raton, 129–148

22 Anaya AL, Hernandez-Bautista BE, Jimenez-Estrada M, Velasco-Ibarra L (1992) Phenylacetic
   acid as a phytotoxic compound of corn pollen. *J Chem Ecol* 18: 897–905

23 Boucher DH (1979) La nodulacion del frijol en policultivo: el efecto de la distancia entre las plan-
   tas de frijol y maiz. *Agric Trop (CSAT)* 1: 276–283

24 Bethlenfalvay GJ, Reyes-Solis MG, Camel SB, Ferrera-Cerrato R (1991) Nutrient transfer
   between the root zones of soybean and maize plants connected by a common mycorrhizal inoculum. *Physiol Plant* 82: 423–432

25 Gliessman SR (1982) Nitrogen cycling in several traditional agroecosystems in the humid tropical lowlands of southeastern Mexico. *Plant Soil* 67: 105–117

26 Risch S (1980) The population dynamics of several herbivorous beetles in a tropical agroecosystem: the effect of intercropping corn, beans and squash in Costa Rica. *J Appl Ecol* 17: 593–612

27 Letourneau DK (1986) Associational resistance in squash monocultures and polycultures in tropical Mexico. *Env Ent* 15: 285–292

28 Liebman M, Ohno T (1998) Crop rotation and legume residue effects on weed emergence and
   growth: implications for weed management. *In*: JL Hatfield, JL Buhler, BA Stewart (eds):
   Integrated weed and soil management, Ann Arbor Press, Chelsea, 181–221

29 USDA (1938) *Soils and Men*, Yearbook of Agriculture, U.S. Department of Agriculture,
   Washington, DC

30 Lal R, Regnier E, Exkert DJ, Edwards WM, Hammond R (1991) Expectations of cover crops for

sustainable agriculture. *In*: WL Hargrove (ed): *Cover crops for clean water*. Soil and Water Conservation Society, Iowa, 1–14

31 Gliessman SR (1989) Allelopathy and agricultural sustainability. *In*: CH Chou, GR Waller (eds): *Phytochemical ecology: allelochemicals, mycotoxins, and insect pheromones and allomones*. Institute of Botany, Academia Sinica Monograph Series No. 9, Taipai, Taiwan, 69–80

32 Lovett JV, Hoult AHC (1995) Allelopathy and self-defense in barley. *In*: Inderjit, KMM Dakshini, FA Einhellig (eds): *Allelopathy: organisms, processes, and applications*, American Chemical Society Symposium Series 582, Washington, DC, 170–183

33 Putnam AR, DeFrank J (1983) Use of phytotoxic plant residues for selective weed control. *Crop Prot* 2: 173–181

34 Barnes JP, Putnam AR, Burke BA (1986) Allelopathic activity of rye (*Secale cereale* L.). *In*: AR Putnam, CS Tang (eds): *The science of allelopathy*. John Wiley and Sons, New York, 271–286

35 Paulus J (1994) Ecological aspects of orchard floor management in apple agroecosystems of the central California coast. Ph.D. dissertation in Biology, University of California, Santa Cruz

36 Brown PD, Morra MJ (1995) Glucosinolate-containing plant tissues as bioherbicides. *J Agric Food Chem* 43: 3070–3074

37 Vaughn SF, Boydston RA (1997) Volatile allelochemicals released by crucifer green manures. *J Chem Ecol* 23: 2107–2116

38 Jimenez-Osornio JJ, Gliessman SR (1987) Allelopathic interference in a wild mustard (*Brassica campestris* L.) and broccoli (*Brassica oleracea* L. var. *italica*) intercrop agroecosystem. *In*: GR Waller (ed): *Allelochemicals: role in agriculture and forestry*. American Chemical Society Symposium Series 330: 262–274

39 Jimenez-Osornio JJ (1984) Interactions in a wild mustard (*Brassica campestris* L.) and broccoli (*Brassica oleracea* L. var. *italica*) intercrop agroecosystem. MA Dissertation, Biology, University of California, Santa Cruz

40 Brown PD, Morra MJ, McCaffrey JP, Auld DL, Williams L (1991) Allelochemicals produced during glucosinolate degradation in soil. *J Chem Ecol* 17: 2021–2034

41 Altieri MA (1994) *Biodiversity and pest management in agroecosystems*. Food Products Press, New York

42 Gliessman SR, Altieri MA (1982) Polyculture cropping has advantages. *Calif Agric* 36: 14–17

43 Rosado-May FJ (1991) Ecological role of wild mustard (*Brassica kaber* (D.C.) L.C.Wheeler) in the management of soil-pathogenic fungi and nematodes in a corn agroecosystem. PhD Dissertation, Biology, University of California, Santa Cruz

Chemical Ecology of Plants: Allelopathy in Aquatic and Terrestrial Ecosystems
ed. by Inderjit and Azim U. Mallik
© 2002 Birkhäuser Verlag/Switzerland

Allelochemicals phytotoxicity in explaining weed invasiveness and their function as herbicide analogues

Inderjit[1] and Prasanta C. Bhowmik[2]

[1] Department of Botany, University of Delhi, Delhi 110007, India
[2] Department of Plant and Soil Sciences, Stockbridge Hall, University of Massachusetts, Amherst, MS 01003-7245, USA

Introduction

Allelopathy is defined as the effect of one plant (including microorganisms) on the growth of another plant through the release of chemical compounds into the environment [1]. This definition includes both stimulatory and inhibitory effects. However, some ecologists prefer to include only inhibitory activities [2]. Any discussion on the definition of the term "allelopathy" is beyond the scope of this chapter, and we will focus only on inhibitory aspects of allelopathy.

The area occupied by weeds is increasing at an alarming rate. For example, land infested with weeds was 6.25 million hectares in 1985, increased to 22.5 million hectares in 1995, and may have increased by 47.5 million hectares by now [3]. The aggressive nature of the spread of weeds could be due to their efficient competitive abilities for resources, to allelopathic interference, or to both. Generally, perennial weeds are considered more troublesome than annual weeds [4–6]. Annual weeds, however, can cause significant damage to crops in terms of yield loss. Common lamb's-quarter (*Chenopodium album* L.), for example, can grow as tall as corn (*Zea mays* L.) (Fig. 1) and tomato (*Lycopersicon esculentum* Mill.) [7] and can compete with the crop and cause significant yield loss. Common lamb's-quarter has developed resistance to triazine herbicides since 1974 [8, 9]. Currently there are 216 herbicide-resistant weed biotypes in 45 countries [10].

Inderjit and Keating [4] listed 112 weed species that are reported to have potential allelopathic activities. They suggested the need for further field research to select the species exhibiting allelopathic interference in field situations. The objectives of this chapter are to discuss the potential role of allelochemicals in understanding (1) weed invasiveness (negative aspects of allelopathy) and (2) the potential for using allelochemicals for weed management (positive aspects of allelopathy).

Figure 1. General view of a corn field infested with common lamb's-quarters.

Allelochemicals and invasiveness

Because of their invasiveness, weed species have great potential to spread and
establish at places other than their natural habitat. Such weeds are not only
problematic from an agricultural standpoint, but they also pose a great threat
to the plant diversity of nearby flora. An invasive plant is an introduced species that becomes established and subsequently spreads outside its natural
range. Certain introduced species become invasive and reduce native plant
diversity. An obvious consequence of weed invasion is that of reduced plant
diversity. *Lantana camara* L., for example, is spreading in and around New
Delhi Ridge, India, and continues to dominate other plant species. This leads
to an adverse effect on the plant diversity of the New Delhi Ridge. Allelopathy
is suggested as a possible competitive strategy of *Lantana* in dry rainforest
ecotones, warm temperate rainforests, and wet sclerophyll forests [11]. The
conversion of grasslands and savannas to shrublands and woodlands by invasive weeds has become a global problem. *Acacia nilotica* ssp. *indica* has
become one of the most serious tropical rangeland problems in Australia [12].

There is a distinction between the attributes of successful weed species in
arable land ecosystems and those of invasive species. Some perennial weeds
possess strong competitive abilities in terms of dense root systems and allelopathic potential. For example, *Pluchea lanceolata* (DC.) C. B. Clarke (a noxious weed in semiarid regions of India) is spreading into both cultivated and
uncultivated fields [13]. In addition to the damage to crop yields, *Pluchea*
influences plant diversity, altering the roles of competition, interference, and

substratum degradation in ecosystem process. Boundaries of cultivated fields that often provide a variety of habitats for plant species have been degraded because of the intensification of agriculture [14, 15]. Edges of arable fields have suffered reduced species richness and invasion by annual grasses in UK, posing weed-management problems for growers. Smith et al. [15] suggested that management of uncropped edges of arable fields to enhance plant diversity does not promote weed occurrence in adjacent fields. These authors suggested that the use of herbicides in arable field edges promotes invasive annual weeds. Encouraging grasses and wildflower mixtures by managing existing vegetation or sowing a new sward can control certain annual invasive weeds [15]. Because disturbance in agroecosystems is frequent and regular in comparison to natural ecosystems, diversity is difficult to maintain.

Centaurea maculosa Lam. (spotted knapweed) originated in Eurasia but is an invasive weed in North America. This weed has taken over millions of hectares of natural Palouse prairie land, thus adversely affecting plant diversity [16]. Ridenour and Callaway [17] investigated the relative importance of allelopathy in explaining the interference potential of spotted knapweed with respect to the native bunchgrass Idaho fescue (*Festuca idahoensis* Elmer.). It is important to examine the relative importance of allelopathic interference of weed. Root exudates of spotted knapweed were checked for allelopathic suppression of Idaho fescue, and activated charcoal was used to manipulate the allelopathic effects. Ridenour and Callaway [17] planted individual Idaho fescue in pots filled with pure silica-sand. Half of the pots received active carbon and had either conspecifics or spotted knapweed. All pots received sufficient amounts of water and nutrients. The authors found evidence for allelopathic suppression of Idaho fescue and a negative correlation between abundance of spotted knapweed and abundance of Idaho fescue. A significant decrease in root growth of Idaho fescue was observed whether it was in contact with spotted knapweed or not (Idaho fescue roots grew in the rhizosphere zone of spotted knapweed but did not have physical contact) (Fig. 2a, b). The observed decrease in root growth of Idaho fescue, however, was less in presence of activated charcoal for both contact and non-contact roots. In one experiment, these authors found that Idaho fescue grown with knapweed were 85% shorter compared to situation when activated carbon was used. These findings support the contention that allelopathy is involved in the invasiveness of spotted knapweed.

An invasive Eurasian forb (*Centaurea diffusa* Lam.), a noxious weed in North America, has strong negative effects on three native bunchgrass species (*Festuca ovina* L., *Koeleria laerssenii*, and *Agropyron cristatum* (L.) Gaertn.) [18]. Such inhibitory effects, however, were absent in closely related grass species from the native communities of *C. diffusa*. Callaway and Aschehoug [18] proposed that North American grasses have not adapted to allelochemicals released by the introduced forbs, while Eurasian species have. This would suggest that a plant could grow with neighbors in its native habitat without having strong allelopathic effects, whereas it may interfere seriously through the release of allelochemicals with its neighboring species in an alien habitat.

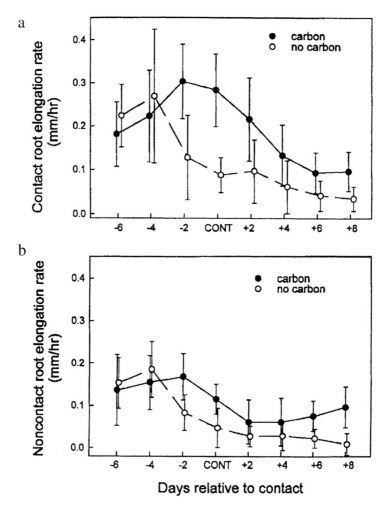

Figure 2 (a) Elongation rates of Idaho fescue roots that made physical contact with spotted knapweed roots in root observation chambers ("*contact roots*"), with or without activated carbon, from 6 days before until 8 days after contact. Elongation rates of all roots were converted to mm/hour and standardized in time by aligning their days of contact at "*day 0*". Error bars represent two standard errors. Source: Ridenour and Callaway [17]. Reproduced with permission from Springer-Verlag. (b) Elongation rates of Idaho fescue roots that grew into spotted knapweed rhizospheres but did not make physical contact with spotted knapweed ("*noncontact roots*"), with or without activated carbon, from 6 days before until 8 days after contact. Elongation rates of all roots were converted to mm/hour and standardized in time by aligning their days of contact at "*day 0*". Day 0 for noncontact roots was taken to be average day of contact for sister contact roots of approximately the same age on the same plant. Error bars represent two standard errors. Source: Ridenour and Callaway [17]. Reproduced with permission from Springer-Verlag.

Mugwort (*Artemisia vulgaris* L.) is a problematic invasive weed in the eastern United States [19]. Owing to its vigorous growth, mugwort is very competitive and pure stands of mugwort often can be seen. Mugwort dramatically reduces biodiversity, and infestation also results in increased production expense in nursery crops, reduced crop quality, and much higher maintenance expenses in turf, landscapes, and other urban settings. Inderjit and Foy [20] assessed the nature of interference by mugwort under controlled conditions. They examined the effect of soil amended with mugwort leaf material on seedling growth of red clover (*Trifolium repens* L.). In general, addition of mugwort leachates to soil resulted in chemical changes in the soil, including changes in available water-soluble phenolics, and suppressed red clover seedling growth when compared to that in non-amended soils. N and P fertilization in unsterilized soil amended with mugwort leachate could not eliminate these negative effects on red clover growth, but the addition of charcoal eliminated the leachate effects. This suggests that mugwort allelochemicals play a role in interference with red clover growth. Inderjit et al. [21] analyzed mugwort-infested and mugwort-free soils for selected soil properties and compared them to those of soil amended with mugwort leachate. However, no consistent trend was observed.

Mugwort produces flavonoids such as tricine, jaceosidine, eupafolin, chrysoeriol, diometin, homoeriodictyol, isorhamnetin, apigenin, eriodictyol, luteolin, luteolin 7-glucoside, kaempferol 3-glucoside, kaempferol 7-glucoside, kaempferol 3-rhamnoside, kaempferol 3-rutinoside, quercetin 3-glucoside, quercetin 3-galactoside, quercetin, rutin, and vitexin [22]. Given the diversity of secondary products in aerial portions, mugwort's allelochemicals may play a role in its invasiveness. In addition to plant-plant interference, mugwort allelochemicals may influence other ecosystem components, which in turn drive interactions that determine the community structure [23]. Mugwort allelochemicals may have more far-ranging effects than simply altering the success of other plants and may influence plant diversity of local flora through their influence on resource competition, nutrient dynamics, microbial ecology, and soil chemistry. In nature, several mechanisms usually operate in combination to control vegetation pattern [24–26]. There is a need to investigate the role of mugwort's allelochemicals in relation to its substratum ecology in order to understand its role in competitive strategies.

Allelopathy, however, may not always be a valid explanation to argue the invasiveness of a species. Murphy and Aarssen [27], for example, reported the allelopathic potential of pollen from *Phleum pratense* L. This plant, however, does not appear to be invasive because it does not have other characteristics required to identify a species as invasive [28]. Such characteristics include adaptation or opportunistic abilities to survive in changing environments, strong competitiveness or high rates of successful reproduction, and the type of life-history strategy ("k" or "r" type) used for reproduction. There is a need for a systematic search for a species that might combine the characteristics needed for invasiveness with allelopathic pollen and therefore become a successful invasive plant. In addition to studying the role in crop-growth suppression, attention should be given to the study of ecological roles of allelochemicals in influencing plant diversity and invasiveness.

Allelochemicals play an important role in weed ecology and management. The fact that a weed does not interfere with its neighbors in its native area does not necessarily mean that it would not interfere with its new neighbors in a new environment. More research is needed on the ecology of invasive weeds in order to understand their interference and reproductive strategies.

Allelochemicals and weed suppression

One of the most important components of allelopathy research is to identify putative allelochemicals. Once we know that allelochemicals are an important component of invasive species, actual allelochemicals should be identified. There is a potential for the use of allelochemicals to discover new herbicides [4]. Despite environment and health concerns, herbicides continue to play an important role in integrated weed-management systems. Compared to synthetic herbicides, natural products are more readily degradable in soil [4, 29]. There is a need to discover new herbicides since the number of herbicide-resistant weeds is increasing and conventional synthetic herbicides are no longer effective. The important question is which class of organic compounds (i.e., phenolics, terpenoids, alkaloids, etc.) is a better candidate for natural plant products. The normal range of concentrations tested for allelopathic chemicals is between 10^{-4} and 10^{-7} M. From an environmental safety standpoint, good candidates for natural herbicides should have activity between 10^{-5} and 10^{-7} M [30]. Many phenolic compounds, alkaloids, and quinones, however, have an activity range of 10^{-2} to 10^{-5} M and thus are poor candidates for natural herbicides [30].

Here we discuss two compounds suggested to have potential herbicidal activities, i.e., artemisinin and sorgoleone. Artemisinin $(3R,5aS,6R,8aS,9R,12S,12aR)$-Octahydro-3,6,9-trimethyl-3, 12-epoxy-12H-pyrano[4.3-j]-1,2-benzodioxepin-10(3H)-one), a sesquiterpenoid lactone (Fig. 3), has been shown to inhibit the growth of redroot pigweed (*Amaranthus retroflexus* L.), pitted morning-glory (*Ipomoea lacunosa* L.), annual wormwood (*Artemisia annua* L.), and common purslane (*Portulaca oleracea* L.) [31]. Artimisinin at 33 µM marginally increases the mitotic index of lettuce (*Lactuca sativa* L.) root tips, and chromosomes are less condensed during mitosis. Duke et al. [31] concluded that artemisinin is a selective phytotoxin with herbicidal activity similar to cinmethylin (see also [32]). Although the exact mode of action is still unknown, Dayan et al. [33] suggested that it is somewhat selective and might have a novel mode of action. Attempts have been made to use artimisinin as the basis for discovering new herbicides [34]. Lydon et al. [35] reported that the phytotoxic activities of annual wormwood cannot be explained solely by artemisinin. The dichloromethane extracts of annual wormweed leaves, which contain artemisinin, inhibited seed germination and seedling growth of redroot pigweed more strongly than did similar amounts of artemisinin alone. Furthermore, aqueous extracts without

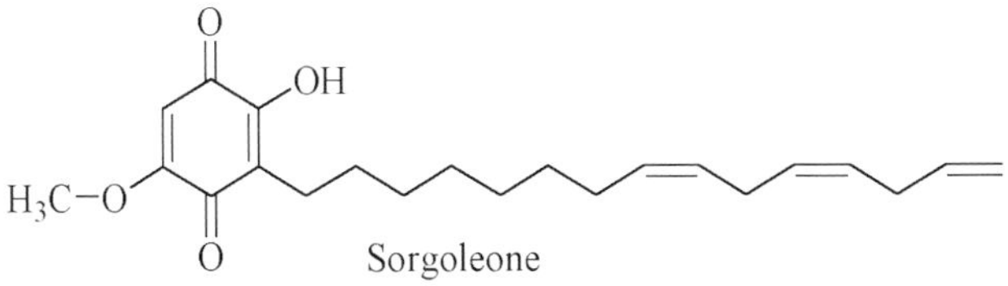

Figure 3. Structure of artemisinin and sorgoleone.

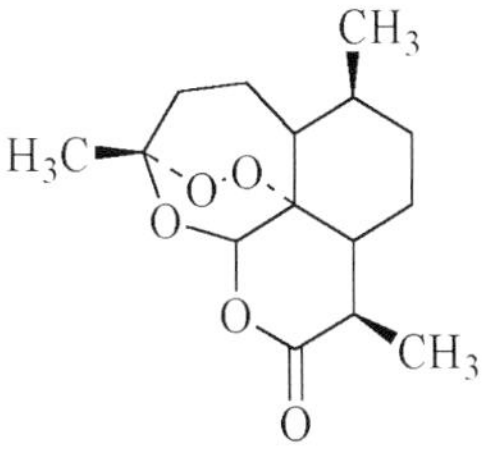

artemisinin had activity similar to that of artemisinin alone. This illustrates the significance of combined action of allelochemicals in mixtures. Einhellig [36] suggested that most allelopathic activities are due to the presence of several compounds in a mixture. The concentration of each compound in a mixture might be significantly less than the concentration of individual compounds needed to cause growth inhibition [37].

Einhellig et al. first reported the phytotoxic activities of sorgoleone (2-hydroxy-5-methoxy-3-[(8'z, 11'z)-8', 11', 14'-pentadecatriene]-*p*-benzo-quinone) (Fig. 3) [36, 38, 39]. Sorgoleone and its analogues are structurally similar to plastoquinone [34]. Sorgoleone inhibited the evolution of O_2 during photosynthesis in potato (*Solanum tuberosum* L.) and in common groundsel (*Senecio vulgaris* L.) [40]. Nimbal et al. [41] carried out a study on sorgoleone using triazine-susceptible potato and redroot pigweed thylakoids. Sorgoleone was a competitive inhibitor of atrazine binding sites. Sorgoleone also inhibited the photosystem II electron-transport reactions [42]. Gattas-Halak et al. [43] reported that sorgoleone influences the cell multiplication cycle by inhibiting the number of cells in prophase, metaphase, and anaphase stages. Sorgoleone is considered as an efficient herbicide because of its potential to inhibit electron transfer between Q_A and Q_B at the reducing site of photosystem II [44]. Duke et al. [34] suggested that sorgoleone is a more potent PS II inhibitor than the commercially available herbicide atrazine [45]. Because *in vivo* activity of sorgoleone does not match its *in vitro* activity, the physico-chemical properties of sorgoleone restrict its use as an herbicide. This is because most pesticide scientists usually attribute this type of discrepancy to unsuitable physico-chemical properties (Dr. Steve Duke, personal communication). Another plausible explanation could be its rapid degradation *in vivo*. Further research is needed to answer this question.

Important considerations

Foliar-applied herbicides may be exuded through roots of weed species and then be absorbed by nearby crop plant. Nicosulfuron is an effective herbicide against johnsongrass [*Sorghum halepense* (L.) Pers.] and quackgrass (*Elytrigia repens* (L.) Beauv) [46–48]. It is considered a good herbicide, as corn is tolerant to the normal field-use rates of nicosulfuron (18 to 105 g active ingredient ha^{-1}) [47–49]. Gubbiga et al. [49] investigated the root exudation of nicosulfuron from nicosulfuron-treated johnsongrass. Johnsongrass was foliar treated with 50 or 100 µg ^{14}C-nicosulfuron per plant. The 23% of ^{14}C-nicosulfuron absorbed by the johnsongrass was found in the medium 30 days after treatment. The subsequent uptake of exuded ^{14}C-nicosulfuron by corn roots was also shown. In a subsequent experiment, these authors found that 10^{-8} M nicosulfuron can cause growth reduction in corn. Therefore, the sensitivity of a crop to a herbicide is an important factor in determining its use as a commercial herbicide.

Allelochemicals from weed species have the potential to be explored as natural herbicides. However, prior to using them as herbicides, the following questions should be addressed.

1. At what minimum concentration does each compound have phytotoxic activity?
2. What is the residence time and fate of the compound in the soil environment?
3. Does the compound influence microbial ecology and physicochemical properties of the soil?
4. What is the mode of action of the compound?
5. Does the compound have any adverse effect on desired crops?

There are certain limitations for using allelochemicals as natural herbicides. High cost and limited efficacy restrict the use of natural herbicides, e.g., maize gluten and pelargonic acid [34]. Natural herbicides may also be toxic to nontarget organisms. A natural compound, alpha-terthienyl, was isolated from the roots of the common marigold (*Tagetes erecta*) and originally patented as an herbicide (Professor Jeff Weidenhamer, personal communication). In addition to its herbicidal activity, alpha-terthienyl was also toxic to nontarget species. This made it impossible for the industry to approve it as an herbicide. Toxicity to the nontarget species is one of the main causes that limit the use of natural compounds as herbicides. Sometimes natural products may have problems like allergy. Sorgoleone, for example, is reported to cause dermatitis (Professor Jeff Weidenhamer, personal communication). Sorgoleone, however, can act as a successful herbicide when released from roots of sorghum. Natural products can be beneficially used as herbicide when they are released from cover crops or weed-suppressing cultivars.

Several phenolic acids (e.g., *p*-hydroxybenzoic, *trans-p*-coumaric, *cis-p*-coumaric, syringic, vanillic, *trans-* and *cis*-ferulic acids, and 2,4-dihydroxy-7-methoxy-1,4-benzoazin-3-one) are identified from wheat seedlings [50]. Allelopathic wheat cultivars are reported to possess higher amounts of these compounds [51]. Although these compounds are reported to be good candidates for herbicidal activity, weed-suppressing wheat cultivars can be used to naturally check the weeds and therefore minimize the use of synthetic herbicides. Inderjit et al. [52] have shown that wheat has the potential to suppress seedling growth of perennial ryegrass (*Lolium perenne*) when sown together. There are no scientific reasons to believe that natural products, when used as an herbicide, can overcome the problem of the development of herbicide-resistant weeds (Professor Jeff Weidenhamer, personal communication).

Acknowledgements
The article was written during the stay of Inderjit as Visiting Professor to the Department of Plant and Soil Sciences, University of Massachusetts. We thank Professors Hans Lambers, Jeff Weidenhamer, Steve Murphy, and Steve Gliessman for their constructive comments. Professor Ray Callaway kindly provided the electronic copy of Figure 2.

References

1 Rice EL (1984) *Allelopathy*. 2nd Edn. Academic Press, Orlando
2 Lambers H, Chapin FS III, Pons TL (1998) *Plant physiological ecology*. Springer-Verlag, New York
3 Otteni L (1998) A national strategy for the management of invasive plants. *In*: S Glenn (ed): *Proceedings of the fifty-second annual meeting of the Northeastern weed science society*. Northeastern Weed Science Society, College Park, MD, 139–146
4 Inderjit, Keating KI (1999) Allelopathy: principles, procedures, processes, and promises for biological control. *Adv Agron* 67: 141–231
5 Bhowmik PC (1994) Biology and control of common milkweed (*Asclepias syriaca*). *Rev Weed Sci* 6: 227–250
6 Bhowmik PC (1997) Weed biology: importance to weed management. *Weed Sci* 45: 349–356
7 Bhowmik PC, Reddy KN (1988) Interference of common lambsquarters (*Chenopodium album*) in transplanted tomato (*Lycopersicon esculentum*). *Weed Technol* 2: 505–508
8 Myers MG, Harvey RG (1993) Triazine-resistant common lambsquarters (*Chenopodium album* L.) control in corn field (*Zea mays* L.). *Weed Technol* 7: 884–889
9 Bandeen JD, Bhowmik PC, Jensen KN (1976) Two strains of lambsquarters with differing tolerance to atrazine. *Abst Weed Sci Soc Am* Abst No. 150
10 Bhowmik PC (2000) Herbicide resistance: a global concern. *Med Fac Landbouww, Univ Gent* 65: 19–30
11 Gentle CB, Duggin JA (1997) Allelopathy as a comparative strategy in persistent thickets of *Lantana camara* L. in three Australian forest communities. *Plant Ecol* 132: 85–95
12 Kriticos D, Brown J, Radford I, Nicholas M (1999) Plant population ecology and biological control: *Acacia nilotica* as a case study. *Biol Control* 16: 230–239
13 Inderjit (1998) Influence of *Pluchea lanceolata* on selected soil properties. *Am J Bot* 85: 64–69
14 Kiss J, Penksza K, Toth F, Kadar F (1997) Evaluation of fields and field margins in nature production capacity with special regard to plant protection. *Agric Ecosyst Environ* 63: 227–232
15 Smith H, Firbank LG, Macdonald DW (1999) Uncropped edges of arable fields managed for biodiversity do not increase weed occurrence in adjacent crops. *Biol Conserv* 89: 107–111
16 Muir AD, Majak W (1983) Allelopathic potential of diffuse knapweed (*Centaurea diffusa*) extracts. *Can J Plant Sci* 63: 989–996

17 Ridenour WM, Callaway RM (2001) The relative importance of allelopathy in interference: the effect of an invasive weed on a native bunchgrass. *Oecologia* 126: 444–450

18 Callaway RM, Aschehoug ET (2000) Invasive plant *versus* their new and old neighbors: a mechanism for exotic invasion. *Science* 290: 521–523

19 Holm L, Doll J, Holm E, Pancho J, Herberger J (1997) *World weeds: natural histories and distribution*. John Wiley and Sons, New York

20 Inderjit, Foy CL (1999) Nature of the interference potential of mugwort (*Artemisia vulgaris*). *Weed Technol* 13: 176–182

21 Inderjit, Kaur M, Foy CL (2001) On significance of field studies in allelopathy. *Weed Technol* 15: 792–797

22 Lee SJ, Chung HY, Maier GG, Wood AR, Dixon RA, Mabry TJ (1998) Estrogenic flavonoids from *Artemisia vulgaris* L. *J Agric Food Chem* 40: 3325–3329

23 Wardle DA, Nilsson MC, Gallet C, Zackrisson O (1998) An ecosystem level perspective of allelopathy. *Biol Rev* 73: 305–319

24 Callaway RM (1995) Positive interactions among plants. *Bot Rev* 61: 306–349

25 Chapin FS III, Walker LR, Fastie CL, Sharman LC (1994) Mechanisms of primary succession following deglaciation at Glacier Bay, Alaska. *Ecol Monogr* 64: 149–175

26 Inderjit, Del Moral R (1997) Is separating allelopathy from resource competition realistic? *Bot Rev* 63: 221–230

27 Murphy SD, Aarssen LW (1995) Allelopathic pollen of *Phleum pratense* reduces seed set in *Elytrigia repens* in the field. *Can J Bot* 73: 1417–1422

28 Murphy SD (1999) Pollen allelopathy. *In*: Inderjit, KMM Dakshini, CL Foy (eds): *Principles and practices in plant ecology: allelochemical interactions*. CRC Press, Boca Raton, 129–148

29 Duke SO (1988) Glyphosate. *In*: PC Kearney, DD Kaufman (eds): *Herbicide: chemistry, degradation and mode of action*. Vol. 3. Marcel Dekker, New York, 1–70

30 Macias FA (1995) Allelopathy in search for natural herbicide models. *In*: Inderjit, KMM Dakshini, FA Einhellig (eds): *Allelopathy: organisms, processes, and applications*. American Chemical Society, Washington, DC, 310–329

31 Duke SO, Vaughn KC, Croom EM, Elsohly HN (1987) Artemisinin, a constituents of annual wormwood (*Artemisia annua*) is a selective phytotoxin. *Weed Sci* 35: 499–505

32 Bhowmik PC (1988) Cinmethylin for weed control in soybeans, *Glycine max*. *Weed Sci* 36: 678–682

33 Dayan FE, Hernandez A, Allen SN, Moraces RM, Vroman JA, Avery MA, Duke SO (1999) Comparative phytotoxicity of artimisinin and several sesquiterpene analogues. *Phytochemistry* 50: 607–614

34 Duke SO, Scheffler BE, Dayan FE (2001) Allelochemicals as herbicides. *In*: NP Bonjoch, MJ Reigosa (eds): *First European OECD allelopathy symposium: physiological aspects of allelopathy*. GAMESAL, SA, Vigo, 47–59

35 Lydon J, Teasdale JR, Chen PK (1997) Allelopathic activity of annual wormwood (*Artemisia annua*) and its role of artimisinin. *Weed Sci* 45: 807–811

36 Einhellig FA (1995) Allelopathy: current status and future goals. *In*: Inderjit, KMM Dakshini, FA Einhellig (eds): *Allelopathy: organisms, processes and applications*. American Chemical Society, Washington, DC, 1–24

37 Blum U (1996) Allelopathic interactions involving phenolic acids. *J Nematol* 28: 259–267

38 Einhellig FA, Souza IF (1992) Phytotoxicity of sorgoleone found in grain sorghum root exudates. *J Chem Ecol* 18: 1–11

39 Einhellig FA, Rasmussen JA, Hejl AH, Souza IF (1993) Effects of root exudate sorgoleone on photosynthesis. *J Chem Ecol* 19: 369–375

40 Nimbal CI, Pedersen JF, Yerkes CN, Weston LA, Weller SC (1996) Phytotoxicity and distribution of sorgoleone in grain sorghum germplasm. *J Agric Food Chem* 44: 1343–1347

41 Nimbal CI, Yerkes CN, Weston LA, Weller SC (1996) Herbicidal activity and site of action of the natural product sorgoleone. *Pest Chem Physiol* 54: 73–83

42 Gonzalez VM, Kazimir J, Nimbal C, Weston LA, Cheniae GM (1997) Inhibition of a photosystem II electron transfer reaction by the natural product sorgoleone. *J Agric Food Chem* 45: 1415–1421

43 Gattas-Hallak AM, Davide LC, Souza IF (1999) Effect of sorghum (*Sorghum bicolor* L.) root exudates on the cell cycle of the bean plant (*Phaseolus vulgaris* L.) roots. *Gen Mol Biol* 22: 95–99

44 Czarnota MA, Paul RN, Dayan FE, Weston LA (2001) Further studies on the mode of action,

localization of production, chemical nature, and activity of sorgoleone: a potent PS II inhibitor produced in *Sorghum* spp root exudates. *Weed Technol* 15: 813–825

45 Streibig JC, Dayan FE, Rimando AM, Duke SO (1999) Joint action of natural and synthetic photosystem II inhibitors. *Pestic Sci* 55: 137–146

46 Bhowmik PC, O'Toole BM, Andaloro J (1992) Effects of nicosulfuron on quackgrass (*Elytrigia repens*) control in corn (*Zea mays*). *Weed Technol* 6: 52–56

47 Kapusta G, Krausz RF (1992) Interaction of terbufos and nicosulfuron on corn (*Zea mays*). *Weed Technol* 6: 999–1003

48 Camacho RF, Moshier LJ, Morishta DW, Devlin DL (1991) Rhizome johnsongrass (*Sorghum halepense*) control in corn (*Zea mays*) with primisulfuron and nicosulfuron. *Weed Technol* 5: 789–794

49 Gubbiga NG, Worsham AD, Corbin FT (1996) Root/rhizome exudation of nicosulfuron from treated johnsongrass (*Sorghum halepense*) and possible implication for corn (*Zea mays*). *Weed Sci* 44: 455–460

50 Wu H, Haig T, Pratley J, Lemerle D, An M (2000) Distribution and exudation of allelochemicals in wheat *Triticum aestivum*. *J Chem Ecol* 26: 2141–2154

51 Wu H, Pratley J, Lemerle D, Haig T, Verbeek b (1998) Differential allelopathic potential among wheat accessions to annual ryegrass. *In*: DL Michalk, JE Prately (eds): *Proceedings, 9th Australian Agronomy Conference*. Wagga Wagga, Australia, Australian Society of Agronomy, 567–571

52 Inderjit, Olofsdotter M, Streibig JC (2001) Wheat (*Triticum aestivum*) interference with seedling growth of perennial ryegrass (*Lolium perenne*): influence of density and age. *Weed Technol* 15: 807–812

Chemical Ecology of Plants: Allelopathy in Aquatic and Terrestrial Ecosystems
ed. by Inderjit and Azim U. Mallik
© 2002 Birkhäuser Verlag/Switzerland

Shift in allelochemical functioning with selected abiotic stress factors

Inderjit[1] and Harsh Nayyar[2]

[1] Department of Botany, University of Delhi, Delhi 110007, India
[2] Department of Botany, Panjab University, Chandigarh 160014, India

Introduction

In order to survive and thrive in the dynamic environment, plants have evolved some complicated strategies. One such strategy involves the release of secondary metabolites by a plant into the environment to out-compete its neighbors. The phenomenon is known as allelopathy. The term "allelochemical" is often used for secondary metabolites having a role in plant defense, interference (allelopathy), herbivory, etc. The terms "allelochemicals" and "secondary substances" are generally used as synonyms [1]. All secondary metabolites may not function as an allelochemical. There are several instances where secondary substances may have a role in primary metabolism [2] or where allelochemicals may act as nutrients [3]. Any discussion of the definition of allelochemical is beyond the scope of this chapter. We will use the term "allelochemical" to denote secondary metabolites having some role in plant-plant interference. Furthermore, allelochemicals may have a variety of ecological functions, but we will focus on their role in plant interference, i.e., allelopathy.

Several workers express their dissatisfaction over the way allelopathy research is done [4–8]. It has been suggested that allelopathy cannot be separated from other mechanisms of interferences such as resource competition, nutrient immobilization, and abiotic and biotic soil factors [9]. Jeffrey Weidenhamer (personal communication), however, feels that allelopathy can be separated from resource competition [10, 11]. We agree that allelopathy can be separated from resource competition under controlled conditions. But, can controlled-conditioned experiments explain field situation in totality? If separation from other mechanisms of plant interference by keeping other factors constant proves allelopathy, how can we rule out the probable role of these factors in determining community structure? How is it then meaningful to discuss the importance of climate, habitat, substratum, mycorrhizae, microbial ecology, and nutrient dynamics in determining allelopathy? We cannot rule out ecological complexity; we can only prove that allelopathy may be a probable factor under a given situation. Inderjit and Weiner [12] argued against the strict plant-plant approach to understand allelopathy and suggested that research in this field can be furthered when allelopathy is investigated in the context of soil chemical ecology. A plant cannot be separated from its environment, which includes substratum and climatic variations. After the release of allelochemicals into the environment, their fate and expression in terms of phytotoxicity is determined by (1) substratum factors and (2) abiotic and biotic stress factors. The role of substratum factors in determining allelopathic expression has been discussed previously [12–17].

Plant interaction with several abiotic and biotic stresses further adds to the complexity of natural systems. The term "stress" is generally used for an adverse effect. Nilsen and Orcutt [18, p. 4] defined stress as, "…. *a set of conditions that cause aberrant change in physiological processes, resulting eventually in an injury*". A common approach at the ecosystem- and whole-plant-level is to consider any situation stressful where the external constraints limit the rate of dry matter production of all or part of vegetation below its genetic potential [19]. Enhanced production of secondary metabolites in the plant under stress has been argued previously [20]. The subsequent release of a myriad of allelochemicals that is due to stress, and their eventual participation in allelopathic interaction, however, is difficult to demonstrate in field situations. Although both abiotic and biotic stress factors are important in terms of the performance of allelochemicals in the ecosystem, we will focus on selected abiotic stress factors. It is argued in the literature that abiotic stress factors generally cause an increase in allelochemicals [20–23], but reports on their release in the substratum and ultimate ecological role are comparatively few. In this chapter, we will discuss allelochemicals and plant stress from an ecological perspective, i.e., effect of stress factor(s) on the release of allelochemicals and their influence on plant growth and ecosystem factors. We raise and discuss some questions on allelochemical functioning under the influence of abiotic stresses. The objective of this article is to present an evaluation of the following questions:

1. Does any increase in the levels of allelochemicals in the plant under abiotic stress cause an increase in the amount of allelochemicals released into the environment?
2. In what situation(s) can allelochemicals initiate abiotic stress on the plant?
3. How is ecological functioning of allelochemicals related to abiotic stress factors?
4. Is it ecologically relevant to argue for the importance of single stresses in the complex ecosystem?
5. What are methodological impasses in studying allelochemical performance in relation to abiotic stress factors?
6. How is the importance of a specific stress factor in a given situation or environment determined?
7. Can abiotic stresses predispose target plants to allelochemical efficacy?

We will discuss each question by taking examples from different types of abiotic stresses such as habitat/climate, drought, irradiation, temperature, and nutrient deficiency.

Does any increase in the levels of allelochemicals in the plant under abiotic stress result in an increase in the amount of allelochemicals released?

Stress factors are generally reported to promote root exudation [24]. The amount of a particular allelochemical may increase with one kind of stress and decrease with another kind of stress [9, 25]. Little information exists on the release of allelochemicals under stress conditions, probably because of difficulty in demonstrating the same under natural conditions and, more so, whether they induce allelopathy or not. Phosphorous deficiency induced the release of more phenolics by the roots as compared to phosphorous-sufficient sunflower (*Helianthus annuus*) plants [26]. Under high-temperature conditions, the exudates released by trichomes of velvetleaf (*Abutilon theophrasti*) were twice as toxic as those from plants growing under normal temperatures [27]. Purple nutsedge (*Cyperus rotundus*) released toxic substances into its rhizosphere under drought stress [23]. Reduced yield of sweet corn (*Zea mays*) under low soil moisture conditions was attributed to higher allelopathic effect of nutsedge [28].

Pluchea lanceolata is a noxious weed of both cultivated and uncultivated land in the semiarid regions of northwest India [29]. The weed is known to possess allelopathic activities and is reported to adversely affect growth and establishment of crop species [30–34]. Uncultivated *Pluchea* soils, compared to cultivated *Pluchea* soils, are considered under abiotic stress for nutrients and water. From an allelopathy standpoint, release of allelochemicals into the environment is considered important. It was found that levels of water-soluble phenolics were higher in cultivated soils when compared to uncultivated soils. Although weed plants from uncultivated areas have higher phenolics, their release was higher in cultivated soils infested with the weed. While some fields in the region are cultivated once a year, others are cultivated regularly or twice a year. Inderjit and Dakshini [35] reported that *Pluchea*-infested soil, cultivated twice a year, had higher amounts of phenolics when compared to soils cultivated once a year. *Pluchea* soils cultivated once a year had higher amounts of Ca^{2+} compared to soils cultivated twice a year. High levels of a flavonoid (quercitrin) were found in *Pluchea* soils cultivated twice a year, followed by soil cultivated once a year, and were bound least in uncultivated soils [36]. Agricultural practices carried out in the region (i.e., incorporation of damaged weed parts during plowing into soil) are a probable cause of high levels of phenolics in regularly cultivated *Pluchea* soils. Compared to uncultivated soils, cultivated soils are under regular disturbance as a result of tillage. Though uncultivated soil suffers from water-deficit conditions, high levels of phenolics were recorded in cultivated soils. Inderjit and Gross [37] discussed the effects of different water regimes on the levels of phenolics in *Pluchea* soils. *Pluchea* soil (1 m^2 plot) was irrigated to maintain (1) 10% moisture (control), (2) 20% moisture, and (3) 40% moisture. Soil with maximum irrigation (40%) had the highest values of water-soluble phenolics, followed by soil with 20% moisture, and least in control soil. Since there was no difference in the levels of phenolics of *Pluchea* plants and their soil at the start of the experiment, increased values could be due to variation in soil moisture levels. To ascertain that enhanced release of phenolics in regularly cultivated soils is from weed parts and not from the crop, several controls were used, e.g., regularly cultivated fields for the same crop but with and without the weed. Phenolic content was determined while farmers prepare their fields for the cultivation, i.e., prior to sowing of crop seeds [32].

Any increase in the accumulation of secondary substances in plants under stress, however, cannot be generalized. For example, the concentration of benzoic and cinnamic acid derivatives in wheat (*Triticum aestivum*) decreased under drought stress [38]. It is also possible that two different compounds in the same plant respond differently to drought stress. Water-deficit conditions increased the levels of α-pinene in loblolly pine (*Pinus taeda*) and decreased the levels of other monoterpenes, e.g., β-pinene, camphene, myrene, and limonene [39]. Physiological responses of allelochemicals to stress conditions may vary due to several intrinsic and extrinsic factors [20], which are unlikely to be common for plants experiencing no stress. Any response of increase/decrease in the levels of allelochemicals and their subsequent release due to a particular stress should be studied only if the plant is actually subjected to that particular stress under natural settings. For example, drought stress may be an appropriate factor in semiarid and arid regions but may not have any ecological relevance in wetlands.

Research is needed to find the relationship between enhanced production of allelochemicals in the plant under stress and higher release (if any) of allelochemicals in the substratum. Very little information is available on the quality of allelochemicals released under stress. The biochemical routes employed by plants under a stress situation for production of different allelochemicals are yet to be determined. Figure 1 illustrates the various pathways for the synthesis of secondary metabolites. Which particular route is initiated by a specific stress is not clear, and whether similar or different types of chemicals are produced under all stresses is also a matter of investigation. Phenylalanine lyase (PAL) and chalcone synthase (CS) are the key enzymes of shikimate pathway [40, p. 278] and are likely to be the target sites for most of the stresses. Enhanced release could also be due to elevated production of allelochemicals as a part of the defense strategy or due to leakage/injury at the cellular level (Fig. 1). This aspect needs further research.

In what situation(s) can allelochemicals initiate abiotic stress on the plant?

Allelochemicals may influence the availability of mineral ions. For example, phenolic compounds may act as a source of labile C for soil microbes [41, 42].

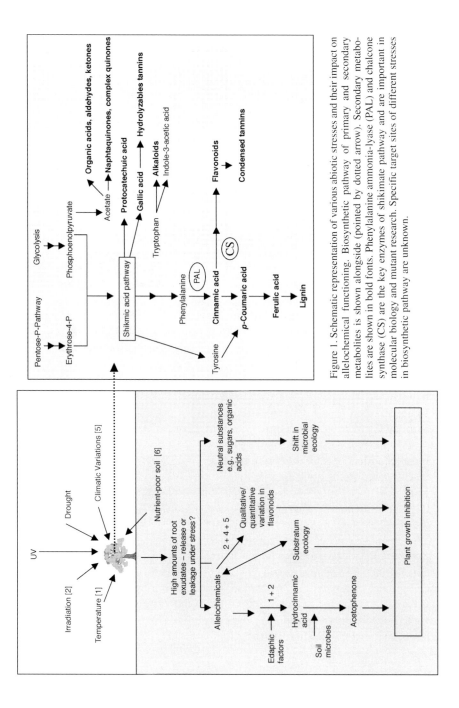

Figure 1. Schematic representation of various abiotic stresses and their impact on allelochemical functioning. Biosynthetic pathway of primary and secondary metabolites is shown alongside (pointed by dotted arrow). Secondary metabolites are shown in bold fonts. Phenylalanine ammonia-lyase (PAL) and chalcone synthase (CS) are the key enzymes of shikimate pathway and are important in molecular biology and mutant research. Specific target sites of different stresses in biosynthetic pathway are unknown.

The release of these molecules may result in an increase in microbial activity, which may influence allelochemical functioning [16, 41, 42]. This results in a shift in microbial ecology, which may result in temporary depletion of nitrogen and/or phosphorus. Soil microbes are reported to preferentially retain soil NO_3^- compared to NH_4^+ [43]. Any deficiency in nitrate level, resulting from allelochemicals, may create nutrient stress to nitrophilous species. Phenolic acids produced by soil microorganisms may have different phytotoxic potential compared to original phenolic acids contributed by the plant [41]. Phenolic acids in free and reversibly bound form are readily utilized by soil microorganisms under favorable environmental conditions (e.g., temperature and pH) and absence of another source of organic matter (e.g., glucose) [44–46]. Plant phenolics are known to influence nutrient cycling [47]. Phenolics have potential to bind to Al^{3+}, Fe^{3+}, and Mn^{2+}, which results in the release of phosphate otherwise bound to these cations [48, 49]. Polyphenols are reported to alter the fate of inorganic and organic soil nutrients [50]. They can form complexes with nutrient ions [47, 51]. In a study by Northup et al. [17] in pygmy forests having low soil pH, polyphenols released from leaf litter of *Pinus muricata* were shown to form complexes with Al^{3+}, which increases their toxicity. Higher phosphate levels in the substratum may enhance growth of ericoid mycorrhizae, e.g., in boreal forests [52]. However, excessively high amounts of phosphorus may inhibit mycorrhizae by reducing carbon flow to roots [53]. Inderjit and Mallik [54] investigated the effects of catechol and protocatechuic, *p*-coumaric, *p*-hydroxybenzoic, and ferulic acids on pH, organic N, organic matter, total phenolics, and selected inorganic ions of the mineral soils from boreal forests. Total phenolic content of soil amended with phenolic compounds (except catechol) was increased compared to unamended control soil. In general, there was a decrease in the levels of organic N, Mn^{2+}, and Fe^{3+} in the mineral soil (B+ horizon) amended with phenolic acids. Soil amended with an equimolar mixture of the above phenolics had lower values for organic N, phosphate, Mn^{2+}, Zn^{2+}, Al^{3+}, Fe^{3+}, and Ca^{2+} and higher values for total phenolics compared to the control.

Phenolic compounds such as caffeic and ferulic acids, myricetin, tannins, and tannin-derivative compounds inhibit oxidation of NH_4^+ to NO_2^- by *Nitrosomonas* [55, 56]. Bremner and McCarty [57, 58], however, disagreed with the above findings and reported that terpenoids and phenolics influence the immobilization of ammonium N by soil microorganisms.

It is evident from the above discussion that allelochemicals released into the environment by a plant may influence the resources (e.g., water and nutrients) available in the soil for target and donor plants. The donor plant may have evolved different mechanisms to adapt to abiotic stresses arising from allelochemicals released by it, while the target plant may lack this ability and face damage from abiotic stress factors or allelochemicals or both. Further research is needed to understand the interactive effect of allelochemicals and abiotic stress factors on target and donor plants.

How is ecological functioning of allelochemicals related to abiotic stress factors?

The functioning of allelochemicals is influenced by (1) climatic factors, (2) abiotic and biotic soil factors, (3) nutrient dynamics, and (4) plant growth. Any qualitative and quantitative variation in allelochemical profile may be influenced by either of the above four ecological factors.

Chaves and Escudero [59] reported that the qualitative and quantitative variations in the flavonoid content of *Cistus ladanifer* were influenced by temperature, light, and water stress. These authors considered that climatic factors are primarily responsible for quantitative (ultravoilet and hydric stress) and qualitative (temperature and hydric stress) variations in flavonoid synthesis. Pacheco et al. [60] studied seasonal variation of flavonoids in *Robinsonia evenia* and reported that the endogenous content of flavone apigenin 7-O-diglycoside was higher in the month of November and that of quercetin was higher during the month of February. Similarly, Chaves et al. [61] observed that flavonoids abundant during spring are apigenin and apigenin-4-(O)methyl and that their concentrations fall in summer. During summer, the higher amounts of kaempferol-3,7-di(O)methyl were observed. Seasonal variations are shown to determine the quality and quantity of flavonoid produced by *Cistus ladnifer* [61]. Dakshini et al. [14] reported that surface soils (0–12 cm) infested with *Pluchea* had higher values for total phenolics during the month of October, and subsoil (12–20 cm) had higher values for total phenolics during the month of July. Since the month of July and August are monsoon months in northern India, *Pluchea* soils sampled in post-rain season (i.e., early October) have higher levels of water-soluble phenolics, likely because of higher leaching.

Phenolics can be transformed by abiotic and biotic catalytic processes [15, 62]. Abiotic catalysts include primary minerals, layer silicates, metal oxides, hydroxides, oxyhydroxides, and poorly crystalline aluminosilicates. Through oxidative polymerization, ring cleavage, decarboxylation, and/or dealkylation, abiotic catalysts can transform phenolics to humic macromolecules. Oxidoreductases, for example, polyphenol oxidases and peroxidases act as biotic catalysts for phenolic transformation and cause polymerization during humic formation. Catechol, pyrogallol, orcinol, ferulic acid, and syringic acid are deemed to be the most important substrates for the polymerization reactions [15]. Abiotic stress factors may play an important role in influencing the toxicity of allelochemicals released into the substratum. *Ceratiola ericoides*, a perennial shrub, produces an inactive dihydrochalcone ceratiolin [63]. Owing to specific climatic (light and heat) and edaphic (pH) substratum conditions, ceratiolin is transformed into the toxic compound hydrocinnamic acid [64]. Hydrocinnamic acid further undergoes microbial degradation to form acetophenone [63]. Both hydrocinnamic acid and acetophenone are reported to inhibit germination and growth of little bluestem grass (*Schizachyrium scoparium*) [64].

Generally, a nutrient-poor substratum is likely to enhance the production of allelochemicals in the plant under stress [20]. *Kalmia angustifolia*, an ericaceous shrub, is a major cause of conifer regeneration failure in boreal forests [65]. *Kalmia* is associated with nutrient-poor microsites [66]. It has been reported that *Kalmia* soils from B+ horizon have lower values for nitrogen, phosphorus, iron, and manganese. *Kalmia* is adapted to nutrient-poor habitats, but such *Kalmia*-dominated, nutrient-poor habitats may create stress on black spruce (*Picea mariana*). Although we are not aware of any study illustrating an increase in allelochemical release from *Kalmia* plants with higher nutrient-deficient conditions, increased levels of allelochemicals in nutrient-deficient *Kalmia* soils can be hypothesized. This is an example where nutrient-poor conditions create stress only on one plant (i.e., black spruce) and enhanced production/release of allelochemical by the donor plant (i.e., *Kalmia*) create additional stress on black spruce. More research is needed to better understand the complex *Kalmia*-black spruce ecosystem. Significance of nutrient deficiencies in the expression of allelochemical interference has been suggested [11]. Harper [67] suggested that any inhibition after the addition of plant debris/allelochemicals is due to temporary depletion of nitrogen rather than allelochemicals. Observed growth inhibition due to allelochemicals is reported to be eliminated after the addition of N and P fertilization [68]. Phosphorous, for example, could reduce the inhibitory effects of *Solidago canadensis* on *Acer saccharum* [69]. This, however, is not always the case. The phytotoxic effects of mugwort (*Artemisia vulgaris*) leachate-amended soil on seedling growth of red clover (*Trifolium repens*) could not be eliminated after the addition of NP fertilizers [70].

Temperature influences the concentration of hydroxamic acids in wheat (*Triticum aestivum*) seedlings [71]. Temperature and availability of nitrates tend to increase gramine content in barley [72, 73]. Hordenine production increased in barley plants growing under low-temperature conditions. The balance of hordenine and gramine production changes in response to environmental changes [74]. Bhowmik and Doll [75] investigated the allelopathic effects of residues of redroot pigweed (*Amaranthus retroflexus*) and yellow foxtail (*Setaria glauca*) on corn (*Zea mays*) and soybean (*Glycine max*) at various temperature and photosynthetic photon flux densities (PPFD). They found that moderate PPFD (380–570 $\mu E/m^2/sec$) and temperature (30–20°C) reduced the allelopathic effects of weed residues on crops. Research is needed to find out the qualitative and quantitative variation in the production and release of allelochemicals from the donor plant under various abiotic stresses.

Recently, Inderjit and Asakawa [68] studied the effect of hairy vetch (*Vicia villosa*) on radish (*Raphanus satius*) in relation to full and reduced light intensity. These authors found that root growth of radish was suppressed under both normal (130 $\mu E/m^2/s$) and reduced (9.7 $\mu E/m^2/s$) light regimes. In reduced light regimes, however, higher values of phenolics in medium used to grow either monoculture of hairy vetch or mixed culture of hairy vetch and radish were observed when compared to those recorded in full light. Although shade generally enhances the levels of allelochemicals [9, 20], this cannot be generalized. Bhowmik and Doll [75] found that PPFD influences the allelopathic potential of residues of certain weeds such as redroot pigweed and yellow fox-tail on corn and soybean. It was found that redroot pigweed and yellow foxtail residues had low allelopathic effects on corn grown under moderate PPFD (380–570 $\mu E/m^2/sec$) when compared to 760 $\mu E/m^2/sec$. Clearcutting and fire in boreal forests allow sufficient light, which induces higher production of allelochemicals in the ericaceous plants, e.g., *Kalmia angustifolia* [76]. In bar-ley (*Hordeum vulgare*) lines growing under increased light intensity and increased temperature at outdoor conditions, gramine production was enhanced upto 100% in some lines [74] compared to a growth chamber.

The stress response of any plant is stage-specific. Accordingly, the produc-tion of allelochemicals also differs. Young tissues having active metabolism may react strongly toward environmental stress and may produce higher con-centrations of allelochemicals. The growing tissues may produce more sec-ondary metabolites than the older ones. It may also be hypothesized that this could be a defense mechanism to protect growing points of the plant from damage by herbivores. The reproductive stage is considered to be most prone to damage by any ecological change [77, 78]. Leaves of *Citrus aurantium* at their earlier stages of development showed enhanced production of naringenin and neohesperidin compared to leaves at later stages [79].

Is it ecologically logical to argue single stress in a complex ecosystem?

It is difficult to simulate nature. One plausible reason of difficulties in stress-related allelochemical studies is that stress factors operate in parallel; their synergistic interaction may better explain the stress-induced responses [9, 20]. Any increase in the levels of allelochemicals in the plant under stress and their subsequent release is likely due to substratum ecology, biotic influences, abi-otic and biotic stress factors, and other unknown factors [13, 20]. It is possible to design an experiment isolating the effect of a single factor. Whether this ever happens in complex field situation is, however, questionable. This makes it dif-ficult to prove stress-related allelopathy in nature because the intricacies of macro- and micro-climate are too complex to be resolved under natural condi-tions.

Balakumar et al. [80] found that the concentration of total phenolics increases by 19% with hydric stress, by 58% with UV irradiation, and by 63% with the combination of both stresses. While summarizing their research on *Cistus ladnifer*, Chaves and Escudero [59] suggested that climatic factors are responsible for quantitative (resulting from UV irradiation and hydric stress) and qualitative (as in case of hydric and temperature stress) variation in the flavonoid synthesis. More research is needed to find out the individual and interactive effects of abiotic stress or the expression of allelochemicals. Chaves and Escudero [59], for example, reported that higher temperature induces the activity of the enzyme 7-methyltransferase, which results in higher accumula-tion of flavonoids, methylated at the same position in the *C. ladnifer* exudate.

Gross [81] discussed the significance of joint interaction of several abiotic stress factors (e.g., low levels of nutrients and light) in interference success of milfoil (*Myriophyllum spicatum*) in the aquatic ecosystem (Fig. 2). Milfoil is reported to release several polyphenols, and the main allelochemical reported is tellimagrandin II [82]. Gross et al. [83] studied the effect of tellimagrandin II on alkaline phosphatase activity (APA). It was found that milfoil polyphenols released in the medium inhibit APA of selected cyanobacteria, chlorophytes, diatoms, and some epiphytes. Milfoil, however, derives its phosphorus from sediments and rarely has P-limited conditions [81]. Milfoil has high biomass during mid-summer when phytoplankton in freshwater lakes have P-limited conditions. Prolific growth of milfoil in P- and N-limited conditions leads to higher production of polyphenols, which causes allelopathic suppression of epiphytes and phytoplanktons. On the other hand, higher eutrophication leads to luxuriant growth of epiphytes, and water surface is covered with filamentous algae. Reduced light results in reduced production of allelochemicals in milfoil, resulting in prolific growth of epiphytes and phytoplanktons (Fig. 2). Nutrient (N and P) deficiency and light regime, therefore, control the amount of polyphenols (particularly tellimagrandin II) released by the milfoil and thus its allelopathic interference to epiphytes and phytoplanktons. Any response in qualitative and/or quantitative variation in allelochemicals with these stress factors may not be simple and linear. For example, N-limited conditions result in higher content of tellimagrandin II in milfoil but do not influence the total

Myriophyllum spicatum in shallow water

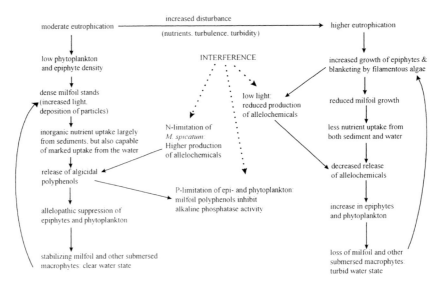

Figure 2. Schematic diagram showing an interaction of allelopathy and resource competition in milfoil stands. Source: Gross [81]. Reproduced with permission from CRC Press.

phenolic content [81]. Thus, allelochemical response changes depend upon the environmental conditions, and it becomes more important to find out which allelochemicals are actually playing a vital role in the natural ecosystem.

What are methodological impasses in studying allelochemical functioning in relation to abiotic stress factors?

While defining stress, Callow [84] opined that stress may be level- and sub-ject-dependent. Environmental stresses may have an impact at the cellular or molecular level, which may not be expressed at the organism or population level because of homeostasis. Continuous stress may result in developing tol-erance, and the tolerant genotypes in a population may not experience stress. Another problem associated with methodology is the selection of allelochem-icals. It is a difficult task to decide about the class of organic compound like-ly to respond to stress factors. It may be possible that two different groups of organic compounds respond to two different types of abiotic stress factors (Fig. 1). A strong collaboration between plant physiologist and natural product chemist is desired. Another riddle is to know whether the enhanced root exu-dation from a plant under stress has implications at the population or ecosys-tem level. Enhanced root exudation may influence soil microbial ecology, mycorrhizae, and nutrient dynamics, and any effect on the ecosystem factor may ultimately have an effect on the community structure [13]. In most of the stress-related studies, the effect on a plant species (i.e., population level) is investigated, while ecosystem components remain neglected.

The selection of an appropriate mutant for a particular allelochemical can help in better understanding the mechanism of stress-induced interference in secondary metabolism. Because of appreciable advancement in the area of molecular and biotechnological tools, it is now possible to locate a specific gene that is responsible for the production of a particular allelochemical [85]. Much of these riddles can be solved by using mutants of some plants with identified and explored allelochemicals, e.g., barley or wheat, or by exploring this possibility in *Arabidopsis* . The mutants lacking or overproducing a spe-cific secondary metabolite involved in allelopathy can be developed and stud-ied for their interaction with a particular abiotic stress. This would provide an insight about the control points in the biosynthetic pathways of allelochemi-cals that respond to specific stress factors. Phenylalanine ammonia lyase (PAL) could be targeted for this purpose. Being a key enzyme of shikimate pathway, its manipulation (repression, derepression, or overexpression) could provide answers to some of the questions about production and functioning of allelochemicals in relation to stress factors. The use of such mutants or exper-iments involving radiolabeled compounds can provide a better insight into the functioning of secondary metabolism under stress conditions. The suggested molecular methods, however, are not without pitfalls. Allelopathic activities result from joint action of several compounds rather than one compound [19].

Jeffrey Weidenhamer (personal communication) raised the following questions: (1) Can we really be certain that a mutant that lacks an enzyme produces a particular allelochemical? (2) Can we really be certain that it is only the amount of that particular allelochemical that is impacted and that this controls the outcome of the experiment? (3) What about the amount of other metabolites that are altered because of the absence of that enzyme? It therefore becomes important to first identify the allelochemicals actually controlling allelopathic expression of a particular plant.

Kleiner et al. [86] have provided evidence for turnover of secondary metabolism using ^{14}C in *Populus tremuloides*. By using this approach, these workers were able to assign the amount of photosynthate into different functions such as growth, maintenance, and defense. Radiotracer studies also have provided information in the past on the relative metabolic costs, turnover, and proportion of photosynthates partitioned to allelochemicals [87–89]. Such studies are indispensable in studying the production of secondary substances, their release and fate in soil, their uptake by target plants and their metabolism thereafter to assure their functioning as allelochemicals. Labeled studies can be helpful in finding a particular pathway of biosynthesis of secondary substances produced and released in response to a specific stress. At present, sophisticated methods, such as ^{14}C-labeled radiotracer technique [90], are available to study transport and fate of allelochemicals in soil. Microbial degradation often limits the studies of the fate of allelochemicals. To trace different pathways during degradation, labeled compounds can be used. These, however, should be ring-labeled compounds, not side-chain labeled. Side chains are easily lost because of microbial activity. Synthesis of ring-labeled compounds, however, is very expensive (Professor Udo Blum, personal communication).

Generally the effect of a single compound is investigated as a response to a stress factor. In nature, however, allelopathic interactions are explained by more than one allelochemical. Equimolar concentrations should not be used to study the joint action of allelochemicals. Reference models, e.g., the Additive Dose Model (ADM) and the Multiplicative Survival Model (MSM), can be used to study the joint action of allelochemicals. When two or more allelochemicals have similar modes of action in the receiver plant, the ADM should be used. In the event that the allelochemicals exert their effects independently of each other, the MSM should be employed [91]. Joint action of allelochemicals is either additive or partially antagonistic. In the event that the number of open membrane sites is higher than the total number of molecules surrounding a root membrane, the effects are additive for the mixture, e.g., at low concentrations. On the other hand, if the number of open membrane sites is lower than the number of molecules, the more competitive molecule will capture more sites and the second molecule will capture fewer sites, resulting in antagonistic activities [92]. Because phenolic acids have essentially similar sites of action, their joint action could never be synergistic. In stress-related studies, different types of molecules released should be assessed. After knowing their mechanism of action, an appropriate model should be used to study their phytotoxicity.

The significance of studies exploring the relative importance of allelo-chemicals and direct interference of abiotic factors in growth reduction should also be realized. For example, hairy vetch is a winter annual crop in a no-till cropping system. The potential of hairy vetch to suppress weeds has been argued [93]; however, the possibility of allelopathic interference was ruled out [68]. Teasdale [94] investigated the relative importance of light and allelopa-thy on the establishment of certain weed species (common lamb's-quarters, *Chenopodium album*; Green foxtail, *Setaria viridis*; velvetleaf, *Abutilon theophrasti*; and corn) in soil having surface residues of hairy vetch. Under full light, hairy vetch residues (0–616 g m^{-2}) had no effect on any of the species except common lamb's-quarter, where significant reduction in its establish-ment was observed. However, under shade (9% light transmittance), hairy vetch residues inhibited velvetleaf, green foxtail, and common lamb's-quar-ters. Approximately similar levels of light were maintained with shade cloth and under hairy vetch residue (308 g m^{-2}). Interestingly, there were no signif-icant differences in percent establishment of selected weed species in both sit-uations (Tab. 1). This strongly rules out any possibility of allelopathy and sig-nifies the importance of light in the weed-suppressing potential of hairy vetch. This study emphasizes the significance of abiotic factors such as temperature and light in experiments on plant allelochemical interference. While designing an experiment on stress-induced allelochemicals, such control studies should be carried out to find out the direct interference of abiotic factors as compared to their effect through enhanced levels of allelochemicals.

Table 1. Comparison of weed establishment under shade cloth *versus* hairy vetch residue at similar light levels[a]

Species	Weed establishment	
	Hairy vetch residue[b]	Shade cloth[c]
	%	
Abutilon theophrasti	73.6	72.6
Setaria viridis	66.3	57.9
Chenopodium album	14.6	11.2

[a] No pair of values in the same row were significantly different according to an LSD test (P = 0.05).
[b] Treatment without shade cloth and with 308 gm^{-2} of hairy vetch residue. Light transmittance under residue averaged 8%.
[c] Treatment with shade cloth and without hairy vetch residue. Light transmittance under shade cloth averaged 9%.
Source: Teasdale [94]. Reproduced with the permission of Weed Science Society of America (WSSA).

How can the importance of a specific stress factor be determined in a given situation or environment?

The significance of a particular stress in a given habitat can be evaluated by studying its ecology. For example, nutrient-related stresses are important in habitats dominated by ericaceous shrubs such as *Kalmia angustifolia, Ledum groenlandicum, and Calluna vulgaris*. Other kinds of abiotic stress factors important in such habitat are clearcutting and fire in the boreal forests [76]. Increasing soil acidity and iron pan formation is a result of long-term occupancy of *Kalmia* in boreal forests. Water stress may not be an important stress in such ecosystems, but it is the main abiotic stress in arid and semiarid regions. In aquatic environments, nutrients, light, and temperature are important abiotic stress factors. It is important to find out whether a plant actually experiences a particular kind of abiotic stress in a natural system. Studying any stress under controlled conditions will generate some response, but these data might not be meaningful in terms of explaining field situations.

Can abiotic stresses predispose target plants to allelochemicals?

Sorghum (*Sorghum* spp.) and soybean (*Glycine max*) were observed to show more damage to ferulic acid application while growing at supraoptimal temperatures [95]. Likewise, soybean seedlings grown at a temperature of 34°C were inhibited more with 100 μM ferulic acid than those grown at 23°C, even though the temperature difference itself did not affect the growth. Einhellig [22] observed that water stress of any degree lowered the ferulic-acid-inhibition threshold on germination and seedling growth of sorghum. Residues of redroot pigweed and yellow foxtail caused more inhibition to corn growing under low temperature and low photosynthetic photon flux density conditions. Several promoters (e.g., nitrate), inhibitors (e.g., phenolic acids), and neutral substances (e.g., glucose) interact in the soil environment to influence phytotoxicity [44]. Non-inhibitory levels of glucose enhanced the inhibitory activity of *p*-coumaric acid on morning glory (*Ipomoea hederacea*) seedling biomass [45]. Phenolic acids are often related to stress studies. It is therefore important to investigate the performance of stress-released allelochemicals in relation to neutral substances and promoters present in the substratum.

Conclusion and prospects

Qualitative and quantitative variation in allelochemical concentration resulting from several abiotic stresses is convincingly demonstrated. The response of allelochemical(s) to stress factors depends upon (1) type of abiotic stress, (2) substratum factors, and (3) kind of allelochemicals under investigation. Furthermore, the production of allelochemicals in the plant under stress and their subsequent release in the environment may differ with different kinds of abiotic stresses. The question now is, why would a plant spend energy to synthesize secondary metabolites and release them into the environment at the cost of primary metabolism when it is under stress? Stress hampers primary metabolism and initiates the cycle of secondary metabolism. Wolfe et al. [96]

stated, "Plants that manufacture constitutive defense toxicants are at risk of self-toxicity and also use metabolic energy producing these compounds that may not be needed." At present we do not have any experimental evidence, but we strongly believe that plants would not waste energy to produce secondary metabolites that have no physiological or ecological function. Recently, Kleiner et al. [86] reported, that when using ^{14}C, plants contribute a significant amount of their photosynthate into production of defense molecules like phenolic glycosides and tannins. Analysis of phenolic fractions showed that total phenolics were 2 times and condensed tannins were 1.7 times greater in sink than in source leaves. It was also evident from their study that secondary metabolites may not be synthesized afresh from photosynthesis but could come from stored molecules and get partitioned into different organs under stress. Another question that arises is, how do stressed plants avoid self-toxicity from enhanced levels of secondary metabolites? Stressed plants may have a mechanism for storing toxic metabolites in different forms. For example, the marine alga *Emiliania huxleyi* produces acrylate and avoids self-toxicity of acrylate by storing it as dimethylsulfoniopropionate.

As is evident from the above discussion, stress-induced production and release (quantitative and/or qualitative) of allelochemicals depend on the climatic and environmental factors. Extreme seasonal or environmental conditions (e.g., temperature, water stress, toxic levels of certain inorganic ions) can damage cell membranes ([18] pp 50–82) and organelles, particularly vacuoles, resulting in the leakage of secondary metabolites. There are different biochemical routes to trigger the production of secondary substances, and to pinpoint a specific route in response to different stresses is challenging. Whether a plant under stress produces and releases secondary metabolites as part of its defense strategy or whether it is due to leakage/injury at the cellular level is difficult to answer. It is also extremely difficult, if not impossible, to predict the behavior of stressed-induced allelochemicals under the dynamic soil environment.

The release of secondary substances implicated as allelochemicals may serve another additional purpose under stress conditions. For example, in addition to herbivore defense, enhanced release of allelochemicals may also influence soil nitrogen status through their influence on nodule formation in leguminous plants. The potential role of root exudates in stimulating nodule formation is well known [97]. Flavonoids may be an important component of root exudates [51, 98, 99] and have a significant role in legume-*Rhizobium* symbiosis [100]. When inoculated with *Rhizobium* spp., root exudate has been shown to release various flavonoids, which initiate nodule formation in N-fixing legumes [101, 102]. It would be interesting to find the correlation, if any, among stress-induced qualitative and quantitative variation in flavonoid content, nodule formation abilities of legume plant, and soil nitrogen content. Brassinosteroids, implicated as allelochemicals, are reported to have antistress effects on plants [103]. *Arabidopsis* mutants for brassinosteroids biosynthesis corroborate their role as regulators of stress-related genes [104].

Although there are several uncertainties yet to be resolved in stress-related allelochemical research, modern tools in plant physiology and molecular biology can be of help to resolve some of the methodological impasses.

Acknowledgements
We sincerely thank Professors Erik T. Nilsen, David Orcutt, and Jeff Weidenhamer for their constructive comments to improve the quality of the manuscript. The help of Miss Harleen Kaur and Monika Saini in proofreading the manuscript is gratefully acknowledged.

References

1 Berenbaum MR (1995) Turnabout is fair play: secondary roles for primary compounds. *J Chem Ecol* 21: 925–940
2 Seigler DS (1977) Primary roles for secondary compounds. *Biochem Syst Ecol* 5: 195–199
3 Rosenthal GA, Bell EA (1979) Naturally occurring, toxic non-protein amino acids, *In*: GA Rosenthal, DH Janzen (eds): *Herbivores, their interaction with secondary plant metabolites.* Academic Press, New York, 353–385
4 Connell JH (1990) Apparent *versus* "real" competition in plants. *In*: JB Grace, D Tilman (eds): *Perspectives in plant competition.* Academic Press, New York, 9–26
5 Williamson GB (1990) Allelopathy, Koch's postulates and the neck riddle. *In*: JB Grace, D Tilman (eds): *Perspectives in plant competition.* Academic Press, New York, 143–162
6 Inderjit, Dakshini KMM (1995) On laboratory bioassays in allelopathy. *Bot Rev* 61: 28–44
7 Weidenhamer JD (1996) Distinguishing resource competition and chemical interference: overcoming the methodological impasse. *Agron J* 88: 866–875
8 Romeo JT, Weidenhamer JD (1998) Bioassays for allelopathy in terrestrial plants. *In*: KF Haynes, JG Millar (eds): *Methods in chemical ecology. Vol. 2. Bioassay methods.* Kluwer Academic Publishing, Norvell, 179–211
9 Inderjit, Del Moral R (1997) Is separating resource competition from allelopathy realistic? *Bot Rev* 63: 221–230
10 Weidenhamer JD, Hartnett DC, Romeo JT (1989) Density-dependent phytotoxicity: distinguishing resource competition and allelopathic interference in plants. *J Appl Ecol* 26: 613–624
11 Nilsson M-C (1994) Separation of allelopathy and resource competition by the boreal dwarf shrub *Empetrum hermaphroditum* Hagerup. *Oecologia* 98: 1–7
12 Inderjit, Weiner J (2001) Plant allelochemical interference or soil chemical ecology? *Perspect Plant Ecol Evol Syst* 4: 3–12
13 Inderjit (2001) Soils: environmental effect on allelochemical activity. *Agron J* 93: 79–84
14 Dakshini KMM, Foy CL, Inderjit (1999) Allelopathy: one component in a multifaceted approach in ecology. *In*: Inderjit, KMM Dakhini, CL Foy (eds): *Principles and practices in plant ecology: allelochemical interactions.* CRC Press, Boca Raton, 3–14
15 Huang PM, Wang MC, Wang MK (1999) Catalytic transformation of phenolic compounds in soils. *In*: Inderjit, KMM Dakhini, CL Foy (eds): *Principles and practices in plant ecology: allelochemical interactions.* CRC Press, Boca Raton, 287–306
16 Inderjit, Dakshini KMM (1999) Bioassay for allelopathy: interactions of soil organic and inorganic constituents. *In*: Inderjit, KMM Dakhini, CL Foy (eds): *Principles and practices in plant ecology: allelochemical interactions.* CRC Press, Boca Raton, 35–44
17 Northup RR, Dahlgren RA, Aide TM, Zimmerman JK (1999) Effect of plant polyphenols on nutrient cycling and implications for community structure. *In*: Inderjit, KMM Dakhini, CL Foy (eds): *Principles and practices in plant ecology: allelochemical interactions.* CRC Press, Boca Raton, 369–380
18 Nilsen ET, Orcutt DM (1996) *Physiology of plant under stress: abiotic factors.* John Wiley, New York
19 Grime JP (1979) *Plant strategies and vegetation processes.* John Wiley, Chichester
20 Einhellig FA (1999) An integrated view of allelochemicals amid multiple stresses. *In*: Inderjit, KMM Dakhini, CL Foy (eds): *Principles and practices in plant ecology: allelochemical interac-*

tions. CRC Press, Boca Raton, 479–494

21 Einhellig FA (1995) Allelopathy: current status and future goals. *In*: Inderjit, KMM Dakshini, FA Einhellig (eds): *Allelopathy: organisms, processes and applications.* American Chemical Society, Washington, DC, 1–24

22 Einhellig FA (1989) Interactive effects of allelochemicals and environmental stress. *In*: CH Chou, GR Waller (eds): *Phytochemical ecology: allelochemicals, mycotoxins, and insect pheromones and allomones.* Institute of Botany, Academia Sinica Monograph Series No. 9, Taipei, ROC, 101–116

23 Tang CS, Cai WF, Kohl K, Nishimoto RK (1995) Plant stress and allelopathy. *In*: Inderjit, KMM Dakshini, CL Foy (eds): *Allelopathy: organisms, processes, and applications.* ACS Symposium Series 582, American Chemical Society, Washington, DC, 142–157

24 Curl EA, Truelove B (1986) *The rhizosphere.* Springer-Verlag, Berlin

25 Gershenzon J (1984) Changes in the levels of plant secondary metabolites under water stress and nutrient stress. *Recent Adv Phytochem* 18: 273–320

26 Koeppe DE, Rohrbaugh LM, Wender SH (1976) The relationship of varying U.V. intensities on the concentrations and leaching of phenolics from sunflowers grown under varying phosphate nutrient conditions. *Can J Bot* 54: 593–599

27 Sterling TM, Houtz RL, Putnam AR (1987) Phytotoxic exudates from velvetleaf (*Abutilon theophrasti*) glandular trichomes. *Am J Bot* 74: 543–550

28 Ardi (1986) Interference between sweet corn (*Zea mays* L.) and purple nutsedge (*Cyperus rotundus* L.) at different irrigation levels. Masters thesis, University of Hawaii, Honolulu

29 Inderjit, Foy CL, Dakshini KMM (1998) *Pluchea lanceolata*: a noxious weed. *Weed Technol* 12: 190–193

30 Inderjit (1998) Influence of *Pluchea lanceolata* on selected soil properties. *Am J Bot* 85: 64–69

31 Inderjit, Dakshini KMM (1990) The nature of interference potential of *Pluchea lanceolata* (DC) C. B, Clarke (Asteraceae). *Plant Soil* 122: 298–302

32 Inderjit, Dakshini KMM (1992) Interference potential of *Pluchea lanceolata* (Asteraceae): growth and physiological responses of asparagus bean, *Vigna unguiculata* var. sesquipedalis. *Am J Bot* 79: 977–981

33 Inderjit, Dakshini KMM (1994) Allelopathic effects of *Pluchea lanceolata* (Asteraceae) on characteristics of four soils and growth of mustard and tomato. *Am J Bot* 81: 799–804

34 Inderjit, Dakshini KMM (1994) Allelopathic potential of phenolics from the roots of *Pluchea lanceolata*. *Physiol Plant* 92: 571–576

35 Inderjit, Dakshini KMM (1996) Allelopathic potential of *Pluchea lanceolata*: comparative study of cultivated fields. *Weed Sci* 44: 393–396

36 Inderjit, Dakshini KMM (1995) Quercetin and quercitrin from *Pluchea lanceolata* and their effects on growth of asparagus bean. *In*: Inderjit, KMM Dakshini, FA Einhellig (eds): *Allelopathy: organisms, processes and applications.* ACS Symposium Series 582. American Chemical Society. Washington, DC, 86–95

37 Inderjit, Gross EM (2001) Plant phenolics: potential role in aquatic and terrestrial ecosystems. *In*: S Martens, D Treutter, G Forkmann (eds): *Polyphenol 2000.* INRA, Germany, 206–234

38 Tsai SD, Todd GW (1972) Phenolic compounds of wheat leaves under drought stress. *Phyton* 30: 67–75

39 Gilmore AR (1977) Effects of soil moisture stress on monoterpenes in loblolly pine. *J Chem Ecol* 3: 667–676

40 Hopkins WG (1999) Introduction to plant physiology. John Wiley and Sons, New York

41 Blum U, Shafer SR, Lehman ME (1999) Evidence for inhibitory allelopathic interactions involving phenolic acids in field soils: concepts *versus* experimental model. *Crit Rev Plant Sci* 18: 673–693

42 Schmidt SK, Ley RE (1999) Microbial competition and soil structure limit the expression of phytochemicals in nature. *In*: Inderjit, KMM Dakhini, CL Foy (eds): *Principles and practices in plant ecology: allelochemical interactions.* CRC Press, Boca Raton, 339–351

43 Stark JM, Hart SC (1997) High rates of nitrification and nitrate turnover in undisturbed coniferous forests. *Nature* 385: 61–64

44 Blum U, Gerig TM, Worsham AD, Holappa LD, King LD (1993) Modification of allelopathic effects of p-coumaric acid on morning glory seedling biomass by glucose, methionine, and nitrate. *J Chem Ecol* 19: 2791–2811

45 Pue KJ, Blum U, Gerig TM, Shafer SR (1995) Mechanism by which noninhibitory concentrations

of glucose increase inhibitory activity of p-coumaric acid on morning-glory seedling biomass accumulation. *J Chem Ecol* 21: 833–847

46 Blum U (1998) Effects of microbial utilization of phenolic acids and their phenolic acid breakdown products on allelopathic interactions. *J Chem Ecol* 24: 685–708

47 Appel HM (1993) Phenolics in ecological interactions: the importance of oxidation. *J Chem Ecol* 19: 1521–1552

48 Kafkafi U, Bar-Yosef B, Rosenberg R, Sposito G (1988) Phosphorus adsorption by kaolinite and montmorillonite: II. Organic anion competition. *Soil Sci Soc Am J* 52: 1585–1589

49 Tan K, Binger A (1986) Effect of humic acid on aluminum toxicity in corn plants. *Soil Sci* 141: 20–25

50 Hättenschwiler S, Vitousek PM (2000) The role of polyphenols in terrestrial ecosystem nutrient cycling. *Trends Ecol Evol* 15: 238–243

51 Inderjit (1996) Plant phenolics in allelopathy. *Bot Rev* 62: 186–202

52 Sanders FE, Tinker PB (1973) Phosphate flow into mycorrhizal roots. *Pestic Sci* 4: 385–395

53 Inderjit, Mallik AU (1996) The nature of interference potential of *Kalmia angustifolia*. *Can J For Res* 26: 1899–1904

54 Inderjit, Mallik AU (1997) Effect of phenolic compounds on selected soil properties. *For Ecol Manage* 92: 11–18

55 Rice EL, Pancholy SK (1973) Inhibition of nitrification by climax ecosystems. II. Additional evidence and possible role of tannins. *Am J Bot* 60: 691–702

56 Rice EL, Pancholy SK (1974) Inhibition of nitrification by climax ecosystems. III. Inhibitors other tannins. *Am J Bot* 61: 1095–1103

57 Bremner JM, McCarty GW (1988) Effects of terpenoids on nitrification in soil. *Soil Sci Soc Am J* 52: 1630–1633

58 Bremner JM, McCarty GW (1990) Reply to "Comments on "Effects of terpenoids on nitrification in soil." *Soil Sci Soc Am J* 54: 297–298

59 Chaves N, Escudero JC (1999) Variation of flavonoid synthesis induced by ecological factors. *In*: Inderjit, KMM Dakhini, CL Foy (eds): *Principles and practices in plant ecology: allelochemical interactions*. CRC Press, Boca Raton, 267–285

60 Pacheco P, Crawford DJ, Stuessy TF, Silva M (1985) Flavonoid evolution in *Robinsonia* (Compositae) of Juan Fernandez islands. *Am J Bot* 72: 989–998

61 Chaves N, Escudero jc Gutierrez-Merino C (1993) Seasonal variation of exudate of *Cistus* ladanifer. *J Chem Ecol* 19: 2577–2591

62 Inderjit, Cheng HH, Nishimura H (1999) Plant phenolics and terpenoids: transformation, degradation, and potential for allelopathic interactions. *In*: Inderjit, KMM Dakhini, CL Foy (eds): *Principles and practices in plant ecology: allelochemical interactions*. CRC Press, Boca Raton, 255–266

63 Fischer NH, Williamson GB, Weidenhamer JD, Richardson DR (1994) In search of allelopathy in Florida scrub: the role of allelopathy. *J Chem Ecol* 20: 1355–1380

64 Tanrisever N, Fronczek FR, Fischer NH, Williamson GB (1987) Ceratiolin and other flavonoids from *Ceratiola ericoides*. *Phytochemistry* 26: 175–179

65 Mallik AU (1994) Autecological response of *Kalmia angustifolia* to forest types and disturbance regimes. *For Ecol Manage* 65: 231–249

66 Inderjit, Mallik AU (1999) Nutrient status of black spruce (*Picea mariana*) forest soils dominated by *Kalmia angustifolia*. *Acta Oeco* 20: 87–92

67 Harper JL (1977) *Population biology of plants*. Academic Press, London

68 Inderjit, Asakawa C (2001) Nature of interference potential of hairy vetch (*Vicia villosa* Roth) to radish (*Raphanus sativus* L.): does allelopathy play any role? *Crop Prot* 20: 261–265

69 Fisher R, Woods RA, Glavicic MR (1978) Allelopathic effects of goldenrod and aster on young sugar maple. *Can J For Res* 8: 1–9

70 Inderjit, Foy CL (1999) Nature of the interference potential of mugwort (*Artemisia vulgaris*). *Weed Technol* 13: 176–182

71 Gianoli E, Niemeyer HM (1997) Environmental effects on the accumulation of hydroxamic acids in wheat seedlings: the importance of plant growth rate. *J Chem Ecol* 23: 543–551

72 Corcuera LJ (1993) Biochemical basis of the resistance of the barley to aphids. *Phytochemistry* 33: 741–747

73 Hanson AD, Ditz KM, Singletray GW, Leland TJ (1983) Gramine accumulation in leaves of barley grown under high temperature stress. *Plant Physiol* 71: 896–904

74 Lovett JV, Hoult AH, Christen O (1994) Biological active secondary metabolites of barley. IV. Hordenine production by different barley lines. *J Chem Ecol* 20: 1945–1954

75 Bhowmik PC, Doll JD (1983) Growth analysis of corn and soybean response to allelopathic effects of weed residues at various temperatures and photosynthetic photon flux densities. *J Chem Ecol* 9: 1263–1280

76 Mallik AU (1995) Conversion of temperate forests into heaths: role of ecosystem disturbance and regeneration strategies of three ericaceous plants. *Environ Manage* 19: 675–684

77 Gibson LR, Paulsen GM (1999) Yield components of wheat grown under high temperature stress during reproductive growth. *Crop Sci* 39: 1841–1846

78 O'Toole JC, Chang TT (1979) Drought resistance in cereals – rice: a case study. *In*: H Mussell, RC Staples (eds): *Stress physiology of crop plants*. John Wiley, New York, 374–405

79 Castillo J, Benavente O, del Rio JA (1992) Naringin and neohesperidin levels during development of leaves, flower buds and fruits of *Citrus aurantium*. *Plant Physiol* 99: 67–73

80 Balakumar T, Vincent VH, Paliwal K (1993) On the interaction of UV-B radiation (280–315 nm) with water stress in crop plants. *Physiol Plant* 87: 217–222

81 Gross E (1999) Allelopathy in benthic and littoral areas: case studies on allelochemicals from benthic cyanobacteria and submerged macrophytes. *In*: Inderjit, KMM Dakhini, CL Foy (eds): *Principles and practices in plant ecology: allelochemical interactions*. CRC Press, Boca Raton, 179–199

82 Gross EM, Wolk CP, Juttner F (1991) Fischerellin, a new allelochemical from the freshwater cyanobacteria *Fischerella muscicola*. *J Phycol* 27: 686–692

83 Gross EM, Meyer H, Schilling G (1996) Release and ecological impact of algicidal hydrolyzable polyphenols in *Myriophyllum spicatum*. *Phytochemistry* 41: 133–138

84 Callow P (1999) *Blackwell's concise encyclopedia of ecology*. Blackwell, Oxford

85 Duke SO, Scheffler BE, Dayan FE (2001) Allelochemicals as herbicides. *In*: NP Bonjoch, MJ Reigosa (eds): *First European OECD allelopathy symposium: physiological aspects of allelopathy*. GAMESAL, SA, Vigo, 47–59

86 Kleiner KW, Raffa KF, Dickson RE (1999) Partitioning of ^{14}C-labeled photosynthate to allelochemicals and primary metabolites in source and sink leaves of aspen: evidence for secondary metabolites turnover. *Oecologia* 119: 408–418

87 Mihaliak CA, Gerhenzon J, Croteau R (1991) Lack of rapid nonterpene turnover in rooted plants: implication for theories of plant chemical defense. *Oecologia* 87: 373–376

88 Mooney HA, Chu C (1974) Seasonal carbon allocation in *Heteromeles arbutifolia*, a California evergreen shrub. *Oecologia* 14: 295–306

89 Gershenzon J (1994) Metabolic costs of terpenoids accumulation in higher plants. *J Chem Ecol* 20: 1281–1328

90 Cheng HH (1995) Characterization of the mechanisms of allelopathy: modeling and experimental approaches. *In*: Inderjit, KMM Dakshini, FA Einhellig (eds): *Allelopathy: organisms, processes and applications*. American Chemical Society, Washington, DC, 132–141

91 Morse PM (1978) Some comments on the assessment of joint action in herbicide mixtures. *Weed Sci* 26: 58–71

92 Inderjit, Streibig JC, Olofsdotter M (2002) Joint action of phenolic acid mixtures and its significance in allelopathy research. *Physiol Plant* 114: 422–428

93 Abdul-Baki AA, Teasdale JR (1997) *Sustainable production of fresh-market tomatoes and other summer vegetable with organic mulches*. U.S. Department of Agriculture, Agricultural Research Service, Farmer's Bulletin No. 2279: 3–23

94 Teasdale JR (1993) Interaction of light, soil moisture, and temperature with weed suppression by hairy vetch residue. *Weed Sci* 41: 46–51

95 Einhellig FA, Eckrich PC (1984) Interactions of temperature and ferulic acid stress on grain sorghum and soybean. *J Chem Ecol* 10: 161–170

96 Wolfe GV, Steinke MS, Kirst GO (1997) Grazing-activated chemical defense in a unicellular marine alga. *Nature* 387: 894–897

97 Long SR (1989) *Rhizobium*-legume nodulation: life together in the underground. *Cell* 56: 203–214

98 Harborne JB (1988) *Flavonoids: advances and research*. Chapman and Hall, London

99 Harborne JB (1994) *Flavonoids: advances in research since 1986*. Chapman and Hall, London

100 Flores HE, Weber C, Puffett J (1996) Underground plant metabolism: the biosynthetic potential of roots. *In*: Y Waisel, A Eshel, U Kafkafi (eds): *Plant roots: the hidden half, Edn II*. Marcel

Dekker, New York, 931–956

101 Maxwell CA, Hartwig UA, Joseph CM, Phillips DA (1989) A chalcone and two related flavonoids released from alfalfa roots induce nod genes of *Rhizobium meliloti*. *Plant Physiol* 91: 842–847

102 Scheidemann P, Wetzel A (1997) Identification and characterization of flavonoids in the root exudate of *Robinia pseudoacaia*. *Trees* 11: 316–321

103 Dhaubhadel S, Chaudhary S, Dobinson KF, Priti K (1999) Treatment with 24-epibrasinolide, a brassinosteroid, increases the basic thermo tolerance of *Brassica napus* and tomato seedlings. *Plant Mol Biol* 40: 333–342

104 Szekers M, Koncz C (1998) Biochemical and genetical analysis of brassinosteroids metabolism and function in *Arabidopsis*. *Plant Physiol Biochem* 36: 145–155

Pitfalls in interpretation of allelochemical data in ecological studies: implications for plant-herbivore and allelopathic research

Julia Koricheva[1] and Anna Shevtsova[2]

[1] Section of Ecology, Department of Biology, University of Turku, FIN-20014 Turku, Finland,
[2] Department of Forest Vegetation Ecology, Swedish University of Agricultural Sciences, S-90183 Umeå, Sweden

Introduction

Until the middle of the twentieth century, plant secondary compounds were considered to be metabolic waste products and were largely ignored by ecologists. Interest in these compounds exploded in the 1960s and 1970s, when the importance of secondary metabolites in plant interactions with the biotic and abiotic environment was recognized [1–3], and the term "allelochemic" was coined for compounds with primarily ecological (as opposed to physiological) functions [3]. Attention was directed in particular to the role of secondary metabolites in plant interactions with herbivores [4] and with other plant species (allelopathy [5]) and in the regulation of decomposition and nutrient cycling [6–8]. Nowadays, allelochemical assays of plants or plant parts form an integral part of ecological studies, and standard analytical procedures for quantifying major classes of allelochemicals and identifying individual compounds have been developed (e.g., [9]).

Ecological studies of plant allelochemicals pursue two general aims: (1) examination of the mechanisms behind variation in plant allelochemical composition (genetic, ontogenetic, and environmental variation; changes in resource acquisition, partitioning, and allocation) and (2) elucidation of the possible consequences of this variation for other organisms (herbivores, pathogens, other plants) and ecosystem processes (e.g., decomposition and nutrient cycling). Depending on the purpose of the study, the amount of a particular compound present in plant tissues can be expressed either in absolute terms, i.e., as content per leaf, shoot, or plant, or in relative terms, i.e., as a concentration[1]. In gener-

---

[1] The terms "concentration" and "content" are sometimes used interchangeably in the ecological literature, causing considerable confusion [10]. In this chapter, we follow the terminology suggested by the editors of Oecologia (http://www.ecophys.biology.utah.edu/Oecologia/suggestions.html) and define content as the total amount of a compound present in a specific amount of plant tissues (e.g., mg per leaf, shoot, whole plant) and concentration as the amount of a compound present in a unit amount of plant tissue (e.g., mg/g or % w/w). Note that in chemistry the term "concentration" indicates how much solute is contained in a given volume or mass of solution or solvent and is usually expressed as molarity, normality, or titer (e.g., [11]).

al, the use of absolute values is recommended when the mechanisms behind these changes are under investigation, whereas relative (standardized) values may be more useful for questions concerning the consequences of changes in response variables [12]. Because ecologists are interested in both the mechanisms and the consequences of variation in plant allelochemical composition, concentrations and contents might be expected to be equally common as forms of allelochemical data presentation in ecological studies. For instance, concentrations are more relevant estimates of plant quality for herbivores, since the latter usually consume only a fraction of the total plant biomass and are therefore likely to be affected by the amount of defensive compounds present in each gram of biomass eaten (i.e., concentration) rather than by the total content of defensive compounds per plant or plant part. On the other hand, the absolute content of allelochemicals per plant or plant part should be the variable of primary interest in plant allocation studies because plants allocate and produce molecules (quantities) rather than concentrations of allelochemicals [13, 14]. In practice, however, the majority of ecological studies presents the results of allelochemical analyses as concentrations [14].

Although rarely stated explicitly, there appear to be three main reasons for the preference of concentrations over content in presenting and analyzing allelochemical data in ecological studies: (1) the tradition in biochemistry to express the results in relative terms per unit of mass or volume, (2) the prevalence of a "consumer perspective" (e.g., the consequences of variation in allelochemical composition for herbivores) over "phytocentric perspective" (causes of variation in plant allelochemical composition) in chemical ecology, and (3) the belief that concentrations adjust for differences in biomass, thus allowing comparisons of patterns among plants or treatments. It is easily forgotten that allelochemical concentrations represent a ratio between the absolute content of a compound present in a certain amount of plant material and the biomass of this material. Variation in allelochemical concentrations may therefore be a consequence of changes in plant growth and biomass accumulation rather than of shifts in allelochemical production and defense allocation. This fact has long been recognized in plant nutrient studies, and the need to take into account changes in plant biomass and total nutrient pools (content) when interpreting changes in nutrient concentrations has been repeatedly emphasized [15, 16]. Similarly, the fallacy of indices such as water percentage and body fat have been discussed by physiologists [17, 18]. In contrast, ecologists working with plant allelochemicals appear to be much less aware of the potential pitfalls in using allelochemical concentrations [14].

In the present chapter, we discuss some common problems in the use of allelochemical concentrations in ecological studies, particularly in those addressing plant-herbivore and allelopathy theories, and ways in which the analysis and interpretation of allelochemical data can be improved to provide a direct answer to the specific questions behind the study. We want to emphasize that our aim is not to convince ecologists to abandon the use of concentrations altogether, but rather to restrict it to the situations when it is appropriate.

Fallacies of concentrations

As a ratio between the content of the compound and plant biomass, concentrations share the statistical drawbacks of ratios—such as widening of the sampling variation and non-normal distribution [19, 20]. These statistical properties of ratios can lead to a loss of sensitivity in statistical tests [21], making biological models framed in terms of ratios untestable [22]. In addition, under some circumstances the use of ratios might actually increase the correlation between the original variables, giving rise to so-called "spurious correlations" [19, 23, 24]. We will not discuss the general statistical shortcomings of ratios in detail, since ample literature exists on the subject [18, 19, 22, 25]. Instead, we focus on the biological implications of these statistical problems, in particular for the interpretation of variation in allelochemical concentrations.

A critical implicit assumption underlying the frequent use of allelochemical concentrations in ecological studies is that concentrations adjust for differences in biomass, thus allowing comparisons among plants, treatments, etc. The use of concentrations to eliminate the effect of plant size is analogous to the use of ratios for scaling data in taxonomy [19], physiology [18], plant reproductive ecology [26, 27], and insect nutritive ecology [28]. Studies in the above fields have shown that ratios eliminate the effect of the denominator variable on the numerator only under two restrictive conditions: (1) when the relationship between numerator and denominator is linear and (2) when the function passes through the origin, i.e., the relationship is isometric [18, 26] (Fig. 1A). Size dependence occurs when either or both of these conditions are violated (i.e., the relationship is curvilinear and/or there is a nonzero intercept).

While the assumptions of isometry have seldom been directly tested in allelochemical studies, there is good reason to believe that allelochemical concentrations often fail to exclude the effects of plant biomass on content and are therefore size-dependent. The evidence comes from three sources: (1) studies on the kinetics of accumulation of allelochemicals and growth in cell cultures, (2) allometric analysis of allelochemical production in intact plants, and (3) correlations between allelochemical concentrations and plant biomass.

Studies of the dynamics of secondary metabolite accumulation in cell cultures (reviewed in [29, 30]) indicate that allelochemical production is often a nonlinear function of growth. Most of the allelochemicals seem to fall into two categories: those produced during the active growth phase and undergoing rapid turnover (Fig. 1B) and those produced during the stationary phase, with less obvious turnover or none (Fig. 1C). Note that in both cases the relationship between allelochemical content and growth is nonlinear, and in the latter case a significant intercept may emerge (Fig. 1C). These product-growth pat-

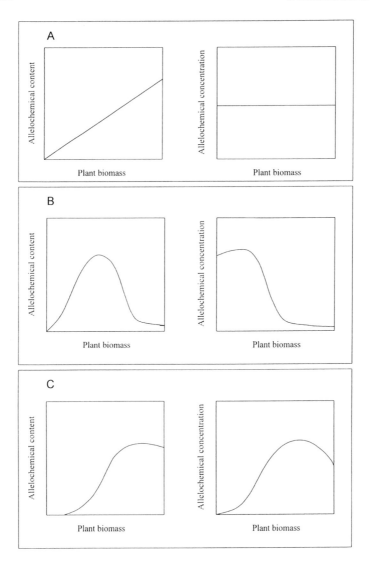

Figure 1. Changes in contents and concentrations of allelochemicals in relation to growth as affected by the timing of synthesis of an allelochemical. (A) synthesis parallels growth; (B) synthesis occurs during the active growth phase and is followed by catabolism and/or translocation of an allelochemical; (C) synthesis takes place during the stationary phase, and no substantial turnover or retranslocation occurs.

terns are also well corroborated by analyses of the time-course of the expression of enzymes responsible for the synthesis of allelochemicals [29, 30].

Evidence that the relationship between allelochemical content and plant biomass is nonlinear also comes from a number of studies of the dynamics of allelochemical production in intact plants. For instance, analysis of the ln-ln regression line between nicotine content and biomass of wild tobacco revealed that slopes differed significantly from 1.0 in both control and damaged plants [31, 32]. Because the slope of a log-log regression line represents the exponent of the untransformed relationship (see e.g., Fig. 2 in [27]), this indicates that the untransformed relationship between nicotine and biomass accumulation in these studies was nonlinear. Again, furanocoumarin production in wild parsnip seedlings was not commensurate with biomass accretion, and divergent rates of biomass and furanocoumarin accumulation resulted in fluctuations in furanocoumarin concentrations [33]. Finally, the relationship between 2,4-dihydroxy-1,4-benzoxazolin-2-one (DIBOA) content and leaf biomass in rye was better described by a hyperbolic or logistic function than by a linear one, and the intercept of the linear regression was significantly different from 0 [13]. Thus, both of the assumptions of isometry—linearity and zero intercept—appear to be violated in the relationship between allelochemical content and plant biomass.

Finally, evidence that concentrations often fail to eliminate the effect of plant biomass on content comes from frequently reported correlations between allelochemical concentrations and plant biomass (Tab. 1). Significant positive or negative correlations for at least some of the compounds measured or treatments applied were found in eight out of nine studies (Tab. 1), indicating that size dependence is common for various types of allelochemicals and various plant species. Negative correlations between allelochemical content and plant biomass are likely to be characteristic of allelochemicals synthesized early in plant ontogeny (Fig. 1B), whereas positive correlations will be obtained for the compounds synthesized once the active growth phase is over (Fig. 1C). Indeed, for alkaloids and some phenolic glucosides, which are mainly synthesized in young tissues [34, 35], most of the reported correlations were negative, while positive correlations often have been reported for condensed tannins and their precursor (+)-catechin, known to be synthesized later in plant development (Tab. 1).

Having demonstrated that allelochemical concentrations are often size-dependent, we will examine the ways in which this fact complicates the interpretation of the results of ecological studies.

Misuses of allelochemical concentrations in ecological studies

Allelochemical concentrations as indicators of changes in allelochemical production

Changes observed in allelochemical concentrations often are interpreted as changes in allelochemical synthesis and are used as evidence in support of hypotheses predicting patterns in allelochemical production. For instance, reduced concentrations of carbon-based secondary compounds (CBSCs) in fertilized or shaded plants are often cited as support for the carbon-nutrient

Table 1. Reported correlations between concentrations of allelochemicals and plant biomass production

Plant species	Allelochemical	Pearson's r (P value)	Reference
Ascophyllum nodosum	Phlorotannins	−0.1 ... −0.77 (0.60) (0.01)	[129]
Betula pendula	salidroside proanthocyanidins (+)-catechin myricitrin quercitrin	−0.31 (<0.01) −0.46 (<0.001) −0.36 (<0.001) −0.27 (<0.01) −0.43 (<0.001)	[87]
Diplacus aurantiacus	phenolic resin	−0.05 (>0.05)	[130]
Lycopersicon esculentum	rutin	0.26 ... 0.65 (>0.05) (<0.01)	[90]
	chlorogenic acid	0.14 ... 0.40 (>0.05) (<0.05)	
Lotus corniculatus	condensed tannins	0.24 ... 0.87 (0.47) (0.01)	[64]
	cyanogenic glycosides	−0.64 ... 0.84 (0.09) (0.04)	
Nicotiana tabacum	nicotine	−0.69 (<0.05)	[131]
Plantago lanceolata	iridoid glycosides	NS ... 0.73 (>0.05) (<0.05)	[132]
Salix myrsinifolia	salicortin salicin proanthocyanidins (+)-catechin	NS (>0.05) −0.58 (0.0001) 0.24 (0.014) 0.41 (0.0001)	[133]
Senecio jacobaea	pyrrolizidine alkaloids	−0.09 ... −0.57 (>0.05)(<0.05)	[88]

balance hypothesis [36], which predicts that the production of CBSCs will be reduced under conditions where carbohydrates are limited and allocated to plant growth. Similarly, increased concentrations of allelochemicals in response to defoliation often are interpreted as an indication of increased allelochemical synthesis, as predicted by the induced defense hypothesis [37, 38]. However, fertilization, defoliation, and other abiotic and biotic factors affect plant biomass accumulation. For instance, fertilization with limiting nutrients usually greatly increases plant biomass, whereas defoliation often causes reduction in leaf size (e.g., [39]). Changes in the concentrations of CBSCs in fertilized or defoliated plants may therefore simply reflect the distribution of the same amount of the allelochemical in a larger or smaller biomass.

In order to discriminate between these different mechanisms underlying changes in allelochemical concentrations, changes in content and plant bio-

Table 2. Interpretation of shifts in allelochemical concentration, content and plant biomass (after [14] with modifications)

Changes in concentration	Changes in content	Changes in biomass	Interpretation
↑	↑↑	↑	Excess synthesis
0	↑/↓	↑/↓	Steady state
↓	0/↑	↑/↑↑	Dilution effect
↓	↓/↓↓	0/↓	Reduced synthesis
↑	0/↓	↓/↓↓	Concentration effect

↑ = increase; ↓ = decrease; 0 = no changes.
Number of arrows indicates the magnitude of change.

mass have to be considered as well. Five possible outcomes can be distinguished ([14]; Tab. 2).

1. An increase in concentration accompanied by corresponding changes in content and biomass indicates that the allelochemical synthesis rate is higher than the biomass accumulation rate (excess synthesis).
2. A situation where the concentration does not change, while both content and biomass either increase or decrease, indicates that allelochemical synthesis rates match changes in biomass accumulation (steady state).
3. A decrease in concentration accompanied by an increase in content and biomass indicates growth dilution.
4. A decrease in concentration together with a decrease in content and a decrease or no change in biomass indicates that allelochemical synthesis is reduced more than biomass accumulation.
5. An increase in concentration accompanied by reduced content and biomass indicates that growth is reduced more than allelochemical synthesis (concentration effect).

To get an idea of the relative frequency of these different outcomes, we compiled the findings of the handful of studies that have reported environmentally induced changes in plant allelochemical composition in terms of both concentrations and contents (Tab. 3). The results clearly demonstrate that as determinants of changes in allelochemical concentrations, changes in plant biomass are at least as important as changes in allelochemical production (Tab. 3). Dilution of allelochemical concentrations by growth has been observed in response to fertilization [40, 41], increased light [42], and elevated CO_2 [43], while increases in concentrations that are due to reduced plant biomass (concentration effects) were caused by nutrient limitations [40], shading [44], and drought [42]. In several cases, changes in the rates of allelochemical production matched the rates of biomass accumulation, so that no significant changes in allelochemical concentrations occurred (Tab. 3). Such steady state situations appear to be common in plants subjected to elevated CO_2 (e.g., [41, 45]).

Table 3. Examples of environmentally induced changes in plant allelochemical composition and their interpretation

Plant species	Treatment	Allelochemical	Changes in concentration	Changes in content	Changes in biomass	Interpretation[1]	Reference
Betula pendula	UV-B	Flavonoids	↑	↑↑	↑	Excess synthesis	[43]
		Condensed tannins	0	↑	↑	Steady state	
	Elevated CO$_2$	Flavonoids	→	0	0/↑	Dilution effect	
		Condensed tannins	↑	↑	0/↑	Excess synthesis	
Delphinium barbeyi	Shading	Alkaloids	↑	0	→	Concentration effect	[44]
Erysimum cheiranthoides	Low N	Cardenolides	↑	0	→	Concentration effect	[40]
		Glucosinolates	→	↓↓	→	Reduced synthesis	
	High N	Cardenolides	0	↑	↑	Steady state	
		Glucosinolates	→	0	↑	Dilution effect	
Mentha piperita	Elevated CO$_2$	Mono- and sesquiterpenes	0	↑	↑	Steady state	[45]
Pinus taeda	High N	Condensed tannins	→	↑	↑↑	Dilution effect	[41]
	Elevated CO$_2$		0/↑	↑	↑	Steady state/excess synthesis	
Tabernaemontana pachysiphon	High light	Monoterpene indole alkaloids	→	0	↑	Dilution effect	[42]
	Fertilization		↑	↑↑	↑	Excess synthesis	
	Drought		↑	→	↓↓	Concentration effect	

[1] Shifts in allelochemical concentrations, contents, and plant biomass are interpreted as in Table 2 following [14].

At first sight, several of the patterns revealed in the studies compiled in Table 3 are consistent with the predictions of the CNB hypothesis: reduced concentrations of condensed tannins in response to fertilization [41], increased concentrations of alkaloids in response to shading [44], and increased concentrations of C-based cardenolides in response to reduced N supply [40]. However, a more careful examination of these patterns reveals that changes in concentrations were not accompanied by corresponding changes in allelochemical production, as predicted by the CNB hypothesis. In fact, in each of the above cases, changes in allelochemical production (as revealed by content) were either not significant or pointed in the opposite direction from changes in concentrations (Tab. 3). On the other hand, the lack of changes in concentrations of C-based compounds in response to elevated CO_2 reported in some studies (e.g., [41, 44]) has been interpreted as an indication that the CNB hypothesis fails to predict plant responses to increased carbon availability [46, 47]. However, as is clear from Table 3, responses in terms of content revealed increased production of C-based compounds in response to elevated CO_2, as predicted by the CNB hypothesis (see also [48]).

With regard to existing tests of the induced defense theory, it is impossible at present to estimate the proportion of studies in which reported increases in allelochemical concentrations were due to reduced plant biomass (concentration effect) rather than to increased synthesis of allelochemicals, since very few studies have reported induced changes in terms of both concentrations and total content per leaf or per plant (but, see [39, 49]). The same problem also may affect studies of induced mechanical defenses (trichomes, thorns, spines, etc). As noted by Karban and Baldwin [38], in the majority of studies it is not clear whether the induced increase in trichome density (an analogue of allelochemical concentrations) is the result of increases in the total number of trichomes per leaf (an analogue of content) and/or a decrease in leaf area. This matter clearly warrants more research.

Another potential limitation on the use of concentrations as indicators of synthesis is the dynamic nature of some allelochemicals. For compounds undergoing rapid catabolism, translocation, or volatilization, static concentrations may considerably underestimate the rate of production [50]. For instance, Litvak and Monson [51] found that simulated herbivory resulted in reduced concentrations of monoterpenes in needles of ponderosa pine, lodgepole pine, and white fir. Had monoterpene concentrations been the only parameter measured in this study, the authors could have concluded that their results were at odds with the predictions of the induced defense hypothesis and indicated that no induction was taking place. Fortunately, however, monoterpene cyclase activity (a key determinant of the synthesis rate of monoterpenes) and monoterpene emission rates were also measured, and both parameters were found to increase in response to wounding [51]. The authors therefore concluded that monoterpene synthesis was significantly induced in the wounded needles but that large increases in the rate of monoterpene volatilization resulted in an actual decrease in monoterpene concentrations. Usually, however, the quantities of allelochemicals lost through processes of volatilization, leaching, and root exudation represent only a small fraction of their total content in the plant [48, 52, 53]. With regard to metabolic turnover, earlier studies conducted mainly on excised plants or plant parts have demonstrated high turnover rates for several types of allelochemicals such as monoterpenes, phenolic glycosides, and alkaloids (references in [50, 52]). However, more recent studies on intact plants and under more physiologically realistic conditions have found much lower rates of turnover (reviewed in [52]). The dynamic nature of allelochemicals thus probably causes fewer problems in interpretation of changes in concentrations than do shifts in plant biomass.

Allelochemical concentrations as measures of plant defensive investments

Another common implicit assumption in ecological studies is that concentrations of allelochemicals describe the amount of resources allocated to chemical defense. This assumption is incorporated into most of the hypotheses predicting inter- and intraspecific patterns of plant resource allocation to secondary metabolite production. For instance, the plant apparency hypothesis [54] predicts that apparent plants will allocate more resources to chemical defense than will unapparent plants and that the former will therefore exhibit higher concentrations of allelochemicals than the latter. Similarly, the optimal defense hypothesis [55] postulates that defense allocation is a function of defense costs, the value of the plant tissue/organ, and the risk of herbivory and predicts that concentrations of allelochemicals will be higher in those plant tissues that are most valuable to the plant and most vulnerable to herbivory. Finally, the resource availability hypothesis [56] assumes that defense investments are determined by inherent plant growth rates, which in turn are functions of plant resource availability, and predicts that allelochemical concentrations will be higher in slow-growing plant species adapted to resource-limited environments as compared to fast-growing species from resource-rich environments. Because the predictions of these hypotheses are formulated in terms of allelochemical concentrations, it is not surprising that studies testing the hypotheses have focused largely on comparing allelochemical concentrations among and within plants. For instance, the higher concentrations of defensive compounds commonly observed in juvenile plants and young leaves as compared to mature plants and leaves, or in reproductive tissues as compared to vegetative organs (reviewed in [57]), often have been interpreted as the indication of higher plant resource allocation to defense of the most valuable plant tissues, in accordance with the predictions of the optimal defense hypothesis.

However, plants produce and allocate molecules (quantities) rather than concentrations of allelochemicals. The latter are merely the result of the distribution of a given amount of compound in a given amount of plant biomass. Allelochemical concentrations can therefore be used as operational measures of defensive investments only when the passive dilution or concentration of allelochemicals by changes in plant biomass is negligible [58]. When large changes in plant biomass occur, the concentrations of allelochemicals will reflect biomass changes rather than defensive investments. For instance, the reduction in the concentrations of some allelochemical compounds with leaf age may often be the consequence of growth dilution rather than reduced defense allocation because total content of allelochemicals per leaf usually increases during leaf maturation (e.g., [34, 59]). For this reason, total contents rather than concentrations should be used as the currency in studies addressing plant defense allocation [13]. This practice is commonly used in plant nutrient studies that examine the variation in a plant's total pool of nutrients under various conditions (e.g., [60]).

Recent meta-analysis by Koricheva [61] also has revealed that allelochemical concentrations are poor predictors of fitness costs of chemical defenses. Even the production of compounds that occur at very low concentrations in plants (<1%) was often found to impose substantial costs in terms of growth and reproduction (e.g., [57, 62–64]), presumably because of significant investments of resources into the storage, transport, and maintenance of these compounds [52, 57].

Allelochemical concentrations as indicators of plant allelopathic potential

In allelopathic research, allelochemical concentrations in a source plant are sometimes assumed to reflect its allelopathic (phytotoxic) potential. For instance, increases in allelochemical concentrations under various forms of abiotic and biotic stress (e.g., UV radiation, mineral deficiencies, water stress, temperature, disease infections, and herbivores) are often interpreted as an indication of increased plant allelopathic potential (references in [5, 65, 66]). However, changes in the concentrations of allelochemicals in a source plant may have no direct implications for the effects on target plants or microorganisms, for several reasons. First of all, as discussed above, allelochemical concentrations are poor predictors of allelochemical production. If the increase in allelochemical concentrations in environmentally stressed plants is due to reduced growth (e.g., [67, 68]), the plant biomass and the total amount of allelochemicals produced by a source plant may actually decrease. It is noteworthy that although a great deal of research has been done on the effect of various environmental stresses on the concentrations of allelopathic compounds in source plants (reviewed by [5, 65, 66]), the concomitant changes in the plant biomass and contents of allelochemicals have almost never been considered (but, see [69, 70]). However, a number of researchers have recognized that the increase in leaf size with age [59] or mass loss during leaf senescence [71] may to some extent explain the observed variation in concentration of allelochemicals between young and mature leaves or between living and dead plant organs, respectively.

In addition to biomass accumulation by individual plants or plant organs, the density and productivity of a source plant at the community level are also important determinants of the amounts of allelochemicals potentially entering the ecosystem. For instance, a significant increase in the concentration of phenols in shoots of *Vaccinium myrtillus* L. in response to clearcutting was associated with a decrease in the number and production of shoots per m^2 [72], which would actually result in a decrease of phenol content in *V. myrtillus* per unit of soil surface. It must also be considered that, in addition to production, the amount of an allelopathic compound added to the environment is a function of its release rate through leaching, volatilization, root exudation, or litter decay. Release rates are not necessarily correlated with allelochemical concentrations [48] but may instead depend on environmental conditions (e.g., [48, 65]). For example, phosphorus deficiency has been demonstrated to increase the leaching of phenolic compounds from intact roots and dried plant material [67], while both light and temperature are involved in controlling terpene emissions from *Quercus ilex* (e.g., [73, 74]).

Once allelochemicals are released into the environment, their allelopathic expression in the soil depends on a number of factors: the presence of the allelochemicals in biologically active concentrations, their persistence and fate in the soil, their flux rates (input rates and output rates, the latter including microbial and non-microbial degradation and transformation, uptake by plants, and leaching), the relative toxicity of degraded and transformed products, and possible synergistic or antagonistic interactions between different allelochemicals [5, 75–78]. Again, many abiotic and biotic environmental factors are suggested as determining the activity of allelopathic compounds in soil and the magnitude and duration of their phytotoxicity (e.g., [5, 77]. For instance, nutrient-rich conditions were found to favor the microbial degradation of ferulic acid added experimentally to the soil, while under nutrient-poor conditions, ferulic acid was able to accumulate temporarily [79].

Finally, environmental factors may also affect the sensitivity of a target plant or microogranism to a given amount of the specific allelopathic agent [80, 81]. For instance, several studies have demonstrated that effects of allelopathic compounds are more severe under low fertility, while raising nutrient levels can alleviate some of the allelochemical effects [82, 83]. The mechanisms whereby fertilization negates the action of allelochemicals are not totally clear. Several lines of evidence have demonstrated that allelochemical compounds may inhibit mineral uptake by target plants [65] as well as interfere indirectly with mineral absorption by inhibiting mycorrhizal symbionts [84, 85]. The inhibition of nutrient uptake is more pronounced at low pH [86], suggesting that improving infertile and acidic conditions by fertilization can alleviate the negative effect of allelochemicals on the target plant's nutrient uptake.

In summary, by themselves the concentrations of allelopathic compounds in the source plant are poor predictors of allelopathic potential because the latter is a function of numerous factors, including allelochemical production by a source plant, concomitant changes in plant biomass accumulation, release of allelochemicals, changes in the productivity of the source plant at the community level, availability and activity of allelochemical compounds and their byproducts in the soil, and sensitivity of a the target organism to the given amount of allelochemical compound.

Spurious correlations and covariations

The major assumption behind many theories of plant chemical defense is that the production of allelochemicals is costly because plant resources are limited and that allocation to defense leads to reduced allocation to other functions such as growth and reproduction [46, 57]. Correlations between allelochemical concentrations and plant biomass are usually regarded as manifestations of growth-defense tradeoffs and are interpreted in terms of costs associated with the synthesis of allelochemicals. For instance, negative correlations between allelochemical concentrations and plant biomass are often taken as evidence that the synthesis of these compounds is costly in terms of growth, whereas positive correlations are usually considered to be an indication that allelochemicals are so cheap that their production parallels biomass accumulation [64, 87, 88]. However, as discussed above, significant correlations between allelochemical concentrations and plant biomass may arise because allelochemical production is allometrically (rather than isometrically) related to plant biomass accumulation (Fig. 1).

If significant correlations between allelochemical concentrations and plant biomass arise because allelochemical production is restricted to a particular phase of plant growth (Fig. 1), the exact opposite interpretation of cost is possible. Negative correlations between allelochemical concentrations and plant biomass occur when the peak in the synthesis of the allelochemicals coincides with active plant growth (Fig. 1B). Such overlap in compound synthesis and growth may indicate that the production of these compounds is cheap enough to allow the plant to combine both functions (allelochemical production and growth). On the other hand, positive correlations between allelochemical concentrations and plant biomass may be due to the fact that the synthesis of allelochemicals starts after the active growth phase (Fig. 1C), which may indicate that their production is costly and cannot be combined with growth [89]. At the whole-plant level, significant correlations between allelochemical concentrations and plant biomass also may arise because of an uneven distribution of allelochemicals in various plant organs and ontogenetic changes in allometric relationships between plant parts. For instance, the increased proportion of stem and root biomass in larger plants might dilute whole-plant concentrations of allelochemicals concentrated mainly in leaves [90].

Moreover, negative correlations between allelochemical concentrations and plant biomass may arise simply because plant biomass is a denominator in the ratio defining concentration. The expected correlation of a ratio with its own denominator is as high as -0.706 when there is equivalent variability in each variable and the numerator and the denominator are uncorrelated [19]; this value is higher than most of the reported correlations between allelochemical concentrations and plant biomass (Tab. 1). Interestingly, some of these correlations disappear or change their sign when independent measures of plant growth and reproduction are used instead of the biomass of the plant part in which the allelochemicals have been measured. For instance, concentrations of pyrrolizidine alkaloids in ragwort (% fresh weight) were negatively correlated with plant fresh weight, but not with plant relative growth rate (RGR) (estimated by the slope of the linear regression of ln fresh weight plotted against time) [88]. On the other hand, concentrations of condensed tannins in shoots of *Lotus corniculatus* were positively correlated with shoot biomass but tended to be negatively correlated with fruit biomass [64]. Significant correlations between allelochemical concentrations and plant biomass may thus be spurious and should not be automatically interpreted as an indication of growth-defense tradeoffs.

Similar pitfalls await researchers in the assessment of correlations between concentrations of different compounds, e.g., in order to test whether concentrations of CBSCs are positively correlated with concentrations of carbohydrates as predicted by the CNB hypothesis [41]. When concentrations are expressed per unit total dry weight, the total composition will always add up to a concentration of 100%. Therefore, as Poorter and Villar [91] have pointed out, concentrations of different plant compounds would be expected to covary in a negative way as a result of a change in one compound only. These "spurious covariations" should be particularly strong for the compounds present in plants at relatively high concentrations. For example, an increase in the concentration of compound A from 100 to 250 mg/g will automatically imply a 13% decrease in concentrations of all other compounds [91].

Ontogenetic drift

Another complication involved in the interpretation of changes in allelochemical concentrations in response to manipulations of abiotic and biotic factors is that otherwise identical plants growing in different environments often do not have the same developmental rate and thus will be ontogenetically dissimilar when sampled at a given age [92]. Changes in allelochemical concentrations observed in response to manipulations in environmental conditions may therefore reflect ontogenetic variation induced by the treatments rather than specific functional responses. For instance, an increase in the concentrations of hydrolyzable tannins in the leaves of defoliated birch trees may be the result of a delay in the ontogenetic development of the leaves of the defoliated trees rather than of a specific chemical response to defoliation [93]. Similarly, the foliar concentrations of phenolic glycosides and salicin in fertilized eastern cottonwood saplings were affected by changes in leaf development induced by fertilization but not by the rates of N addition [94].

Alternative approaches

We tried to demonstrate above that relying solely on concentration data in eco-logical studies often may lead to serious misinterpretations, particularly where the mechanisms behind variation in allelochemical concentrations are con-cerned. While many alternative approaches to the interpretation of phenotypic variation in plant chemistry have been developed, particularly by researchers working with plant nutrients, these methods so far have been seldom applied in ecological studies of plant allelochemicals. We therefore briefly review these methods below.

As in any research topic, the choice of method for the analysis and presen-tation of allelochemical data depends critically on the goals of the study (Tab. 4). Expressing the results in terms of concentrations is valid when the purpose of the study is to assess the quality of the plant as a food for herbi-vores, since the effects of allelochemicals on herbivores (particularly general-ist ones) are often concentration- or dosage-dependent (e.g., [95, 96]). Therefore, Gianoli et al. [13] proposed that studies aiming at the characteriza-tion of optimal patterns of plant defense should use allelochemical concentra-tions as a currency. However, plant quality is often determined by the concen-trations of several individual compounds and by the interactions between them. Thus, carefully conducted bioassays with herbivores [97, 98], isobolo-graphic analysis for testing for interactions among allelochemicals [99], and multivariate techniques such as multiple regression and path analysis [100, 101] may be more appropriate for establishing which of the set of compounds is most important in determining herbivore preference and performance and the resulting plant damage. Moreover, herbivores may be affected not only by concentrations of allelochemicals within the plant tissues or on the plant sur-face but also by various volatile compounds released by the plant in response to insect feeding. A recent study by Kessler and Baldwin [102], for instance, has shown that herbivore-induced volatile allelochemicals released from wild tobacco plants affect herbivores both directly (by reducing oviposition rates) and indirectly (by attracting natural enemies), resulting in more than 90% reduction of herbivore numbers. Thus, the emission rates of allelochemicals may sometimes be an even more important determinant of herbivore perform-ance than the concentrations of allelochemicals within the plant.

Allelochemical concentrations are even poorer predictors of plant allelo-pathic potential because, as discussed above, allelopathic effects on target neighboring plants depend not only on allelochemical production by the source plant but also on the amounts of allelochemicals released into the environment, the availability and activity of allelochemical agents in the soil, and the uptake rate and susceptibility of the target plant. Therefore, in addition to the concen-tration and content in the source plant, such methods as quantifying the allelo-chemical at the release rate (e.g., in leaf soaks and washes [103] and in litter, snowmelt, and throughfall [104]) and accounting for the biomass and content of allelochemicals in the source plant at the community level [70] are essential

Table 4. Classification of methods of allelochemical data analysis

Purpose of study or analysis	Suggested method	Examples
To examine the importance of variation in plant allelochemical composition for herbivores		
• Effects of individual compounds	Bioassays with herbivores	[97, 98]
• Relative importance of several compounds	Multiple regression, path analysis	[100, 101]
• Interactions among compounds	Isobolographic analysis	[99]
• Importance of plant volatiles	Analysis of herbivore-induced volatile emission	[51, 102]
To examine changes in allelopathic potential:		
• Allelochemical production and release by source plant	Quantification in a source plant while accounting for changes in growth	[70]
	Assessment of allelochemical release rate	[103, 104]
	Accounting for source plant productivity at the community level	[70]
	Phytotoxicity bioassays	[113, 134]
• Availability and activity of allelochemicals in soils solution	Concentration in soil solution	[104, 107]
	Assessment of allelochemical flux rates (dynamic availability)	[109, 110]
	Translocation of ^{14}C-labeled chemicals [111]	
• Uptake by and sensitivity of target plant	Plant growth experiments on joint effect of allelochemicals and stress	[83]
To examine changes in allelochemical production	Isotope pulse-chase techniques Enzyme activities	[49, 52, 114] [51, 115]
To examine changes in defense allocation under various conditions	Allometric analysis of the relationship between allelo-chemical content and plant biomass	[31, 118]
To elucidate the mechanisms behind changes in allelochemical concentrations	Graphical vector analysis	[14, 43]
To remove the confounding effect of plant biomass on allelochemical data	Analysis of covariance	[41, 122, 123]
To minimize potential dilution or concentration effects	Calculation of concentrations on X-free (e.g., TNC-free) basis	[124, 125]
	Expression of results of area rather than mass basis	[126, 127]
	Calculation of concentration on fresh- rather than dry-weight basis	[125]

for validating that variation in the allelochemical synthesis by the source plant is related to the changes in the amount of phytotoxic compounds supplied to the environment. An alternative to the direct analysis of allelochemical content and release is testing for phytotoxicity activity of plant extracts or leachates (by means of allelopathic bioassays and greenhouse and field experiments), which can be used as a relative measure of the allelopathic potential of the source plant (see [105] for examples of appropriate bioassays).

The phytotoxicity of allelochemicals in the soil is usually analyzed by measuring allelochemical concentrations in the soil (e.g., [106, 107]). Williamson and Weidenhammer [75] and Weidenhammer [76] also strongly recommend assessing the flux rates (dynamic availability) of allelochemicals, e.g., by utilizing different adsorbent materials (e.g., [108]). In this respect, a very promising method is the extraction of organic compounds by means of polyester capsules containing non-ionic carbonaceous resins [109, 110]. This approach allows the non-destructive *in situ* extraction of root exudates and rhizosphere secondary metabolites; it also probably most accurately represents the toxicity actually experienced by the root system of target plants, since it accounts for the dynamic nature of allelochemicals being continuously added to and removed from the soil solution.

Finally, if the allelopathic response (uptake by and sensitivity of target plant) is under investigation, the appropriate methods will include following the translocation of ^{14}C radiolabeled chemicals in order to confirm and to quantify the uptake [111] and plant growth experiments to test for the joint effect of phytotoxic compound and environmental stresses (e.g., [83]).

Given the complexity of allelochemical action in plant-plant interactions, it is difficult if not impossible to conduct a study accounting for all aspects of variation in allelopathic potential. Logically constructed and carefully conducted studies are therefore needed to link different aspects of allelopathic interaction: from the production of allelochemical compounds by the source plant to their effect on the target organism. To date, there are several examples of systems in which the production, release, and soil fate of allelochemicals have been investigated in extensive studies. Among them are the Florida scrub (reviewed in [76, 105]), sub-alpine bilberry-spruce forest (e.g., [71, 106]), and boreal forests dominated by the ericacious dwarf shrub *Empetrum hermaphroditum* (crowberry). In the case of the crowberry, a specific phytotoxic compound, batatasin-III, was identified [112] and quantified in different leaf cohorts on the source plant (e.g., [107, 113]) and in litter, snowmelt, and throughfall, thus clarifying the transport from the source plant to the soil, as well as in associated soils, demonstrating the persistence of batatasin-III in the soil [107, 110]. This information provides the solid base needed for further investigation of the importance of different environmental factors for variation in allelochemical production and possible changes in allelopathic potential.

If the aim of the study is to follow the dynamics of allelochemical production, techniques such as isotope pulse-labeling are more appropriate [53, 114]. The same method also can be used to detect the rates of allelochemical turnover and volatilization [49, 53]. Measurements of enzyme activities also may help to shed the light on changes in allelochemical production (e.g., [51, 115]), although enzyme activities may be poor indicators of the time course of allelochemical accumulation if the latter is regulated by factors other than enzyme activities, e.g., substrate availability (e.g., [43]).

Studies addressing plant defense allocation may use absolute content of allelochemicals per plant or plant part as currency [13]. Another tool that can be recommended for studies aiming at examination of plant allocation patterns is allometric analysis. This technique has a strong tradition in zoology [116]; more recently, it has been proposed in plant ecology as an alternative to ratios in assessing plant reproductive effort [26, 27] and root/shoot allocation [117]. The use of allometric analysis in allelochemical studies has been pioneered by Baldwin et al., who assessed allocation to nicotine production in tobacco plants [31, 118]. Allocation to nicotine was defined in these studies as the slope of the allometric relationship between the whole-plant nicotine level and the nicotine-free biomass, while the intercept was used as an indicator of the onset time of nicotine production. The relationship found in undamaged plants served as a null model, against which allocation patterns in damage-induced plants were compared.

Allometric analysis is particularly useful for comparisons among treatments that may potentially affect plant growth patterns because it allows comparisons at a given biomass rather than at a given age, thus taking care of the problem of ontogenetic drift [92, 117]. Coleman et al. [119], for instance, have demonstrated that two annual plant species grown at an elevated CO_2-level had lower N concentrations compared to control plants of the same age. The difference between the treatments, however, disappeared when allometric relationships between total plant nitrogen content and plant biomass were compared, indicating that the reduction in nitrogen concentrations at an elevated CO_2-level is the product of accelerated growth rather than of increased nitrogen-use efficiency [119].

Another simple technique that makes it possible to distinguish between shifts in allelochemical production *versus* changes in plant biomass in determining allelochemical concentrations is graphical vector analysis (GVA). This technique is based on the plotting of changes in compound concentrations and contents as well as the biomass of the relevant plant tissue in the form of a vector diagram (Fig. 2). The method was originally developed for diagnosing nutrient limitations in forest stands [16, 120, 121] but has recently been suggested as a tool for the interpretation of ontogenetic and environmental effects on allelochemical composition [14]. Vector analysis is particularly useful for the comparison of results obtained for different types of allelochemicals (Fig. 2), plant species, treatments, and experiments. The first applications of GVA to plant allelochemical studies [14, 41, 43] indicate that as determinants of allelochemical concentrations, changes in plant biomass are as important as shifts in allelochemical synthesis.

Analysis of covariance (ANCOVA) has been recommended as a tool to exclude the confounding effect of body size on the variable of interest in taxonomy [19], physiology [17, 18], and insect nutritive ecology [28]. ANCOVA has several advantages over the use of ratios: it is not based on the restrictive

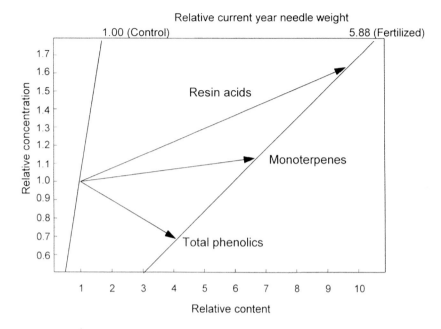

Figure 2. Vector diagram of effects of nitrogen fertilization on the concentration and content of phenolics and terpenoids and on needle biomass of *Pinus sylvestris* (data from [135]). Note that the relative content of all groups of allelochemicals is higher in the needles of fertilized plants as compared to control, indicating increased allocation to synthesis of carbon-based compounds. However, production rates of total phenolics are lower than the rates of biomass accumulation,l resulting in reduction in concentrations (growth dilution). Rates of monoterpene production approximately equal rates of biomass accumulation resulting in stable concentrations (steady state), whereas rate of resin acid production exceeds that of biomass accumulation, causing increase in concentration (cf. Table 2).

assumption of an isometric relationship between denominator and numerator variables, and it provides greater sensitivity and precision as compared to the analysis of ratios, particularly when the coefficient of variation of the denominator is greater than that of the numerator [17, 21, 28]. Several recent studies of plant allelochemicals have used plant biomass or growth rate as a covariate [41, 122, 123]. Gianoli and Niemeyer [122], for instance, demonstrated that the positive effect of temperature on concentrations of hydroxamic acids (Hx) in wheat disappeared after differences in plant growth rates were accounted for by ANCOVA, indicating that the temperature effect on Hx was mediated through its effect on the plant-growth rate. One factor that might limit the use of ANCOVA in studies on plant chemical composition, however, is that its critical assumption of the homogeneity of slopes among different groups [20] does not always hold for allelochemical data (e.g., [13, 32, 118]).

Finally, if allelochemical concentrations are the variable of primary interest (as is often the case in plant-herbivore studies), a number of approaches could

be used to reduce the potential dilution or concentration effects of plant biomass on allelochemical concentrations (Tab. 4). In cases where treatments are
known to cause large changes in a compound X, concentrations of other compounds can be calculated on X-free basis. For instance, elevated CO_2 is known
to cause the accumulation of total nonstructural carbohydrates (TNC), and
concentrations of other compounds are therefore often calculated on a TNC-
free basis. This approach has been used to examine the effects of elevated CO_2
on the chemical composition of various plant species [124, 125]. The above
studies have found that changes in some compounds (e.g., protein and minerals) and plant types (e.g., grassland species) disappear or become less pronounced when calculated on a TNC-free basis, while changes in other compounds (e.g., soluble phenolics) or plant types (e.g., ruderals) become even
more conspicuous than when calculated on the total-weight basis.

The expression of chemical composition on a unit-area basis also may be an
alternative to conventional expression on a mass basis (e.g., [126, 127]). Foliar
nutrient concentrations expressed as percentage dry weight often show strong
seasonal trends, which may confound the results if the aim of the study is to
assess the effects of non-seasonal environmental factors. Leaf area, on the
other hand, is more sensitive than leaf biomass to environmental changes but
is less subjected to seasonal variation, making it a preferable unit for the comparison of chemical variables. Gholz [126] also found that foliar chemical
composition expressed on an area-unit basis responded less to leaf age than did
mass-based values and correspondingly showed clearer patterns in environmental gradients. In the case of allelochemicals, the expression of concentrations on a unit-area basis is particularly relevant for the compounds that are
present on the plant surface and that may act as oviposition or feeding stimuli
[128]. The importance of distinguishing between internal and surface fractions
of allelochemicals has been demonstrated by Hugentobler and Renwick [40],
who showed that concentrations (on an area basis) of cardenolides on the leaf
surface of a wild crucifer respond differently to nutrient stress as compared to
whole-tissue concentrations (expressed on a mass basis).

Expressing concentrations per unit fresh weight also may be sometimes
advantageous because changes in one compound will have a proportionally
smaller effect on concentrations of other compounds on a fresh-weight basis
as compared to a dry-weight basis [91]. This practice is often used for allelochemicals extracted from fresh plant tissues, e.g., hydroxamic acids [13, 122]
and monoterpenes [135]. This method, however, is valid only if there is no difference in water content between the treatments [13, 91].

Conclusions

The general aim of ecological studies of plant allelochemicals is to examine
the mechanisms underlying variation in plant allelochemical composition and
the consequences of this variation for other organisms and ecosystem process-

es. Up to now, one of the most commonly used approaches to achieve these aims has been the study of patterns of variation in allelochemical concentrations among and within plants. However, while allelochemical concentrations represent a meaningful currency in the context of plant-herbivore interactions, they provide little (or misleading) information about patterns of allelochemical production, defense allocation, and plant allelopathic potential. Different approaches have to be used to reveal these patterns. We therefore urge ecologists to broaden their methodological arsenal by taking greater advantage of the numerous methods of chemical data analysis and interpretation that have been developed in related fields, such as biochemistry, physiology, and plant nutrient ecology. The first applications of these methods in studies of plant allelochemicals (e.g., [13, 14, 31, 32, 41, 43]) indicate that much more information can be derived from the data than from the analysis of patterns of allelochemical concentrations alone.

The fact that changes in plant biomass may be responsible for changes in allelochemical concentrations also has important implications for the building of theoretical models in chemical ecology, since most existing theories are based on the assumption that allelochemical concentrations reflect allelochemical production and defensive investments. New theories should explicitly include the dilution and concentration of allelochemicals by plant biomass as potential mechanisms behind changes in plant allelochemical composition (e.g., [58]).

Acknowledgements
We are grateful to Cristiane Gallet, Erkki Haukioja, Marie-Charlotte Nilsson, and two anonymous referees for constructive comments on earlier versions of this manuscript and to Ellen Valle for checking the English. Research funding from the Academy of Finland (to J.K.) and from the Swedish Council for Forestry and Agricultural Research (to A.S.) is acknowledged.

References

1 Fraenkel GS (1959) The raison d'être of secondary plant substances. *Science* 129: 1466–1470
2 Ehrlich PR, Raven PH (1964) Butterflies and plants: a study in coevolution. *Evolution* 18: 586–608
3 Whittaker RH, Feeny PP (1971) Allelochemics: chemical interactions between species. *Science* 171: 757–770
4 Rosenthal GA, Berenbaum MR (1991) *Herbivores, their interactions with secondary plant metabolites*. 2nd ed. Vol. 1. Academic Press, San Diego
5 Rice EL (1984) *Allelopathy*. 2nd ed. Academic Press, Orlando
6 White CS (1994) Monoterpenes: their effects on ecosystem nutrient cycling. *J Chem Ecol* 20: 1381–1406
7 Northup RR, Dahlgren RA, McColl JG (1998) Polyphenols as regulators of plant-litter-soil interactions in California's pygmy forest: a positive feedback? *Biogeochemistry* 42: 189–220
8 Hättenschwiler S, Vitousek PM (2000) The role of polyphenols in terrestrial ecosystem nutrient cycling. *TREE* 15: 238–243
9 Waterman PG, Mole S (1994) *Analysis of phenolic plant metabolites*. Blackwell Scientific Publications, Oxford
10 Farhoomand MB, Peterson LA (1968) Concentration and content. *Agron J* 60: 708–709
11 Harris DC (1999) *Quantitative chemical analysis*. 5th ed. Freeman and Co., New York

12 Goldberg DE, Scheiner SM (1993) ANOVA and ANCOVA: field competition experiments. *In*: SM Scheiner, J Gurevitch (eds): *Design and analysis of ecological experiments*. Chapman and Hall, New York, 69–93

13 Gianoli E, Rios JM, Niemeyer HM (1999) Within-plant allocation of a chemical defense in *Secale cereale*. Is concentration the appropriate currency of allocation? *Chemoecology* 9: 113–117

14 Koricheva J (1999) Interpreting phenotypic variation in plant allelochemistry: problems with the use of concentrations. *Oecologia* 119: 467–473

15 Jarrell WM, Beverly RB (1981) The dilution effect in plant nutrition studies. *Adv Agron* 34: 197–224

16 Weetman GF (1989) Graphical vector analysis technique for testing stand nutritional status. *In*: WJ Dyck, CA Mees (eds): *Research strategies for long-term site productivity*. Forest Research Institute, New Zealand, 93–109

17 Blem CR (1984) Ratios in avian physiology. *Auk* 101: 153–155

18 Packard GC, Boardman TJ (1988) The misuse of ratios, indices, and percentages in ecophysiological research. *Physiol Zool* 61: 1–9

19 Atchley WR, Gaskins CT, Anderson D (1976) Statistical properties of ratios. I. empirical results. *Syst Zool* 25: 137–148

20 Sokal RR, Rohlf FJ (1995) *Biometry*, 3rd ed. Freeman and Co, New York

21 Anderson DE, Lydic R (1977) On the effect of using ratios in the analysis of variance. *Biobehav Rev* 1: 225–229

22 Jasienski M, Bazzaz FA (1999) The fallacy of ratios and the testability of models in biology. *Oikos* 84: 321–326

23 Kenney BC (1982) Beware of spurious self-correlations! *Water Resour Res* 18: 1041–1048

24 Kronmal RA (1993) Spurious correlation and the fallacy of the ratio standard revisited. *J R Statist Soc A* 156: 379–392

25 Raubenheimer D (1995) Problems with ratio analysis in nutritional studies. *Funct Ecol* 9: 21–29

26 Samson DA, Werk KS (1986) Size-dependent effects in the analysis of reproductive effort in plants. *Am Nat* 127: 667–680

27 Klinkhamer PGL, de Jong TJ, Meelis E (1990) How to test for proportionality in the reproductive effort of plants. *Am Nat* 135: 291–300

28 Raubenheimer D, Simpson SJ (1992) Analysis of covariance: an alternative to nutritional indices. *Entomol Exp Appl* 62: 221–231

29 Sakuta M, Komamine A (1987) Cell growth and accumulation of secondary metabolites. *In*: F Constabel, IK Vasil (eds): *Cell culture and somatic cell genetics of plants, vol. 4. Cell culture in phytochemistry*. Academic Press, San Diego, 97–114

30 Barz W, Beimen A, Dräger B, Jaques U, Otto C, Süper E, Upmeier B (1990) Turnover and storage of secondary products in cell cultures. *In*: BV Charlwood, MJC Rhodes (eds): *Secondary products from plant tissue culture*. Clarendon, Oxford, 79–102

31 Baldwin IT, Karb MJ (1995) Plasticity in allocation of nicotine to reproductive parts in *Nicotiana attenuata*. *J Chem Ecol* 21: 897–909

32 Lynds GY, Baldwin IT (1998) Fire, nitrogen, and defensive plasticity in *Nicotiana attenuata*. *Oecologia* 115: 531–540

33 Lohman DJ, McConnaughay KD (1998) Patterns of defensive chemical production in wild parsnip seedlings (Apiaceae: *Pastinaca sativa* L.). *Chemoecology* 8: 195–200

34 Van Dam NM, Verpoorte R, van der Meijden E (1994) Extreme differences in pyrrolizidine alkaloid levels between leaves of *Cynoglossum officinale*. *Phytochemistry* 37: 1013–1016

35 Julkunen-Tiitto R (1996) Defensive efforts of *Salix myrsinifolia* plantlets in photomixotrophic culture conditions: the effects of sucrose, nitrogen and pH on the phytomass and secondary phenolic accumulation. *Écoscience* 3: 297–303

36 Bryant JP, Chapin FSIII, Klein DR (1983) Carbon/nutrient balance of boreal plants in relation to vertebrate herbivory. *Oikos* 40: 357–368

37 Haukioja E (1990) Induction of defenses in trees. *Annu Rev Entomol* 36: 25–42

38 Karban R, Baldwin IT (1997) *Induced responses to herbivory*. University of Chicago Press, Chicago

39 Bryant JP, Reichardt PB, Clausen TP, Werner RA (1993) Effects of mineral nutrition on delayed inducible resistance in Alaska paper birch. *Ecology* 74: 2072–2084

40 Hugentobler U, Renwick JAA (1995) Effects of plant nutrition on the balance of insect relevant cardenolides and glucosinolates in *Erysimum cheiranthoides*. *Oecologia* 102: 95–101

41 Gebauer RLE, Strain BR, Reynolds JF (1998) The effect of elevated CO_2 and N availability on tissue concentrations and whole plant pools of carbon-based secondary compounds in loblolly pine (*Pinus taeda*). *Oecologia* 113: 29–36

42 Höft M, Verpoorte R, Beck E (1996) Growth and alkaloid contents in leaves of *Tabernaemontana pachysiphon* Stapf (Apocynaceae) as influenced by light intensity, water and nutrient supply. *Oecologia* 107: 160–169

43 Lavola A, Julkunen-Tiitto R, de la Rosa TM, Lehto T, Aphalo PJ (2000) Allocation of carbon to growth and secondary metabolites in birch seedlings under UV-B radiation and CO_2 exposure. *Physiol Plant* 109: 260–267

44 Ralphs MH, Manners GD, Gardner DR (1998) Influence of light and photosynthesis on alkaloid concentration in larkspur. *J Chem Ecol* 24: 167–182

45 Lincoln DE, Couvet D (1989) The effect of carbon supply on allocation to allelochemicals and caterpillar consumption of peppermint. *Oecologia* 78: 112–114

46 Herms DA, Mattson WJ (1992) The dilemma of plants: to grow or defend. *Q Rev Biol* 67: 283–335

47 Lincoln DE, Fajer ED, Johnson RH (1993) Plant-insect herbivore interactions in elevated CO_2 environments. *TREE* 8: 64–68

48 Peñuelas J, Llusià J (1997) Effects of carbon dioxide, water supply, and seasonality on terpene content and emission by *Rosmarinus officinalis*. *J Chem Ecol* 23: 979–993

49 Baldwin IT, Karb MJ, Ohnmeiss TE (1994) Allocation of ^{15}N from nitrate to nicotine: production and turnover of a damage-induced mobile defense. *Ecology* 75: 1703–1713

50 Reichardt PB, Chapin III FS, Bryant JP, Mattes BR, Clausen TP (1991) Carbon/nutrient balance as a predictor of plant defense in Alaskan balsam poplar: potential importance of metabolite turnover. *Oecologia* 88: 401–406

51 Litvak ME, Monson RK (1998) Patterns of induced and constitutive monoterpene production in conifer needles in relation to insect herbivory. *Oecologia* 114: 531–540

52 Gershenzon J (1994) The cost of plant chemical defense against herbivory: a biochemical perspective. *In*: EA Bernays (ed): *Insect-plant interactions*. Vol. 5. CRC Press, Boca Raton, 105–173

53 Gershenzon J, McConkey ME, Croteau RB (2000) Regulation of monoterpene accumulation in leaves of peppermint. *Plant Physiol* 122: 205–213

54 Feeny P (1976) Plant apparency and chemical defense. *Rec Adv Phytochem* 10: 1–40

55 Rhoades DF (1979) Evolution of plant chemical defense against herbivores. *In*: GA Rosenthal, DN Janzen (eds): *Herbivores: their interactions with secondary plant metabolites*. Academic Press, New York, 3–54

56 Coley PD, Bryant JP, Chapin FS (1985) Resource availability and plant antiherbivore defense. *Science* 230: 895–899

57 Zangerl AR, Bazzaz FA (1992) Theory and pattern in plant defense allocation. *In*: RS Fritz, EL Simms (eds): *Plant resistance to herbivores and pathogens: ecology, evolution and genetics*. University of Chicago Press, Chicago, 363–391

58 Jones CG, Hartley SE (1999) A protein competition model of phenolic allocation. *Oikos* 86: 27–44

59 Koeppe DE, Rohrbaugh LM, Rice EL, Wender SH (1970) The effect of age and chilling temperatures on the concentration of scopolin and caffeoylquinic acids in tobacco. *Physiol Plant* 23: 258–266

60 Ohlson M, Malmer N (1990) Total nutrient accumulation and seasonal variation in resource allocation in the bog plant *Rhynchospora alba*. *Oikos* 58: 100–108

61 Koricheva J (2002) Meta-analysis of sources of variation in fitness costs of plant antiherbivore defenses. *Ecology* 83: 176–190

62 Berenbaum MR, Zangerl AR, Nitao JK (1986) Constraints on chemical coevolution: wild parsnips and parsnip webworm. *Evolution* 40: 1215–1228

63 Baldwin IT, Sims CL, Kean SE (1990) The reproductive consequences associated with inducible alkaloidal responses in wild tobacco. *Ecology* 71: 252–262

64 Briggs MA, Schultz JC (1990) Chemical defense production in *Lotus corniculatus* L. II. Trade-offs among growth, reproduction and defense. *Oecologia* 83: 32–37

65 Devi R, Pellissier F, Prasad MNV (1997) Allelochemicals. *In*: MNV Prasad (ed): *Plant ecophysiology*. John Wiley and Sons, Inc., New York, 253–303

66 Einhellig FA (1999) An integrated view of allelochemicals amid multiply stresses. *In*: Inderjit, KMM Dakshini, CL Foy (eds): *Principles and practices in plant ecology: allelochemical interac-*

tions. CRC Press, Boca Raton, 479–494

67 Koeppe DE, Southwick LM, Bittell JE (1976) The relationship of tissue chlorogenic acid concentrations and leaching of phenolics from sunflowers grown under varying phosphate nutrient conditions. *Can J Bot* 54: 593–599

68 Richardson MD, Bacon CW (1993) Cyclic hydroxamic acid accumulation in corn seedlings exposed to reduced water potentials before, during, and after germination. *J Chem Ecol* 19: 1613–1624

69 Mwaja VN, Masiunas JB, Weston LA (1995) Effects of fertility on biomass, phytotoxicity, and allelochemical content of cereal rye. *J Chem Ecol* 21: 81–96

70 Sène M, Dore T, Gallet C (2001) Relationships between biomass and phenolic production in grain sorghum grown under different conditions. *Agron J* 93: 49–54

71 Gallet C, Lebreton P (1995) Evolution of phenolic patterns in plants and associated litters and humus of a mountain forest ecosystem. *Soil Biol Biochem* 27: 157–165

72 Atlegrim O, Sjöberg K (1996) Response of bilberry (*Vaccinium myrtillus*) to clear-cutting and single-tree selection harvests in uneven-aged boreal *Picea abies* forests. *For Ecol Manage* 86: 39–50

73 Staudt M, Seufert G (1995) Light-dependent emission of monoterpenes by holm oak (*Quercus ilex* L.). *Naturwissenschaften* 82: 89–92

74 Loreto F, Ciccioli P, Cecinato A, Brancaleoni E, Frattoni M, Tricoli D (1996) Influence of environmental factors and air composition on the emission of α-pinene from *Quercus ilex* leaves. *Plant Physiol* 110: 267–275

75 Williamson GB, Weidenhamer JD (1990) Bacterial degradation of juglone: evidence against allelopathy? *J Chem Ecol* 16: 1739–1742

76 Weidenhamer JD (1996) Distinguishing resource competition and chemical interference: Overcoming the methodological impasse. *Agron J* 88: 866–875

77 Inderjit, Cheng HH, Nishimura H (1999) Plant phenolics and terpenoids: transformations, degradation, and potential for allelopathic interactions. *In*: Inderjit, KMM Dakshini, CL Foy (eds): *Principles and practices in plant ecology: allelochemical interactions.* CRC Press, Boca Raton, 255–266

78 Schmidt SK, Ley RE (1999) Microbial competition and soil structure limit the expression of allelochemicals in nature. *In*: Inderjit, KMM Dakshini, CL Foy (eds): *Principles and practices in plant ecology: allelochemical interactions.* CRC Press, Boca Raton, 339–351

79 Blum U, Shafer SR (1988) Microbial populations and phenolic acids in soil. *Soil Biol Biochem* 20: 793–800

80 Einhellig FA, Eckrich PC (1984) Interactions of temperature and ferulic acid stress on grain sorghum and soybeans. *J Chem Ecol* 10: 161–170

81 Schulz M, Friebe A (1999) Detoxification of allelochemicals in higher plants and enzymes involved. *In*: Inderjit, KMM Dakshini, CL Foy (eds): *Principles and practices in plant ecology: allelochemical interactions.* CRC Press, Boca Raton, 383–400

82 Glass AM (1976) The allelopathic potential of phenolic acids associated with the rhizosphere of *Pteridium aquilinum. Can J Bot* 54: 2440–2444

83 Stowe LG, Osborn A (1980) The influence of nitrogen and phosphorus levels on the phytotoxicity of phenolic compounds. *Can J Bot* 58: 1149–1153

84 Nilsson MC, Hogberg P, Zackrisson O, Fengyou W (1993) Allelopathic effects by *Empetrum hermaphroditum* on development and nitrogen uptake by roots and mycorrhizae of *Pinus sylvestris. Can J Bot* 71: 620–628

85 Souto C, Pellissier F, Chiapusio G (2000) Allelopathic effects of humus phenolics on growth and respiration of mycorrhizal fungi. *J Chem Ecol* 26: 2015–2023

86 Harper JR, Balke NE (1981) Characterization of the inhibition of K$^+$ absorption in oat roots by salicylic acid. *Plant Physiol* 68: 1349–1353

87 Lavola A, Julkunen-Tiitto R (1994) The effect of elevated carbon dioxide and fertilization on the primary and secondary metabolites in birch, *Betula pendula* (Roth). *Oecologia* 99: 315–321

88 Vrieling K, van Wijk CAM (1994) Cost assessment of the production of pyrrolizidine alkaloids in ragwort (*Senecio jacobaea* L.). *Oecologia* 97: 541–546

89 Haukioja E, Ossipov V, Koricheva J, Honkanen T, Larsson S, Lempa K (1998) Biosynthetic origin of carbon-based secondary compounds: cause of variable responses of woody plants to fertilization? *Chemoecology* 8: 133–139

90 Wilkens RT (1997) Limitations of evaluating the growth-differentiation balance hypothesis with only two levels of light and water. *Écoscience* 4: 319–326

91 Poorter H, Villar R (1997) The fate of acquired carbon in plants: chemical composition and construction costs. *In*: FA Bazzaz, J Grace (eds): *Plant resource allocation*. Academic Press, San Diego, 39–72

92 Coleman JS, McConnaughay KDM, Ackerly DD (1994) Interpreting phenotypic variation in plants. *TREE* 9: 187–191

93 Kaitaniemi P, Ruohomäki K, Ossipov V, Haukioja E, Pihlaja K (1998) Delayed induced changes in the biochemical composition of host plant leaves during an insect outbreaks. *Oecologia* 116: 182–190

94 Wait DA, Jones CG, Coleman JS (1998) Effects of nitrogen fertilization on leaf chemistry and beetle feeding are mediated by leaf development. *Oikos* 82: 502–514

95 Langenheim JH, Foster CE, McGinley RB (1980) Inhibitory effects of different quantitative compositions of *Hymenaea* leaf resins on a generalist herbivore, *Spodoptera exigua*. *Biochem Syst Ecol* 8: 385–396

96 Lincoln DE, Newton TS, Ehrlich PR, Williams KS (1982) Coevolution of the checkerspot butterfly *Euphydryas chalcedona* and its larval food plant *Diplacus aurantiacus*: larval response to protein and leaf resin. *Oecologia* 52: 216–223

97 Ayres MP, Clausen TP, MacLean SF, Redman AM, Reichardt PB (1997) Diversity of structure and antiherbivore activity in condensed tannins. *Ecology* 78: 1696–1712

98 Hare D (1998) Bioassays methods with terrestrial invertebrates. *In*: Haynes, JG Millar (eds): *Methods in chemical ecology*, vol. 2. b*ioassay methods*. Kluwer Academic Publishing, Norvell, 212–270

99 Nelson AC, Kursar TA (1999) Interactions among plant defense compounds: a method for analysis. *Chemoecology* 9: 81–92

100 Philippi TE (1993) Multiple regression: herbivory. *In*: SM Scheiner, J Gurevitch (eds): *Design and analysis of ecological experiments*. Chapman and Hall, New York, 183–210

101 Matsuki M, MacLean SF (1994) Effects of different leaf traits on growth rates of insect herbivores on willows. *Oecologia* 100: 141–152

102 Kessler A, Baldwin IT (2001) Defensive function of herbivore-induced plant volatile emissions in nature. *Science* 291: 2141–2144

103 Weidenhamer JD, Menelaou M, Macias FA, Fischer NH, Richardson DR, Williamson GB (1994) Allelopathic potential of menthofuran monoterpenes from *Calamintha ashei*. *J Chem Ecol* 20: 3345–3359

104 Wallstedt A, Nilsson MC, Zackrisson O, Odham G (2000) A link in the study of chemical interference exerted by *Empetrum hermaphroditum*: quantification of batatasin-III in soil solution. *J Chem Ecol* 26: 1311–1323

105 Romeo JT, Weidenhamer JD (1998) Bioassays for allelopathy in terrestrial plants. *In*: K Haynes, JG Millar (eds): *Methods in chemical ecology*, vol. 2. *bioassay methods*. Kluwer Academic Publishing, Norvell, 179–211

106 Gallet C, Pellissier F (1997) Phenolic compounds in natural solutions of a coniferous forest. *J Chem Ecol* 23: 2401–2412

107 Gallet C, Nilsson MC, Zackrisson O (1999) Phenolic metabolites of ecological significance in *Empetrum hermaphroditum* leaves and associated humus. *Plant Soil* 210: 1–9

108 Tang CS, Young CC (1982) Collection and identification of allelopathic compounds from the undisturbed root system of bigalta limpograss (*Hemarthria altissima*). *Plant Physiol* 69: 155–160

109 Callaway RM, DeLuca TH, Belliveau WM (1999) Biological-control herbivores may increase competitive ability of the noxious weed *Centaurea maculosa*. *Ecology* 80: 1196–1201

110 Morse CC, Yevdokimov IV, DeLuca TH (2000) *In situ* extraction of rhizosphere organic compounds from contrasting plant communities. *Commun Soil Sci Plant Anal* 31: 725–742

111 Chiapusio G, Pellissier F (2001) Methodological setup to study allelochemical translocation in radish seedlings. *J Chem Ecol* 27: 1701–1712

112 Oden PC, Brandtberg PO, Andersson R, Gref R, Zackrisson O, Nilsson MC (1992) Isolation and characterization of a germination inhibitor from leaves of *Empetrum hermaphroditum* Hagerup. *Scand J For Res* 7: 497–502

113 Nilsson MC, Gallet C, Wallstedt A (1998) Temporal variability of phenolics and batatasin-III in *Empetrum hermaphroditum* leaves over an eight-year period: interpretations of ecological function. *Oikos* 81: 6–16

114 Gershenzon J, Murtagh GJ, Croteau R (1993) Absence of rapid terpene turnover in several

diverse species of terpene-accumulating plants. *Oecologia* 96: 583–592
115 McConkey ME, Gershenzon J, Croteau RB (2000) Developmental regulation of monoterpene biosynthesis in the glandular trichomes of peppermint. *Plant Physiol* 122: 215–223
116 Reiss MJ (1989) *The allometry of growth and reproduction*. Cambridge University Press, Cambridge
117 Farrar J, Gunn S (1998) Allocation: allometry, acclimation—and alchemy? *In*: H Lambers, H Poorter, MMI Van Vuren (eds): *Inherent variation in plant growth. physiological mechanisms and ecological consequences*. Backhuys Publishers, Leiden, 183–198
118 Ohnmeiss TE, Baldwin IT (1994) The allometry of nitrogen allocation to growth and an inducible defense under nitrogen-limiting growth. *Ecology* 75: 995–1002
119 Coleman JS, McConnaughay KDM, Bazzaz FA (1993) Elevated CO_2 and plant nitrogen-use: is reduced tissue nitrogen concentration size-dependent? *Oecologia* 93: 190–200
120 Timmer VR, Stone EL (1978) Comparative foliar analysis of young balsam fir fertilized with nitrogen, phosphorus, potassium, and lime. *Soil Sci Soc Am J* 42: 125–130
121 Haase DL, Rose R (1995) Vector analysis and its use for interpreting plant nutrient shifts in response to silvicultural treatments. *For Sci* 41: 54–66
122 Gianoli E, Niemeyer HM (1997) Environmental effects on the accumulation of hydroxamic acids in wheat seedlings: the importance of plant growth rate. *J Chem Ecol* 23: 543–551
123 Heyworth CJ, Iason GR, Temperton V, Jarvis PG, Duncan AJ (1998) The effect of elevated CO_2 concentration and nutrient supply on carbon-based plant secondary metabolites in *Pinus sylvestris*. *Oecologia* 115: 344–350
124 Körner C, Miglietta F (1994) Long term effects of naturally elevated CO_2 on Mediterranean grassland and forest trees. *Oecologia* 99: 343–351
125 Poorter H, Van Berkel Y, Baxter R, Den Hertog J, Dijkstra P, Gifford RM, Griffin KL, Roumet C, Roy J, Wong SC (1997) The effect of elevated CO_2 on the chemical composition and construction costs of leaves of 27 C_3 species. *Plant Cell Environ* 20: 472–482
126 Gholz HL (1978) Assessing stress in *Rhododendron macrophyllum* through an analysis of leaf physical and chemical characteristics. *Can J Bot* 56: 546–556
127 Smith RB, Waring RH, Perry DA (1981) Interpreting foliar analyses from Douglas-fir as weight per unit of leaf area. *Can J For Res* 11: 593–598
128 Städler E (1986) Oviposition and feeding stimuli in leaf surface waxes. *In*: B Juniper, R Southwood (eds): *Insects and the plant surface*. Edward Arnold Ltd, London, 105–121
129 Pavia H, Toth G, Åberg P (1999) Trade-offs between phlorotannin production and annual growth in natural populations of the brown seaweed *Ascophyllum nodosum*. *J Ecol* 87: 761–771
130 Han K, Lincoln DE (1994) The evolution of carbon allocation to plant secondary metabolites: a genetic analysis of cost in *Diplacus aurantiacus*. *Evolution* 48: 1550–1563
131 Jones GL, Kenyon JM (1961) Measured crop performance, tobacco 1961. *North Carolina Agricult Exp Stat Res Rep* 29
132 Bowers MD, Stamp NE (1992) Chemical variation within and between individuals of *Plantago lanceolata* (Plantaginaceae). *J Chem Ecol* 18: 985–995
133 Julkunen-Tiitto R, Tahvanainen J, Silvola J (1993) Increased CO_2 and nutrient status changes affect phytomass and the production of plant defensive secondary chemicals in *Salix myrsinifolia* (Salisb.). *Oecologia* 95: 495–498
134 Nilsson MC, Zackrisson O, Sterner O, Wallstedt A (2000) Characterisation of the differential interference effects of two boreal dwarf shrub species. *Oecologia* 123: 122–128
135 Kainulainen P, Holopainen J, Palomäki V, Holopainen T (1996) Effects of nitrogen fertilization on secondary chemistry and ectomycorrhizal state of Scots pine seedlings and on growth of gray pine aphid. *J Chem Ecol* 22: 617–636

Biochemical and physiological aspects of pollen allelopathy

Stephen D. Murphy

Department of Environment and Resource Studies, University of Waterloo, Waterloo, Ontario, N2L 3G1, Canada

Introduction

The exudation of allelochemicals from pollen of one species and the inhibitory effect on pollen germination and tube elongation, stigma and style receptivity, ovule development, and seed set in other species are well studied for only a limited number of species (*Hieracium aurantiacum, Hieracium × floribundum, Hieracium pratense, Parthenium hysterophorus, Phleum pratense,* and *Zea mays* var. *chalquiñocónico*). With so few pollen allelopathic species studied, it is not surprising that the biochemistry and physiology of pollen allelopathy is not well established. The pioneering work on *Zea mays* var. *chalquiñocónico* indicated that phenylacetic acid (PAA) is the pollen allelochemical and that it uncouples mitochondria oxidative phosphorylation (between coenzyme Q and cytochrome *c*), altering membrane permeability and halting mitosis. Since PAA is typically a broad-acting growth hormone, it seems likely that a more specialized PAA derivative is the actual allelochemical. Compounds similar to the flavonoid isorhamnetin 3,4-diglucoside in *Phleum pratense* could, in theory, uncouple 6-phosphogluconate dehydrogenase and prevent pollen tube germination. Ironically, the most extensive work on the biochemical and physiological mechanisms of pollen allelopathy has been completed on species in which the effect is less established (*Artemisia vulgaris, Betula verrucosa, Gaillardia* spp., *Ledum* spp., *Urtica dioica*). More than 20 pollen allelochemicals have been identified. These may act as free radicals to alter membrane permeability of heterospecific pollen *via* oxidation-induced changes to the extant enzymes or glycoproteins on the cell walls and membranes. Allelopathic compounds also may enter heterospecific pollen and create a cascade whereby cyclic adenosine monophosphate (cAMP) or cyclic guanosine monophosphate (cGMP) turn off the enzymes involved in pollen tube germination. Nonetheless, there is much more work needed to elucidate all aspects of the physiology and biochemistry of pollen allelopathy. Given that the ecological relevance of pollen allelopathy is established, such work is likely to accelerate over the next five years.

A brief history of pollen allelopathy

While not studied extensively, it is clear that the pollen of a few Anthophyta species have detrimental effects on the vegetative and reproductive functions of many sympatric species [1–21]. The earliest reports actually predate what was thought to be a new phenomenon identified in the 1970s. Roshchina [22, 23] credits I. N. Golubinskii with the first published reports (1946 and 1951 in sources not widely available, i.e., in "Doklandy AN SSR"). The inhibition of pollen germination and reproductive success in plants by allelochemicals in heterospecific pollen did not become known as "pollen allelopathy" until Char [4] published a brief note reporting on Sukhada and Jayachandra's seminal work (they published more extensive articles in 1980 [19, 20]).

Systematic studies of pollen allelopathy began with anecdotal field observations. For example, it was observed that crops and weeds located near *Zea mays* L. var. *chalquiñocónico* Hernández (Poaceae) (in Mexico) and *Parthenium hysterophorus* L. (Asteraceae) (in India) became chlorotic [1–5, 17–19]. Species affected by *Zea mays* L. var. *chalquiñocónico* included *Echinochloa crus-galli* (L.) Beauv. (Poaceae), *Bidens pilosa* L. (Asteraceae), *Rumex crispus* L. (Polygonaceae), *Cassia jalapensis* L. (Fabaceae), and *Amaranthus leucocarpus* Wats. (Amaranthaceae) [1–5, 17–19]. All sympatric species were affected *by Parthenium hysterophorus*, especially the crop *Phaseolus vulgaris* L. [18, 19].

In other cases, "normal" old field succession (in Canada) seemed to be impeded, and species that usually became extirpated persisted (i.e., *Phleum pratense* L. [Poaceae], *Hieracium aurantiacum* L., *Hieracium* × *floribundum* W. & G, and *Hieracium pratense* Tausch. [= *H. caespitosum* Dumort.] [all Asteraceae]) [7–16]. After careful elimination of expected causes (e.g., competition for nutrients, light, water, fungal infections, herbivory, off-target pesticide damage, soil mediated allelopathy, and pollution), the unifying cause of these anecdotal observations turned out to be the existence of sympatric species that produce large quantities of allelopathic pollen [8]. In half of the cases (all except *Hieracium*), these species were wind pollinated; hence, the allelopathic pollen often was easily transported to heterospecific competitors [1–20]. The allelopathic pollen from *Parthenium hysterophorus* and *Zea mays* var. *chalquiñocónico* mainly affected the leaves and roots of competitors and other sympatric species [1–5, 17–19]; allelopathic pollen from *Hieracium* spp. and, especially, *Phleum pratense* targeted the reproductive success of competing species [7–16, 20].

While allelopathic pollen from *Phleum pratense* and *Hieracium* spp. affect all extra-tribal species *in vitro*, the ecological importance of allelopathic pollen in these species is limited to sympatric species that have floral morphologies, pollination mechanisms, and flowering periods similar to the given pollen allelopathic species [6–16]. In terms of ecological importance, allelopathic pollen from *Phleum pratense* mainly affects such Poaceae species as *Bromus inermis* Leyss, *Danthonia compressa* Aust., *Danthonia spicata* (L.) Beauv. ex

Roem, & Schult., and *Elytrigia repens* (L.) Nevski [6–11, 13–16]; allelopath-ic pollen from *Hieracium* species affects several Asteraceae (e.g., *Achillea millefolium* L., *Cirsium arvense* (L.) Scop., *Sonchus arvensis* L., and *Sonchus oleraceous* L.) [7–10].

In all cases, it seems that pollen allelopathy is characterized by its lack of autotoxicity; in contrast, this is a vexing phenomenon in soil-mediated allelopathy because it runs counter to most concepts of natural selection, i.e., why produce a chemical that kills oneself and one's offspring?

None of the people involved in research on pollen allelopathy originally considered this odd phenomenon to be the likely explanation for crop declines or unusual pathways of succession. Indeed, the notion of plants manufacturing and/or sequestering toxins (allelochemicals) in their pollen to use in competi-tion with other species was rather novel and extraordinary [4]. Readers here may want to review the evidence that pollen allelopathy exists. Publications have provided biochemical and physiological evidence [2, 5, 11–13, 17, 19, 21–26], ecological and agricultural evidence [1, 3, 7–10, 14–16, 20], relevant anecdotal evidence [27–30], and evidence related to toxicity in other repro-ductive functions [31–34]. General evolutionary roles of pollen chemicals and physical mechanisms also have been discussed [35, 36]. Table 1 provides an overview of known pollen allelopathic species, their "target" species, and any allelopathic compounds isolated. In this chapter, the biochemical and physio-logical mechanisms of pollen allelopathy will be explored more fully in the context of ecological and evolutionary relevance.

Current state of research on biochemical and physiological aspects of pollen allelopathy

While it seems clear that allelopathic pollen influences (interference) compe-tition between some species, the exact biochemical and physiological mecha-nisms involved are less well known. In part, this reflects the complexity of pollen genetics, biochemistry, and physiology; for such a small, short-lived mass of tissue, pollen has elaborate metabolic processes. Perhaps this com-plexity should not be surprising, as it has the important primary function as the mobile tissue used to deliver the male gametes to a suitable stigma, style, and ovule. Additionally, female tissues can discriminate and eliminate less fit or incompatible male genotypes; hence, there are strong sexual selection pres-sures exerted on pollen. Pollen is also complex because it carries both game-tophytic and sporophytic tissues and metabolites; while the bulk of pollen compounds are the result of *de novo* processes that create the haploid gameto-phyte, there are glycoproteins, lipids, and some pollen chemicals that are directly "loaded" from the male parent. Research on pollen requires an aware-ness that the subject matter is really two generations combined, although the haploid generation dominates.

Table 1. Summary of established pollen allelopathic species, allelopathic compounds, and target species

Pollen allelopathic species	Allelopathic compounds	Target species	References
Dicotyledonae			
Artemisia vulgaris *Betula verrucosa* *Gaillardia* spp. *Ledum* spp. *Urtica dioica*	Carotenoids: • α- and β-carotene • Crocetine • Flavoxanthine Catecholamines: • Dopamine • Noradrenaline Nitrogenous compounds: • Acetylcholine • Histamine • Serotonin Phenolics: • Benzoic acid • Gallic acid • Kaempferol • Quercetin • Rutin • Vanillic acid Polyacetylines: • Capilline Terpenoids: • Artemisinine • Austricine • Azulene • Citral • Cymol • Damsin • Desacacetylinulicine • Gaillardine • Grosshemine • Inulicine • Ledol • Linalool • Parthenin	*Hemerocallis fulva* *Hippaestrum hybridum* *Larix decidua*	[22–26]
Hieracium aurantiacum *Hieracium ×* *floribundum* *Hieracium pratense*	Unknown	Affects all sympatric species, except species of *Hieracium* Ecological impacts severe only for *Sonchus* spp.	[7–10, 12, 20]
Parthenium hysterophorus	Caffeic acid Chlorogenic acid Vanillic acid Parthenin	All other sympatric species Crops, e.g., *Phaseolus vulgaris*	[18, 19]

(Continued on next page)

Table 1. (Continued)

Pollen allelopathic species	Allelopathic compounds	Target species	References
Monocotyledonae			
Phleum pratense	Rhamnetin derivative? (Isorhamnetin 3,4-diglucoside?)	Affects all sympatric species, except species of contribal Agrostidae	[6–11, 13–16]
		Ecological impacts important for other extratribal Poaceae: • *Bromus inermis* • *Danthonia compressa* • *Danthonia spicata* • *Elytrigia repens*	
Zea mays var. *chalquiñocónico*	Phenylacetic acid	*Amaranthus leucocarpus* *Bidens pilosa* *Cassia jalapensis* *Echinochloa crus-galli* *Rumex crispus*	[1–13, 17]

Further muddying the research has been an inability to sort and identify the structure and function of all the pollen chemicals. This complaint can be made for most plant tissues, but in pollen it gets particularly cumbersome because of the speed at which chemicals are produced and catabolized as pollen matures and germinates. Common to most studies of plant biochemistry and physiology, new analytical techniques have become available and, perhaps tardily, have been applied to pollen. As a result, the physiological and biochemical mechanisms of pollen allelopathy are becoming clearer. The question is whether "becoming clearer" means that we understand the mechanisms or that we are marveling at the diversity of chemical structures that are involved in pollen allelopathy. The latter is still true to some extent, a reflection perhaps that the three groups of researchers who have studied pollen allelopathy in the last decade seem interested in pollen allelopathy as one aspect of plant ecology or biochemistry/physiology rather than as the sole focus of our research careers.

This does not mean that pollen allelopathy is unimportant; it means exactly what I often have mentioned in various articles—pollen allelopathy is one interesting aspect of allelopathy, plant competition, and plant ecology, biochemistry, and physiology. However, its true relevance is in the context of how populations, communities, and ecosystems operate. Most people who study allelopathy (in general) appear to share this view and tend to study different aspects of populations, communities, and ecosystems simultaneously. This raises the eternal debate over whether one's research career ought to be narrowed to one specialty (like pollen allelopathy), but I suspect that many scientists are attracted to research questions rather than specific phenomena. The result is that elucidating the physiology and biochemistry of allelopathic pollen has been a slow process. Nonetheless, there has been progress, as this chapter will demonstrate.

Which tissues are affected by pollen allelochemicals?

Logically, pollen allelochemicals should target the tissues where pollen is supposed to alight—pollen, stigmas, and styles of heterospecifics. However, most of the allelopathic species are wind pollinated, and there is a relatively large probability that pollen will alight all over a plant and on the ground surrounding the roots. This means that pollen allelochemicals could affect different tissues. Indeed, the earliest reports of allelopathic pollen found that the leaves of many crop species were covered in pollen from *Parthenium hysterophorus* and suffered necrosis or that the roots of crop species were similarly affected by large ground deposits of pollen (from *Zea mays* var. *chalquiñocónico*). That these tissues suffered damage from pollen is a reflection of pollen's extraordinary characteristics. In the case of *P. hysterophorus*, the pollen is released in tetrads, and, therefore, much more pollen lands on nearby leaves. For *Z. mays* var. *chalquiñocónico*, the pollen is large and dense; while it does get carried by wind, it does not travel much further than several meters, with a substantial amount dropping to the ground and landing in large masses around the roots of companion or inter-cropped species. This ground-mediated pollen allelopathy works only because of the agricultural (cropping) system used.

The more "logical" targets of pollen allelochemicals (e.g., heterospecific pollen, pollen tube, stigma, style) are affected in the cases of *P. hysterophorus* and *Z. mays* var. *chalquiñocónico*, but these targets are most important for pollen allelopathic *Phleum pratense* and *Hieracium* species. Pollen from *P. pratense* and *Hieracium* spp. does not alight in large amounts anywhere except on the stigmas of heterospecifics (and on conspecifics, of course). In the case of *Hieracium*, non-Asteraceae do not receive sufficient pollen to cause allelopathic inhibition of reproduction. *P. pratense*, however, produces so many pollen grains (billions) that its pollen can be found in relatively large quantities (5–15 grains) on stigmas of heterospecific wind-pollinated Poaceae. For sympatric Poaceae like *Bromus inermis* Leyss., *Danthonia compressa* Aust. and *D. spicata* L., and *Elytrigia repens* (L.) Nevski, most of their pollen fails to germinate and few seeds are set from chasmogamous flowers.

Pollen allelochemicals from *P. pratense* and *Hieracium* spp. do not cause stigma or style necrosis, as observed for pollen allelopathic *P. hysterophorus* [18]. In 10 years of research on pollen allelopathy of *P. pratense* and *Hieracium* spp., the conclusion has been that necrosis is either an artifact caused by physical damage during attempts to manipulate the stigmatic surface or normal senescence—"normal" in the sense that the speed of senescence is unaffected by the presence or absence of allelopathic pollen from *Phleum pratense* [6–16]. Granted, allelochemicals could be transmitted from pollen to stigma and cause localized damage (to papillae or trichomes).

Allelochemicals also could penetrate throughout the stigma to cause widespread damage [6–16]. However, the main effect seems to be a lack of successful pollen germination (tube production) [6–16].

The early work of Thomson et al. [20] suggests a mechanism that does not operate in the stigmatic environment. They demonstrated that instead of acting directly at the stigma surface, pollen allelochemicals could inhibit ovule or seed development. Since most pollen allelopathic interactions appear to operate at the pollen stigma surface, inhibiting ovule and/or seed development might be inconsequential, but it might also represent a fail-safe method of pollen allelopathy. To date, no other study has identified this mechanism; however, no other study has been designed to test whether ovule or seed development is affected directly by pollen allelochemicals.

What are the physical mechanisms facilitating pollen allelopathy?

It is relatively easy to test which target tissues are affected *in vitro* because artificial solvents (e.g., distilled water, agar) for pollen allelochemicals are provided. *In situ*, however, the situation is more complex and requires an understanding of the general biochemical, physiological, and cytological aspects of pollen and pollination [8]. *In situ*, pollen allelochemicals need a suitable, naturally occurring solvent if they are to be transmitted effectively. In theory, solvents would not be needed if allelochemicals were transmitted from direct contact between pollen grains, but it is more likely that pollen allelochemicals are transmitted *via* the exudate from stigmas of many species or from the water that is found on stigmas of most species [8]. As reviewed recently [8], this is a complex process. However, if pollen can alight on stigmas, rehydrate (necessary in the case of pollen from Poaceae, for example), and "fuse" with stigmatic tissues to create a bidirectional flow of water and cytoplasmic compounds between pollen and stigma, then pollen allelopathy can occur.

What are the chemical structures and classes of pollen allelochemicals?

It is not difficult to extract chemicals from pollen; this is expected, since pollen is adapted to germinate and exude chemicals. Double-distilled water (for crude extraction) and standard methods like chromatography (ion-exchange and high-performance liquid) work well [6–16]. The main problems in identifying the allelochemicals are related mainly to timing. First, pollen chemicals are extremely ephemeral; therefore the allelochemicals may only be synthesized right before and during pollen rehydration. Analyzing desiccated pollen, pollen that has not rehydrated sufficiently, or pollen that has already rehydrated may cause a researcher to miss detecting the allelochemicals. The second problem is how to perform extraction. It is more convenient to macerate large quantities of pollen. However, this will not only miss the narrow window of rehydrating pollen but also create artifacts, as "wounded" pollen can produce chemicals not normally found. Identifying allelochemicals becomes a slow process as, essentially, a researcher has to gently "milk" massive quantities of rehydrating pollen of any putative allelochemicals, even to create a crude extract. A third problem could be that the existence of so many chemicals in pollen creates difficulties in detecting and discriminating the allelochemicals; however, as techniques have improved, this is becoming less of an impediment (save for the expense).

Innovative work in identifying pollen allelochemicals has been done by research groups led by Ana Luisa Anaya and Victoria Roshchina. Previous to their studies, little was known about pollen allelochemicals, and much of what was known was based on my hypotheses and synthesis of work that indirectly indicated pollen allelopathy. Sukhada and Jayachandra [19] identified three phenolics (caffeic acid, chlorogenic acid, and vanillic acid) and one terpenoid (parthenin) as the pollen allelochemicals from *Parthenium hysterophorus*. Anaya's research group was the first to systematically identify pollen allelo-chemicals [2, 5, 17]. From pollen of *Zea mays* var. *chalquiñocónico*, a highly polar, acidic, phenolic compound (phenylacetic acid) was isolated and con-vincingly identified as causing most of the pollen allelopathic activity [2]. The actual biochemical reactions that cause phenylacetic acid (PAA) to be allelo-pathic are as yet unknown. It seems odd that PAA would be the allelochemi-cal, as it is common to many species and is definitely not unique to pollen or to *Z. mays* var. *chalquiñocónico*. Perhaps it is more likely that a PAA deriva-tive unique to, or produced in larger quantities of, pollen from *Z. mays* var. *chalquiñocónico* is the true allelochemical.

Roshchina's group has gone furthest in examining the biochemistry of pollen allelochemicals by using microspectrofluorometry [22–26] to test sev-eral pollen-allelopathic species (Tab. 1). Their research indicates that many chemicals cause pollen allelopathy, though it may be that one key chemical compound initiates a biochemical and physiological cascade, i.e., produces many different allelochemicals. To use their classification scheme, pollen alle-lochemicals appear to include carotenoids (α- and β-carotene, crocetine, flavoxanthine), catecholamines (dopamine, noradrenaline), nitrogenous com-pounds (acetylcholine, histamine, serotonin), phenolics (benzoic acid, gallic acid, kaempferol, quercetin, rutin, vanillic acid), polyacetylenes (capilline), and terpenoids (artemisinine, austricine, azulene, citral, cymol, damsin, desacacetylinulicine, gaillardine, grosshemine, inulicine, ledol, linalool, parthenin) [23].

This list seems surprisingly long, but, in part, its length reflects the fact that it includes stimulatory compounds (nitrogenous compounds, dopamine, and [some] terpenoids, carotenoids, polyacetylenes) as well as inhibitory com-pounds [23]. Including stimulatory compounds as allelochemicals is perhaps not typical, but it is correct since allelochemicals do include both inhibitory and stimulatory compounds [8]. Regardless of whether chemicals stimulate or inhibit reproduction, it will be interesting to see whether microspectrofluorometry, applied to more species-species interactions, will confirm that there are so many pollen allelochemicals. To reiterate, it is probable that some of the allelopathic activity may originate from derivatives of these compounds that are short-lived (hard to isolate and detect) or that allelopathy results from synergistic reactions between some or all of these chemicals.

My own limited research in identifying pollen allelochemicals tends to support the concept that allelopathy in pollen results from having a large quantity of a common chemical, albeit one that appears to be slightly altered. I emphasize that this work has only been replicated twice and should be considered to advance a hypothesis rather than to provide convincing evidence that the allelochemicals in *Phleum pratense* and *Hieracium* spp. have been identified.

If crude polar extracts from *Phleum pratense* pollen are purified with ion-exchange chromatography, the fraction containing the allelopathic activity is in the acidic fraction containing most of the pollen's pigments. In *P. pratense*, most of these pigments are flavonoids. To extract flavonoids, the procedure is reasonably straightforward[1] and isolates many compounds. Further studies are needed, but it appears that *P. pratense* has some 10 times the amount of the rather common flavonoid isorhamentin. *Hieracium* spp. pollen also has a large amount of this compound. Isorhamnetin is not likely the cause, as other species that have large quantities of this compound are not pollen allelopathic (e.g., *Rhamnus* L. [Rhamnaceae], the genus that gave "rhamnetin" its common name). However, one the many derivatives of isorhamnetin or rhamnetin may be the allelochemical. Both do have known phytotoxic and general biocidal activities (for the most recent evidence, see [37, 38]), and any derivatives may share these biocidal activities.

Interestingly, there is some historical literature that supports the hypothesis that a rhamnetin derivative is responsible. This is because, with respect to *Phleum pratense*, there has long been interest in examining pollen chemistry related to *P. pratense*'s agricultural importance as a forage crop and risk posed as a health hazard (its pollen is seriously allergenic). While much of this literature relates to proteins involved in incompatibility or immunoglobulins involved in allergenicity, ancillary data on secondary metabolites exists. Of particular interest are papers that examined what was originally called "dactylin" (based on its chemical structure, its more contemporary common name would be isorhamnetin, 3,4-diglucoside [40–43]). This compound purportedly has biological activity, but the conclusions were erroneous, i.e., it was mistakenly thought that isorhamnetin, 3,4-diglucoside was an allergen or was involved in sex determination [40–43]. Nonetheless, isorhamnetin 3,4-diglucoside and flavonoids similar to it are allelopathic in the sense that they may act as a deterrent to insects [39]. Unfortunately, direct tests of plant-plant toxicity are unknown. Based solely on this ancillary literature and a couple of experiment runs using isolated flavonoids, there is insufficient evidence to

---

[1] 2 mL methanolic HCl (2 N) can be used, followed by centrifugation (2 min, 15,600 g), hydrolyzation of 150 μL of suspension in an autoclave (15 min, 120 °C), and HPLC analysis.

conclude that rhamnetin or isorhamnetin is involved in pollen allelopathy; this literature merely provides material for a useful hypothesis that needs repeated testing. Perhaps this can be said of all the literature on identifying pollen allelochemicals.

Elucidating the biochemical and physiological mechanisms of pollen allelopathy

With limited knowledge of the biochemical structures of pollen allelochemicals, determining how pollen allelopathy actually functions on a biochemical and physiological basis is challenging to say the least. For example, I have postulated that pollen allelochemicals might interfere with the ATPase that promotes proper adhesion of pollen to stigmatic papillae [8] but have never been able to replicate this effect sufficiently; hence, this also remains a hypothesis. However, others have focused more explicitly on the cellular scale, and interesting work exists regarding how pollen allelochemicals operate.

As the allelochemical identified in pollen from *Zea mays* var. *chalquiñocónico*, phenylacetic acid (PAA) appears to cause three responses in cells. One was to uncouple mitochondria oxidative phosphorylation (between coenzyme Q and cytochrome *c*) [2, 17], perhaps similar to parthenin inducing chlorosis [19]. The other two responses were altering membrane permeability (theoretically halting the vital rehydration and conservation of membrane integrity needed to germinate pollen of Poaceae and other Monocotyledonae) and halting mitosis. As yet, the specific mechanism behind these responses are unclear.

Given that PAA is considered to be an auxin (hormone) and essential to normal plant development, it is remarkable that it would inhibit metabolism and cell growth as an allelochemical. Plant growth regulators that are closely related to endogenous auxins (e.g., herbicides like 2,4-dichlorophenoxyacetic acid) do have "toxicity" effects. This may suggest further that PAA is actually altered and more closely resembles a plant-growth regulator. That the exact mechanism of action of plant-growth regulators like 2,4-D has been difficult to determine helps explain why elucidating the mode of action of an allelochemical based on PAA is also challenging.

Interestingly, the target species affected by pollen allelochemicals from *Zea mays* var. *chalquiñocónico* all were Dicotyledonae. Unlike Dicotyledonae, Monocotyledonae (like *Z. mays* var. *chalquiñocónico*) do not generally suffer toxicity from auxin-like compounds. It is beneficial to use an auxin-like compound to affect dicot competitors and, simultaneously, eliminate any possible autotoxic effects since, as a monocot, *Z. mays* var. *chalquiñocónico* might be impervious. If, however, altered membrane permeability is a second mode of action of PAA, then monocots like *Z. mays* var. *chalquiñocónico* would be more prone to pollen allelopathic effects and autotoxicity would be a greater risk.

Roshchina [23] has gone furthest in examining the mechanism of allelopathic pollen. Based primarily on her work with terpenoid pollen allelochemicals, the process could be quite complex. For terpenoid pollen allelochemicals, Roshchina [23] postulates that the pollen rehydration and exudate do, as I have also suggested [6–9], facilitate physical release of allelochemicals. Taking this much further into the biochemical level than anyone else has, she suggests that, within the allelopathic pollen, terpenoids bind on specific plasmalemma receptors and stimulate pollen tube formation [23]. This means more exudate will be produced to carry pollen allelochemicals out and create an efficient release of pollen allelochemicals from an otherwise useless heterospecific pollen tube.

Once released, pollen allelochemicals have to enter the target tissues. Roshchina's hypothesis is that they may either be transformed to act as free radicals or form free radicals directly [23]. Compounds like free radicals could alter the membrane permeability *via* oxidation-induced changes to the extant enzymes or glycoproteins on the cell walls and membranes [23]. This model would help explain the lack of autotoxicity because these same terpenoids in the allelopathic pollen can act, in theory, as antioxidants, thereby removing free radicals and halting any developing auto-allelopathic effects [23]. Roshchina compares these oxidation reactions to the inhibitory effects of herbicides on pollen germination and believes the modes of action could be similar [23].

Alternatively, the allelochemicals may be similar enough to compounds involved in normal pollen germination to directly enter heterospecific cells. Though not explicitly mentioned, this presumably occurs as the pollen that is heterospecific to the allelopathic species is just beginning to germinate (a stage too small to see under a basic compound light microscope). cAMP or cGMP could be used to turn off the enzymes involved in pollen-tube production in the target pollen. On a more whole-physiological basis, the effect might be altered patterns of respiration, rapid degradation of any stored carbohydrates and lipids, altered use of calcium as a chemotactic signal, or a complete depolarization of the membrane (loss of differential permeability). All of these mechanisms would stop pollen tube growth prematurely. Roshchina [23] based these effects on the theoretical impact of terpenoids (based on their chemical structures), and this work represents an impressive and useful intellectual leap, providing a better theoretical framework. The results of Roshchina's future work in this area will be enlightening.

Given recent advances in the techniques used to study the cytological scale of pollen allelopathy, another mechanism worth investigating might be related to a simpler version of the biochemical cascade that Roshchina suggests. For example, if an allelochemical is a phenol (e.g., flavonoid), several key enzymes in pollen tube production could be affected. One of these is 6-phosphogluconate dehydrogenase, an enzyme that is particularly prevalent in the pollen of most Poaceae (and hence may explain why allelopathic pollen of *Phleum pratense* is so effective against most other Poaceae). Phenols can decouple this enzyme, and if this occurs, it ultimately could prevent inositols from forming pectins for the pollen tube. The lack of a pollen tube is exactly what is observed in pollen allelopathic interactions, i.e., there are no more dramatic effects like lyzing of the pollen grain. If decoupling occurs only with specific phenolic compounds, it would be relatively easy for a pollen allelopathic species to transform this same compound in its own pollen to prevent autotoxicity. If, for example, an isorhamnetin derivative were the allelochemical, demethylation (to form quercetin, for example) or removal of other functional groups like hydroxyls or glucosides might prevent decoupling of 6-phosphogluconate dehydrogenase in conspecific pollen. This is pure speculation, of course, but as progress is made in identifying the allelochemicals and their mechanisms, the reason for the lack of autotoxicity needs to be addressed or at least considered.

The discussion above makes an implicit assumption that should be addressed briefly. It is assumed that the pollen allelochemicals have a direct impact on the target tissues and no further transformation occurs. Allelochemicals produced by some plants are not toxic unless modified by soil microbes [44] and, further, for a plant to prevent autotoxicity requires that the soil microbes facilitating allelopathy be neutralized. This probably does not apply to pollen allelopathy; while the pollen and stigma surfaces are not sterile, they are less subject to microbial colonization than soil. Nonetheless, there is some precedence for a mechanism of activation or deactivation by microbes; for example, yeasts in nectars of Asclepiadaceae influence reproductive success, including pollination [31–33]. It is possible, therefore, that some cases of pollen allelopathy (and defenses against it) are actually mediated by microbial transformation.

Evolutionary significance of pollen allelopathy

While evidence suggests that pollen allelopathy is ecologically important [1, 3, 7–10, 14–16, 45, 46], the explanation of why pollen allelopathy is found in such disparate species is unclear. Save for the fact that three of six pollen allelopathic species are wind pollinated (again, *Parthenium hysterophorus*, *Phleum pratense*, *Zea mays* var. *chalquiñocónico*) and that the other three are from agamospermous species of *Hieracium*, the lack of a taxonomic pattern suggests that pollen allelopathy arose independently several times. This might not be surprising since the basic chemicals apparently used in pollen allelopathy are common to most Anthophyta, i.e., phenolics and terpenoids, produced by fundamental metabolic pathways of shikimate and isoprene synthesis.

Pollen allelochemicals, like most secondary metabolites, are broad-acting (affect many taxa) and have high biological activity (act at low concentrations). To reiterate the basic controversy from an earlier review [8], some argue that broad-acting chemicals should be rare because most organisms do not have the specific site receptors to be affected. Hence, most secondary metabolites are accidental byproducts of metabolism or even toxic waste that needs to be expunged [6, 8, 47]. Others support the idea that chemicals have to be broad-acting in order to affect a range of taxa [6, 8, 48, 49]. If the mechanism of pollen allelopathy resembles free radical effects on membranes, no specific allelochemical receptor would be needed and the effect would be broad-acting. From an evolutionary perspective, this would allow plants that do not have long coevolutionary histories with most of their target species to have effective pollen allelopathy. Selection could favor broad-spectrum compounds that affect individuals of most any species encountered [6–17].

Even if having broad-acting compounds makes evolutionary sense, it does not answer the question of why only a few species use these broad-acting, biologically active, and apparently easy to manufacture allelochemicals. Theoretically, the relatively uncommon existence of pollen allelopathy might suggest that it is not adaptive. If this line of argument is correct, pollen contains a series of metabolic processes that reflect phylogenetic constraints and/or neutral selection. Rather than there being a reason for producing secondary metabolites, pollen could simply inherit the basic template of metabolism from the sporophyte and uses this to fulfill its main purpose of reproduction. The secondary metabolites would be viewed as an accident of nature and produce an interesting phenomenon humans call pollen allelopathy.

Though there is only limited evidence that allelopathic pollen causes selection responses in sympatric species [7–10], there are several problems with the idea that pollen allelopathy is not adaptive. It implies that pollen has resources (energy) to waste and that there is no strong selection pressure on pollen; most evidence suggests that the opposite is true [6, 8]. Additionally, why would only certain species produce sufficient quantities or types of allelochemicals in pollen when the basic pathways would allow most species to do so?

Future research prospects

The issues I have raised here about the chemical structures and classes of pollen allelochemicals; affected tissues; physical, biochemical, and physiological mechanisms; evolutionary significance of pollen allelopathy; and, raised elsewhere [7–10, 50], the applied and commercial aspects of pollen allelopathy (e.g., in weed management) provide a research agenda. This agenda is simply to examine pollen allelopathy as one mechanism of interference competition with other plants, as opposed to focusing on pollen allelopathy, for its own sake, as merely an interesting and unusual phenomenon. To this end, I do not suggest that we need an army of researchers working on allelopathic pollen. This opinion is not borne of cynicism or fear of competition; on the contrary, I have enjoyed my collaboration with many of the authors cited in this chapter. What I recognize is that working on one narrow aspect of nature for its own sake is not always the best path to take. Indeed, most of the questions I raised are ones that have broader applications to allelopathy research (in general) and ecological biochemistry. I argue that further work on pollen allelopathy should strive to address the broader applications. At the end of all of this research, I would like to understand two things: (1) the way in which plant biochemistry and physiology (as in pollen allelopathy) are affected by selection and (2) the relative role of different competitive attributes, like allelopathic pollen, in community ecology. Pollen allelopathy is by its nature an area of research that lends itself to interdisciplinary study. This is needed to understand not only pollen allelopathy but also any ecological and environmental processes.

Acknowledgments
I thank Azim Mallik for his invitation to write this chapter: it has been a privilege. I also thank, in particular, Inderjit and Stephen Gliessman for helpful comments on a previous version of this paper. There are many individuals who have contributed to the development of the ideas surrounding pollen allelopathy. I have been fortunate to have benefited from many comments and conversations; even the briefest of these has been inspirational. I particularly acknowledge Ana Luisa Anaya and Roccio Cruz-Ortega and thank them for our discussions in 1999 at the International Allelopathy Society Congress meeting in Thunder Bay. Victoria Roshchina kindly sent me one of her early papers on her laboratory's excellent work on the cytological aspects of pollen allelopathy. I look forward to having longer discussions. The early papers by Sukhada Kanchan and Jayachandra and James Thomson, Brenda Andrews, and Chris Plowright were seminal in forming the theory behind pollen allelopathy and my own work that began in 1987. Through the years, my former Ph.D. supervisor, Lonnie Aarssen, has been a major reason for beginning and continuing the work. Contributions and comments of colleagues Scott Armbruster, Judie Bronstein, Hank Cutler, Heidi Dobson, Mike Hutchings, and Bev Rathcke have been invaluable. There are so many people who have taken the time to comment critically to this work that the acknowledgements could go on for many more pages, so I conclude by emphasizing that this work on pollen allelopathy reflects a truly collegial and team atmosphere.

References

1 Anaya AL, Ramos L, Hernandez J, Ortega RC (1987) Allelopathy in Mexico. *In*: GR Waller (ed): *Allelochemicals: role in agriculture and forestry*. ACS Symposium Series 330, American Chemical Society Press, Washington, 89–101

2 Anaya AL, Hernandez-Bautista BE, Jimenez-Estrada M, Velasco-Ibarra L (1992) Phenylacetic acid as a phytotoxic compound of corn pollen. *J Chem Ecol* 18: 897–905

3 Anaya AL, Ortega RC, Rodriguez VN (1992) Impact of allelopathy in the traditional management of agroecosystems in Mexico. *In*: SJH Rizvi, V Rizvi (eds): *Allelopathy: basic and applied aspects*. Chapman and Hall, London, 271–302

4 Char MBS (1977) Pollen allelopathy. *Naturwiss* 64: 489–490

5 Jimenez JJ, Schultz K, Anaya AL, Hernandez J, Espejo O (1983) Allelopathic potential of corn pollen. *J Chem Ecol* 9: 1011–1025

6 Murphy SD (1992) The determination of the allelopathic potential of pollen and nectar. *In*: H-F Linskens, JF Jackson (eds): *Modern methods of plant analysis, volume 13 – plant toxin analysis*. Springer-Verlag, New York, 333–357

7 Murphy SD (1999) Is there a role for pollen allelopathy in biological control of weeds? *In*: SS Narwal, P Tauro (eds): *International allelopathy update – volume 2 (basic and applied aspects)*. Science Publishers, Enfield, 204–215

8 Murphy SD (1999) Pollen allelopathy. *In*: Inderjit, KMM Dakshini, CL Foy (eds): *Principles and practices in plant ecology: allelochemical interactions*. CRC Press, Boca Raton, 129–148

9 Murphy SD (2001) Field testing for pollen allelopathy: a review. *J Chem Ecol* 26: 2155–2172

10 Murphy SD (2001) The role of pollen allelopathy in weed ecology. *Weed Technol* 15: 867–872

11 Murphy SD, Aarssen LW (1989) Pollen allelopathy among sympatric grassland species: *in vitro* evidence in *Phleum pratense* L. *New Phytol* 112: 295–305

12 Murphy SD, Aarssen LW (1995) *In vitro* allelopathic effects of pollen from three *Hieracium* spe-

cies (Asteraceae) and pollen transfer to sympatric Fabaceae. *Am J Bot* 82: 37–45

13 Murphy SD, Aarssen LW (1995) Allelopathic pollen extract from *Phleum pratense* L (Poaceae) reduces germination, *in vitro*, of pollen in sympatric species. *Int J Plant Sci* 156: 425–434

14 Murphy SD, Aarssen LW (1995) Allelopathic pollen extract from *Phleum pratense* L (Poaceae) reduces seed set in sympatric species. *Int J Plant Sci* 156: 435–444

15 Murphy SD, Aarssen LW (1995) Allelopathic pollen of *Phleum pratense* reduces seed set in *Elytrigia repens* in the field. *Can J Bot* 73: 1417–1422

16 Murphy SD, Aarssen LW (1996) Partial cleistogamy limits reduction in seed set in *Danthonia compressa* (Poaceae) by allelopathic pollen of *Phleum pratense* (Poaceae). *Écoscience* 3: 205–210

17 Ortega RC, Anaya AL, Ramos L (1988) Effects of allelopathic compounds of corn pollen on respiration and cell division of watermelon. *J Chem Ecol* 14: 71–86

18 Sukhada KD, Jayachandra (1980) Pollen allelopathy – a new phenomenon. *New Phytol* 84: 739–746

19 Sukhada KD, Jayachandra (1980) Allelopathic effects of *Parthenium hysterophorus* L. Part IV. Identification of inhibitors. *Plant Soil* 55: 67–75

20 Thomson JD, Andrews BJ, Plowright RC (1982) The effect of a foreign pollen on ovule development in *Diervilla lonicera* (Caprifoliaceae). *New Phytol* 90: 777–783

21 Viswanathan K, Lakshmanan KK (1984) Phytoallelopathic effects on *in vitro* pollinnial germination of *Calotropis gigantea* R. Br. *Indian J Exp Bot* 22: 544–547

22 Roshchina VV (1999) Mechanisms of cell-cell communication. *In*: SS Narwal, P Tauro (eds): *International allelopathy update – volume 2 (basic and applied aspects)*. Science Publishers, Enfield, 3–26

23 Roshchina VV (2001) Molecular-cellular mechanisms in pollen allelopathy. *Allelopathy J* 8: 11–28

24 Roshchina VV, Melnikova EV (1996) Microspectrofluorometry: a new technique to study pollen allelopathy. *Allelopathy J* 3: 51–58

25 Roshchina VV, Melnikova EV (1998) Allelopathy and plant generative cells: participation of acetylcholine and histamine in signaling interactions of pollen and pistil. *Allelopathy J* 5: 171–182

26 Roshchina VV, Melnikova EV (1999) Microspectrofluorometry of intact secreting cells applied to allelopathy. *In*: Inderjit, KMM Dakshini, CL Foy (eds): *Principles and practices in plant ecology: allelochemical interactions*. CRC Press, Boca Raton, 99–126

27 Hodgkin T, Lyon GD (1983) Germination of *Lilium* pollens on TLC plates and their inhibition by extracts from *Brassica oleracea* tissues. *In*: DL Mulcahy, E Ottaviano (eds): *Pollen: biology and implications for plant breeding*. Elsevier, New York, 343–349

28 Itokawa H, Oshida Y, Ikuta A, Inatomi H, Adachi T (1982) Phenolic plant growth inhibitors from the flowers of *Cucurbita pepo*. *Phytochemistry* 21: 1935–1937

29 Khan MI, Jahan B (1988) Allelopathic potential of senesced anthers of *Bombax ceiba* L., inhibiting germination and growth of lettuce seeds. *Pak J Bot* 20: 205–212

30 Sedgley M, Blessing MA (1982) Foreign pollination of watermelon (*Citrullus lanatus* [Thunb.] Matsum and Nakai). *Bot Gaz* 143: 210–215

31 Eisikowitch D, Kevan PG, Lachance M-A (1990) The nectar-inhibiting yeasts and their effect on pollen germination in common milkweed, *Asclepias syriaca*. *Isr J Bot* 39: 217–225

32 Eisikowitch D, Lachance M-A, Kevan PG, Willis S, Collins-Thompson DL (1990) The effect of the natural assemblage of microorganisms and selected strains of the yeast *Metschinkowia reukauffi* in controlling the germination of pollen of the common milkweed *Asclepias syriaca*. *Can J Bot* 68: 1163–1165

33 Sandu DW, Waraich MK (1985) Yeast associated with pollinating bees and flower nectar. *Microb Ecol* 11: 51–58

34 Tepedino V, Knapp AK, Eickwort GC, Ferguson DC (1989) Death camas (*Zigadenus nuttallii*) in Kansas: Pollen collectors and a florivore. *J Kans Entomol Soc* 62: 411–412

35 Dobson HEM, Bergström G (2000) The ecology and evolution of pollen odors. *Plant Syst Evol* 222: 63–87

36 Martin FW (1970) Pollen germination on foreign stigmas. *Bull Torr Bot Club* 97: 1–6

37 Macias FA, Simonet AM, Galindo JCG, Castellano (1999) Bioactive phenolics and polar compounds from *Melilotus messanensis*. *Phytochemistry* 50: 35–46

38 Yannai S, Day AJ, Williamson G, Rhodes MJC (1998) Characterization of flavonoids as monofunctional or bifunctional inducers of quinone reductase in murine hepatoma cell lines. *Food*

Chem Toxicol 36: 623–630

39 Tobias RB, Larson RL (1991) Characterization of maize pollen flavonoid 3'-0-Methyltransferase activity and its *in vivo* products. *Biochem Physiol Pflanzen* 187: 243–250

40 Inglett GE (1957) Structure of dactylin. *J Org Chem* 22: 189–192

41 Johnson MC, Hampton SF, Schiele AW, Frankel S (1954) A study of allergenic flavonoid glycoside derived from extracts of timothy pollen. *J Allergy* 25: 82–83

42 Hartshorne JN (1958) Pigmentation and sexuality in plants. *Nature* 182: 1382–1383

43 Moore MB, Moore EE (1931) Studies on pollen and pollen extracts. VII. a glucoside from certain grass pollens. *J Am Chem Soc* 53: 2744–2746

44 Wardle DA, Ahmed M, Nicholson DK (1991) Allelopathic influence of nodding thistle (*Carduus nutans* L.) seeds on germination and radicle growth of pasture plants. *NZ J Agr Res* 34: 185–191

45 Aarssen LW (1989) Competitive ability and species coexistence: a "plant's-eye" view. *Oikos* 56: 386–401

46 Aarssen LW (1992) Causes and consequences of variation in competition ability in plant communities. *J Veg Sci* 3: 165–174

47 Jones CG, Firn RD (1991) On the evolution of plant secondary chemistry. *Phil Trans R Soc (Lond) B* 333: 273–280

48 Downum KR, Swain LA, Faleiro LJ (1991) Influence of light on plant allelochemicals: a synergistic defense in higher plants. *Arch Insect Biochem Physiol* 17: 201–211

49 Harborne JB (1990) Constraints on the evolution of biochemical pathways. *Biol J Linn Soc* 39: 135–151

50 Swanton CJ, Murphy SD (1996) Weed science beyond the weeds: the role of integrated weed management (IWM) in agroecosystem health. *Weed Sci* 44: 437–445

Index

Abies grandis 101
abiotic stress 202, 212
Abutilon theophrasti 201, 211
Acacia melanoxylon 91
Acacia nilotica ssp. *indica* 188
C_{15}-acetogenins 21
acetophenone 67, 104, 119
activated carbon 189
acutiphycin 34, 44
Adenostoma fasiculatum 111
Additive Dose Model (ADM) 210
afterlife effect 134, 139
afzelechin 63
agricultural chemicals 15
agricultural pest control 8
agricultural system 112
agroecological approach 174
agroecology 173, 174
agroecosystem, complexity of 183
agroecosystem interactions 181
agroecosystem-level approach 176
agroecosystems, allelopathy in
    intercropped 176
Agropyron cristatum 189
aka-gusare 8
algal allelopathy 40
algal-fungal relationship 7
algal metabolites 3
algal metabolites, secondary 33
algal oxylipin 14
algicidal compound 15
alginic acid 8
alkaloid 11
allelochemical concentration 220,
    221, 223, 224, 228, 229, 231, 232,
    234, 235, 239
allelopathy in ecological theory 125
allelopathy, definition 125
allolaurinterol 21, 23, 27

allometric analysis 221, 234, 236
alpha-terthienyl 194
Alternaria alternata 21, 27
Amaranthus retroflexus 177, 206
amensalism 110
Anabaena doliolum 42
Anabaena flos-aquae 37, 65
Anabaena laxa 16, 17
Anabaena spp. 34
Anabaena variabilis 41
analysis of covariance (ANCOVA)
    234, 236
anatoxin-a 37, 44
angiogenic factor 36
anhydrohapalindole A 44
anhydrohaploxindole 34
anhydrohapaloxindoles B and M 34
Ankistrodesmus 41
annual wormwood (*Artemisia
    annua*) 192
annuionones A–C 80
annuolide B 82
annuolide D 82
annuolide E 82
anthracnose disease 21, 23
anticyanobacterial compound,
    nostocyclamide 41
antifungal chemical defenses 23
antifungal activity 34–36
antifungal activity, hit-rates of 25
antifungal bioassay, miniaturized 26
antifungal natural product 15
anti-inflammatory triterpenes 78
antillatoxin 34, 44
antimicrobial compound 78
antimicrobial disease-control agent
    15
antimicrotubule agents 36
antimitotic activity 35

antineoplastic macrolide 40
antitumor agent lyngbyastatin 35
aphanatoxin 37
Aphanizomenon 37
Aphanizomenon gracile 40
apigenin 191, 205
aplysiatoxin 37, 40, 44
apple production 180
appressorium 9
apramide A 35, 44
apramide G 35, 44
apratoxin A 35, 44
aquatic fungi, association with
   freshwater algae 7
Arabidopsis 209, 213
arachidonic acid 11, 12
arachidonic acid metabolites 12
aromatic diterpene 27
Artemia salina 35
Artemisia california 114
Artemisia vulgaris 190, 206
artemisinin 192
arthropod and disease management,
   integrating weeds 180
ascomycete fungus 7, 9
Ascophyllum nodosum 7
Aspergillus oryzae 21, 27
Aspirin 12
atrazine 193
Aulosira fertilissima 43
autocoid signaling molecule 11
autotoxicity 111
auxin-inhibiting substance 77
Avena fatua 42, 113

B_{12} 100
Bacillus cereus 42
Bacillus subtilis 42
Bangia atropurpurea 10
barbamide 35, 39, 44
bare zone 114, 116–118
batatasin III 133
below-ground competition 121, 122
bioactive compound 76
bioassay 61, 80, 116, 161, 163

bioassay design 116
biodiversity 141
Biomphalaria glambrata 34, 35
biotoxin 33, 37
Bipolaris incurvata 21, 27
bisnorsesquiterpenes 80
black rot, grape 21
black spruce 206
black walnut 149–151
bloom-dominant organism 41
blue-green algae 15
Blumenol A 65
boreal forest 204
Botrytis cinerea 16, 21, 23, 27
Brachionus calyciflorus 41
Brassica campestris 181
Brassica kaber 179
Brassica oleracea 181
brassinosteroid 213
brine shrimp 38
brine shrimp toxicity 35
brown algae 21, 22
brown rot 16
brown rust 16

caffeic acid 67, 204
caffeine 99
5-*trans*-caffeoylshikimic acid 63
Calluna vulgaris 93, 212
calophycin 17, 18
Calothrix 41, 42
Calothrix brevissima 34
calothrixin A 42
camphene 202
camphor 114
canavanine 99
cancer treatment 35
Candida albicans 34, 35
cannabinoid 35
canopy throughfall 122
Carassius auratus 34
carbon/nitrogen ratio 179
carbon-nutrient balance hypothesis
   223, 224
carmabin A 35, 44

carmabin B 35, 44
carrageenin 8
catechol 204, 205
catfish aquaculture 16
cattail 57
cattail mulches 68
Centaurea diffusa 189
Centaurea maculosa 189
Centroceras clavulatum 22
Ceratiola ericoides 104, 118, 119
ceratiolin 104
Ceratophyllum demorsum 43
chalcone synthase (CS) 202
chalcone kukulkanin B 84
chemical defense, cost of 229
chemical diversity, ecological
significance of 40
chemical interference 112
chemical irritant 34
chemotype 38, 40
Chenopodium album 187, 211
Chlorella pyrenodosa 37
Chlorella vulgaris 65
Chondrus crispus 8
chronic toxin concentration 120
chrysoeriol 191
Chrysoma pauciflosculosa 118
cineole 114
cis-p-coumaric acid 195
Cistus ladnifer 205
Citrus aurantium 207
citrus malsecco disease 22
Cladonia 93
^{14}C-labeled radiotracer technique
210
clearcut harvesting 135, 141
climatic factor 205
CNB hypothesis 227
^{14}C-nicosulfuron 194
coastal grassland, southern
California 115
coastal sage scrub 114, 115
coastal sage species 117
Coelastrum microsporum 41, 42
colicin-immune 105

colicin-sensitive 105
Colletotrichum coccodes 21, 27
Colletotrichum gloeosporioides 27
commensalism 110
common groundsel (*Senecio
vulgaris*) 193
common lamb's-quarter
(*Chenopodium album*) 187, 211
common marigold (*Tagetes erecta*)
194
community structure 101
competition 110, 116, 119, 131
competition, effects of 110
competition, light 116
competition, soil nutrients 116, 119
concentration 219, 221–223, 226–
228
concentration and flux rate 113
concentration effect 225, 226, 234,
238
concentration of terpenes 115, 116
constanolactone F 13
corn 187, 206
Cosmarium lundellii 40, 41
p-coumaric acid 101, 204
trans-p-coumaric acid 195
coumarin 67
cover crop 179
cover crop, allelopathic 178
cover-cropping 178
crayfish toxicity 35
crop rotation 178
crop, soil, and weed management,
integration of 178
crop-yield increase 177
crustacea 41
cryptophycin 36, 45
Cucumis sativus 113
Cucurbita pepo 176
curacin A 35, 45
Curvularia lunata 21
cyanobacteria 15, 37–39, 42–45
cyanobacteria, as a model to
demonstrate allelopathy 43
cyanobacteria, mat-forming 38

cyanobacteria, N-fixing 42
cyanobacteria assemblage 39
cyanobacteria blooms 37, 38
cyanobacteria toxins, effect on 42
cyanobacterial allelochemical, algal
    succession 43
cyanobacterial allelochemical,
    bloom formation 43
cyanobacterial allelochemical,
    effects on soil 43
cyanobacterial allelochemical, role
    as bioherbicides 43
cyanobacterial metabolites,
    ecological significance 38
cyanobacterial survival 39
cyanobacterin LU-12 42
cyanobacterin 37, 42, 45
cyanobacterin, affect on aquatic and
    terrestrial angiosperms 42
cyanotoxin 33
cyanotoxin, biotechnological
    utilization 42
cyclic decapeptide 17
cyclic depsi-octapeptide 27
cyclic depsipeptide 36
cyclic undecapeptide 16, 27
cyclooxygenase 12
cylindrocyclophane 45
Cylindrospermum muscicola 20
cymathere ether A 13
Cyperus rotundus 201
Cystoseira sp., parasitized by
    Spaceloma cecidii 9
Cystoseira tamariscifolia 21
cytochrome P450s 12
cytotoxic depsipeptide 35
cytotoxic macrolide 40
cytotoxicity 35, 36

debromoaplysiatoxin 34, 37, 44
defense allocation 220, 228, 234,
    236, 239
degradation process in soil 90
19-*O*-demethylscytophycin C 20
1-deoxy-D-xylulose,5-phosphate 101

deschloroelatol 23
deuteromycete fungi 9
2,3-dibromo-4,5-dihydroxy-benzyl
    alcohol 34
3,5-dibromo-*p*-hydroxybenzyl
    alcohol 34
dichlorofontonamide 34
Dicranopteris 92
Dictopteris undulata 21
20,21-didehydroacutiphycin 34
cis-dihydromatricaria ester 100
2,3-dihydroxybenzoate 103
2,4-dihydroxy-7-methoxy-1,4-
    benzoazin-3-one 195
dilution effect 225, 226
diometin 191
disease cycle 9
disease, soil-borne, inhibition 182
disease-control agent 11
diterpene, linear natural 82
diterpene, natural 82
dodecapeptides 16
dolastatin 36, 45, 47, 49
dolastatin 3 36, 45
dolastatin 10 36, 49
dolastatin 12 36, 47
dolastatin 13 36, 45
dolastatin G 36
Dollabela auricularia 36
L-dopa 99
dragonomide 35, 45
Dysidea herbacea 35
dysidenamide 35, 45
dysidenin 35, 48

ecological relevance, allelopathy
    110, 125
ecological theory, allelopathy 109
economic losses to agriculture 23
ecosystem engineer 134, 139
ectomycorrhizae 139
Ectrogella perforans 8
eicosanoid 11, 12
elatol 23
elatol/deschloroelatol 27

Elodea canadensis 43
Emiliania huxleyi 213
Empetrum hermaphroditum 120, 121
Enediyne-pyrrolidinedione 27
Enterococcus faecalis 41
environmental stress 229, 235
environmental toxicity 15
epicatechin 63
ericoid mycorrhizae 140
eriodictyol 191
Erysiphe graminis 16, 27
Escherichia coli 105
24-ethylcholestan-3β-ol 37
24-ethylcholest-5,22-dien-3β-ol 37
24-ethylcholest-5-en-3β-ol 37
eubacteria 41
Eucalyptus globulus 91
eukaryotic algae 15
eupafolin 191
Eurotium repens 21, 27
eutrophication 42
evolutionary significance,
    allelopathy 111
experimental design, activated
    charcoal 122

feedback effect 89
feedback process 90, 92
feeding deterrent 38, 39
feeding stimulant 39
ferulic acid 204, 205
ferulic acid, *trans*- and *cis*- 195
Festuca idahoensis 189
Festuca ovina 189
fire 118, 119
fire-prevention mechanism,
    allelopathy 120
fire suppression 135, 138
Fischerella 15, 16, 34, 37, 34, 42
Fischerella musciola 15, 16, 37
fischerellin 15, 16, 27, 37, 41, 42, 45
fischerellin A 15, 16, 27, 41, 45
fischerellin B 15, 16, 41, 45
(+)-*Trans*-fitol 83
flavonoid 11, 74, 201, 207, 245

flavonoid isorhamnetin 245
Flea beetle (*Phyllotreta* sp.) 181
fucosterol 21, 22, 27
fungal disease development 9
fungal growth 8
fungal invasion 9
fungal spore 9
fungicide resistance 25
fungus disease cycle 9
Fusarium 18, 21, 23, 27, 182
Fusarium oxysporum 18, 21, 27
Fusarium oxysporum sp. *lycopersici*
    21
Fusarium oxysporum sp.
    mycopersici 27

Gastroclonium clavatum 22
Gaultheria shalon 120
gene, production of allelochemical
    209
generalist herbivore 39
germination inhibition 62
glucosinolates 180, 182
Glycine max 206, 212
goldfish 34
Gracilaria coronopifolia 34
gramine 206
gram-negative bacteria 42
granadadiene 46
granadamide 46
grape, black rot 21
grape downy mildew 19
graphical vector analysis (GVA)
    234, 237
gray mold 16, 23
gray mold rots 23
Green foxtail 211
greenhouse-grown ornamentals 23
greenhouse-grown tomato 25
grenadadiene 35
grenadamide 35
growth dilution 225, 237
Guignardia bidwelli 21, 27

hairy vetch 206

halogenated sesquiterpene 27
hapalindole A 34, 41, 46
12-*epi*-hapalindole E isonitrile 42
hapalindolinone 34, 46
hapalindolinone A 46
hapalindolinone B 46
hapalonamide G, H, and V 34
Hapalosiphon fontinalis 34, 41
Hebeloma crustuliniforme 93
heliannone A 84
heliannuol 80
Helianthus annuus 73, 201
heliespirone 80
hepatotoxin 37
2-heptyl-3-hydroxy-4-quinone 100
herbivore 219, 220, 233, 234
herbivore, defense mechanism 111
herbivore, repelling of 181
herbivore specialization 39
herbivory 117
hermitamides A and B 35
Hieracium aurantiacum 245
Hieracium floribundum 245
Hieracium pratense 245
homodolastatin 3 36, 45
homoeriodictyol 191
hordenine 206
Hordeum vulgare 178, 207
hormothamnin 36, 38
Hormothamnion enteromorphoides
    36, 38
human mycosis 9
humic solution 93
hybriadalactone 13
hydric stress 207
hydrocinnamic acid 104, 119
hydroquinone 21, 27
p-hydroxyacetophenone 104
p-hydroxybenzoic acid 195, 204
3-hydroxyjuglone 103
16β-hydroxykaurene 83
6-hydroxy-7-*O*-methylscytophycin E
    20
2-hydroxymucoic acid semialdehyde
    103

6-hydroxyscytophycin B 20
Hymenoscyphus ericae 93

ichthyotoxicity 34
immunosuppressive activity 35
indanone 36
indol-[2,3-a]-carbazoles 19
indolcarbazoles 34
induced defense hypothesis 224
induced defense theory 227
Ipomoea hederacea 101
iron pan 139
isodysidenin 35, 48
isomalyngamide A 35, 46
isomalyngamide B 35, 46
isopentenyl diphosphate 101
isoprene 101
isorhamnetin 191
isorhamnetin 3-rutinoside-7-
    rhamnoside 65
isozonarol 21, 22

jaceosidine 191
jasmonic acid 12, 13
johnsongrass (*Sorghum halepense*)
    194
Jolyna laminariodes 21
Juglans nigra 103, 149
juglone 103, 152–157, 159, 160

kaempferol 3-glucoside 191
kaempferol 7-glucoside 191
kaempferol 3-rhamnoside 191
kaempferol 3-rutinoside 191
kalkipyrone 35, 46
kalkitoxin 34, 46
Kalmia angustifolia 120, 124, 205,
    212
(–)-kaur-16-en-19-oic acid 83
kaurene 82
kaurenoic acid 99
kelp 9
keystone species 134, 139
Kobresia myosyroides 104
Koeleria laerssenii 189

kororamide 36, 46

Lactuca sativa 112
Laminaria saccharina 9
lanosol 34, 46
Lantana camara 188
Lapidium sativum 113
latex 99
Laurencia 21
Laurenica obtusa 21
Laurencia rigidia 21
laxaphycin 16, 17, 38
laxaphycin A 17
laxaphycin B 17
leaf litter 91
leaf respiration 167, 168
leaf spot of grasses 21
Ledum groenlandicum 212
legume-*Rhizobium* symbiosis 213
Leischmania 100
Lemna gibba 42
Leptosphaeria sp 21
leukotriene 11
Licmorphora 8
limonene 202
linoleic acid 67
α-linolenic acid 11, 67
lipoxygenase 11, 12
litter decomposition 139
Lophocladia lallemandii 22
lunularic acid 34, 46
luteolin 7-glucoside 191
Lycopersicon esculentum 187
lynbyastatin 1 47
lynbyastatin 2 47
Lyngbya majuscula 18, 34, 36–38, 40, 60
lyngbyabellin A 35, 47
lyngbyabellin B 35, 47
lyngbyapeptin A 47
lyngbyastatin 1 36
lyngbyastatin 2 36
lyngbyatoxin 40
lyngbyatoxin A 34, 37, 47

macrograzer 38
majusculamide A 39, 47
majusculamide B 39, 47
majusculamide C 18, 19, 27, 36, 47
malyngamide A 38, 39, 47
malyngamide B 38, 47
malyngamide H 34
malyngamide I 34
malyngamide J 35, 47
malyngamide K 35, 47
malyngamide L 35, 48
malyngolide 34, 38, 39, 48
mariculture, in Japan and Coastal Asia 8
maricultured red algae 8
marine algae, associated with fungi 7
marine diatoms 8
meroditerpenoid 27
mesograzers 38
methionine 101
methoxybifurcarenone 21, 22, 27
(20S)24-methylenlophenol 65
2-C-methyl-D-erythritol,4-phosphate 101
7-methyltransferase 207
methyl tumonate A 35, 50
methyl tumonate B 35, 50
microbial drug-discovery program 9
microcolin A 35, 48
microcolin B 35, 39, 48
microcystin 37
microcystin-LR 48
Microcystis 37, 41, 42
Microcystis aeruginosa 37, 42
microfilament, depolymerization 19
microsystin-LR 43
milfoil 208
mineral soil seedbed 134
mirabilene-A isonitrile 48
mirabilene isonitriles 34, 48
mixed cyanobacteria bloom 38
mixed cyanobacterial assemblages 36
mixed-crop agroecosystem 176

Monilinia fructigena 16
Monoraphidium convolutum 41, 42
morning glory (*Ipomoea hederacea*)
   212
morphine 99
mugwort 190, 206
Multiplicative Survival Model 210
mutualism 110
mycorrhizae 2, 99, 199, 204
mycorrhizal fungi 93
mycorrhizal symbiosis 93
Mycosphaerella ascophylli 7
Mycotypha microspora 21, 27
myrene 202
myricetin 204
Myriophylum spicatum 208

nakienone A 36, 48
nakienone B 36, 48
nakienone C 36, 48
nakitriol 36, 48
Nannochloris 41
native target species, effect on 113
nematodes, plant parasitic 182
neosaxitoxin 37, 49
neurotoxin 37
neutralism 110
nicosulfuron 194
Nigrogen deficiency 42
nitrification 92
nitrogen cycling 92
nitrogen, biological fixation 177
Nitrosomonas 204
N-methyl-D-aspartate (NMDA)
   receptor 35
Nodularia harveyana 41
Nodularia spumigena 37
nodularin 37
nordysidenin 35, 48
norlyngbyastatin 2 36, 47
nostocyclamide, anticyanobacterial
   compound 41
Nostoc 16, 36, 37, 41, 42, 43
Nostoc muscorum 16, 43
Nostocaceae 36

Nostocales 20, 34
nostocyclamide 42
nostocyclophane 48
nutrient deficiency 200
nutrient dynamic 209
nutrient immobilization 2

oomycete 8, 19
oomycete, infection of algae 8
opisthobranch mollusk 36
orcinol 205
Oryza sativa 43
Oscillatoria acutissima 34, 40
Oscillatoria blooms 41
Oscillatoria late-virens 37, 41, 42
Oscillatoria nigroviridis 37
Oscillatoria planktonica 41
oscillatoxin A 37, 44
oxalic acid 99
oxidoreductase 205
oxylipin 14

P-388 lymphocytic leukemic cell 34
Palouse prairie land 189
paracyclophanes 34
parasitic fungi freshwater algae 8
parasitism 110
parrotfish 38
Parthenium hysterophorus 245
pathogen attachment 9
pathogen attack 9
pathogen growth 9
pathogen host recognition 9
pathogen, penetration and invasion
   9
pathogen reproduction 9
pathogen resistance 15
pathogen sensitivity 25
Pediastrum boryanum 40, 41
Pelvetia canaliculata 7
pepper trees 111
perennial ryegrass (*Lolium perenne*)
   195
peroxidation, lipoxygenase-mediated
   11

pesticide cross-resistance 15
Petersenia pollagaster 8
Phalaris minor 105
Phaseolus vulgaris 176
phenol 99
phenolic acid 101, 120, 122
phenolic allelochemical 133
phenolic concentration and nutrient
    availability 121
phenylacetic acid 245
phenylalanine lyase (PAL) 202, 209
Phleum pratense 191, 245
Phoma tracheiphila 22
Phormidium uncinatum 40, 41
phosphorous deficiency 42
photosynthetic electron transport 42
photosynthetic photon flux densities
    206
photosystem II 15, 37, 193
Phragmites australis 43
Phycomelaina laminariae 9
Phyllosticta capitalensis 21, 27
physiological mechanism,
    allelopathy 113
phytoalexin 11, 13, 76
phytohaem-agglutinin 99
Phytophthora cinnamomi 21, 27
Phytophthora infestans 13, 16, 19,
    21, 27
Phytophthora nicotianae 21, 27
phytoplankton 208
phytotoxic property 60
phytotoxicity 105, 210, 230, 234,
    235
phytotoxin 63, 90
phytotoxins from cattails 63
Picea abies 93
Picea mariana 124, 206
β-pinene 202
Pinus clausa 104, 118
Pinus muricata 92, 204
Pinus sylvestris 121
Pinus taeda 202
Pisum sativum 42, 101
pitipeptolide A 35, 35

pitipeptolide B 35, 48
pitted morning glory 192
Plasmopara viticola 19, 27
plastoquinone 193
Plektonema 37
P-limited condition 208
Plocamium hamatum 22
Pluchea lanceolata 83, 188, 201
pollen allelopathy 177, 245
polyculture, of maize, bean, and
    squash 176
polycyclic terpenoids 21
Polygonium convolvulus 42
polyketide macrolide 27
polymerization, oxidative 205
polyphenol 92, 204, 208
population dynamic 91, 92
Populus tremuloides 210
poriferasterol 37, 48
Porphyra 8, 10, 15
Porphyra miniata 8
Porphyra tenera 8
Porphyra yezoensis 8, 10
porphyran 10
potato (*Solanum tuberosum*) 193
potato late blight 19
prenylated quinones 21
Procambarus clarkii 35
proline 99
pro-metabolites 99
prostacyclin 11, 12
prostacyclin I$_2$ (PGI$_2$) 12
prostaglandin 11, 12
prostaglandin endoperoxide synthase
    12
prostaglandin F$_{2\alpha}$ (PGF$_{2\alpha}$) 12
prostaglandin G$_2$(PGG$_2$) 12
prostaglandin H$_2$ (PGH$_2$) 12
proteinase inhibitor 11
protocatechuic acid 204
PS II inhibitor 193
Pseudocercosporella
    herpotrichoides 16
pseudodysidenin 35, 48
Pseudomonas aeruginosa 100

Pseudomonas putida 103
Pyricularia oryzae 16, 27
pyrogallol 205
Pythium aphanidermatum 182
Pythium porphyrae 8, 10, 15

quackgrass (*Elytrigia repens*) 194
quercetin 191
quercetin 3-galactoside 191
quercetin 3-glucoside 191
Quercus chapmanii 118
Quercus germinata 118
Quercus myrtifolia 118
Quercus rubra 121, 122

rabbitfish, juvenile 38
Raphanus satius 206
reactive oxygen intermediates 14
receptor-binding activity 35
red clover 191, 206
red rot disease 8
redroot pigweed (*Amaranthus retroflexus*) 192, 206
resource competition 2
resource limitation to the fitness 113
respiration rate 117
Rhizobium 213
Rhizoctonia solani (root and stem rots) 16, 18, 27, 182
rhizomatous growth 132
Rhizopogon 93
Rhododendron maximum 120–124
Rivularia firma 34
rivularin D$_3$ 34, 48
Robinsonia evenia 205
root and stem rots 18
root exudates 179, 213
root respiration 167, 168
root rots 21
rotifer 41
Rumax crispus 42
rutin 191

salicylic acid 12

Salix brachycarpa 104
Salvia leucophylla 114, 115
Salvinia minima 60
sand pine scrub 118, 120
sandhill community 118
saponin 99
Sargassum sp., parasitized by *Spaceloma cecidii* 9
saxitoxin 37, 49
Scarus schlegeli 38
Scenedesmus acutus 41, 42
Scenedesmus obliquus 41
Scenedesmus quadricauda 41
Schinus 111
Schizachyrium scoparium 104
Schizothrix calcicola 18, 35, 37, 38
schizotrin A 17, 18
Schutzexcrete 100
Sclerotinia sclerotiorum (cottony rot) 16
Sclerotium rolfsii (Southern blight) 18, 27
Sclerotium sclerotiorum 27
Scytonema burmanicum 20
Scytonema hofmanni 34, 37, 42
Scytonema mirabile 20
Scytonema ocellatum 20
Scytonema pseudohofmanni 20
scytonemin A 17, 18, 27
scytophycin 20, 34, 49
scytophycin A 20, 49
scytophycin B 20, 49
sea hares 38
Secale cereale 178
secondary metabolites 10, 15, 99, 199, 200
secondary succession 141
seed decay 21
seed regeneration 135
seedbed quality 134
seedling growth and light intensity 121
self-thinning mechanism 111
Senecio vulgaris 193
sesquiterpene 27, 80

sesquiterpene lactones 81
sesquiterpenoid 21
Setaria viridis 42, 211
shikimate pathway 202
Siganus argenteus 38
Siganus spinus 38
β-sitosterol(5) 65
smoldering combustion 141
soil microbial community 89
soil microflora 89
soil phenolic compounds 120
soil resource availability 119
Solidago altissima 100
Solidago canadensis 206
somamide A 36, 49
somamide B 36, 49
Sorghum 212
sorgoleone 192
soybean 206
Spaceloma cecidii 9
Spirodela polyrhiza 42
Spirulina platensis 41
spotted knapweed 189
Staphylococcus cerevisiae 42
stem break 16
sterol 27
stigmast-4-ene-3,6-dione 65
stigmasterols 65
stilbene 122
strawberries 25
strawberry plant 16
Streptococcus pneumonia 41
Stylocheilus dietary selection 39
Stylocheilus longicauda 39
Stypopodium zonale 21
sunflower, allelochemicals 73
sustainability 175, 183
sustainability, allelopathy 175
sustainable agriculture 175
sustained yield 175
swimmer itch 40
Symploca hydnoides 36
symplostatin 1 36, 49
symplostatin 2 49
Synechocystis 36

Tagetes erecta 194
tanikolide 34, 49
tannin 99, 204, 213
tantazole 34
tantazole A 49
tantazole B 49
taondiol 21, 27
taxifolin 3-arabinoside 82
tellimagrandin II 208
terpene 99, 114–118
terpenes, solubility 118, 119
Thamnocephalus platyurus 41
Thielaviopsis paradoxa 21, 27
thromboxane 12
thromboxane A_2 (TXA_2) 11, 12
tjipanazole 19
tjipanazole A1 20
2D-TLC 25, 38
2D-TLC fingerprint 25
toandiol 22
Tolypothrix 20
Tolypothrix conglutinata var. colorata 20
Tolypothrix tenuis 43
Tolypothrix tjipanasensis 19
tolytoxin 20, 27, 34, 49
tomato wilt 21
total phenolic content 204
toxin 15
toxintransfer, *in situ* pathway 119
toyocamycin 34, 50
transplant of litter and humus 122
transplanted substrate 123
triazine-susceptible potato 193
tricine 191
Trifolium repens 191, 206
Triticum aestivum 42, 43, 202
tubercidin 34, 50
tumonic acid A 35, 50
tumonic acid B 35, 50
tumonic acid C 35, 50
tumor promoter 34
two-dimensional direct bioautography 23

two-dimensional thin-layer chroma-
    tography (2D-TLC) 25, 38
two-dimensional TLC bioautography
    25
Typha domningesis 3
Typha domingensis P. 57

Udoteaceae 21
understory species 91
Uromyces appendiculatus 16, 27
ursalic acid 118
Ustilago violacea 21, 27
UV irradiation 207

Vaccinium myrtillus 93
vascular endothelial growth factor
    (VEGF) 36
vascular wilt 18, 25
vegetation pattern 191
velvetleaf 211
Verticilium alboatrum 21, 27
verticilum wilt 21
Vesicularia dubyana 43
Vicia faba 178
Vicia villosa 206
vitexin 191

walnut 103
water stress 116
weed, suppression of 176
weed-control strategy 178
weed-management strategy 173
weed reduction, allelopathic 179
weeds, management in
    agroecosystems 182
wood savanna-type community 141
wound-inducible defense 11

yanucamide A 38, 50
yanucamide B 38, 50
yellow foxtail (*Setaria glauca*) 206
yield increase, mechanisms 176
yield stimulation 181
ypaoamide 38, 39, 50

Zea mays 42, 176, 187, 201, 206,
    245
Zea mays var. *chalquiñocónico* 245
zonarol 21, 22
zonarol/isozonarol 27